Chemistry of
Heterocyclic Compounds

Chemistry of
Heterocyclic Compounds

Dr. Rakesh Kumar Parashar, Ph.D.

Associate Professor
Department of Chemistry, Kirori Mal College
University of Delhi, Delhi, India

Dr. Beena Negi, Ph.D.

University of Delhi,
Delhi, India

Taylor & Francis
Taylor & Francis Group
Boca Raton   London   New York

CRC is an imprint of the Taylor & Francis Group,
an informa business

Ane Books Pvt. Ltd.

Chemistry of Heterocyclic Compounds
Dr. Rakesh Kumar Parashar and Dr. Beena Negi

© Authors, 2015

Published by

Ane Books Pvt. Ltd.

4821, Parwana Bhawan, 1st Floor, 24 Ansari Road, Darya Ganj,
New Delhi - 110 002, Tel.: +91(011) 23276843-44, Fax: +91(011) 23276863
e-mail: kapoor@anebooks.com, Website: www.anebooks.com

For

CRC Press

Taylor & Francis Group
6000 Broken Sound Parkway, NW, Suite 300
Boca Raton, FL 33487 U.S.A.
Tel : 561 998 2541
Fax : 561 997 7249 or 561 998 2559
Web : www.taylorandfrancis.com

For distribution in rest of the world other than the Indian sub-continent

ISBN: 978-146-6517-13-4

The Cover Picture: The cover picture displays the drug vincristine along with the plant *Catharanthus roseus* (rosy periwinkle) from which it is isolated. In traditional Chinese and Indian medicine the extracts of the roots and shoots of *Catharanthus roseus,* is used for the treatment of various diseases such as Hodgkin's lymphoma, diabetes and malaria. However, it can be dangerous on oral consumption. Vincristine is used in cancer chemotherapy and given by injection into vein.

British Library Cataloguing in Publication Data
A catalogue record for this book is available from the British Library

Printed and bound in India by Replika Press Pvt. Ltd.

Preface

This book on Chemistry of Heterocyclic Compounds is well-written after considerable planning and thought. The requirement of a good book on heterocyclic chemistry has given a practical shape to this book which contains logically sequenced subject matter, consistency, fluency and preciseness in the content. The language used is simple, educative and enlightening. The book should enable the readers to easily grasp the subject matter.

This book on heterocyclic compounds has 14 chapters in all. The first chapter on introduction and nomenclature provides an effective, clear and readable description of heterocyclic compounds. The nomenclature gives the readers a deep insight into the types of heterocyclic compounds, which are important for the readers to understand other chapters of this book. The nomenclature is explained in an understandable manner and well supported by several examples of simple, fused and spiro heterocycles. Second chapter illustrates the importance of heterocyclic compounds as therapeutics and agrochemicals. The subsequent chapters 3 to 13 follows a similar sequence which covers in detail the properties, preparation and reactions of three to seven membered heterocyclic compounds. The chapters 3 and 4 are about three and four membered heterocyclic compounds with one heteroatom. Chapters 5 and 6 include five membered heterocyclic compounds with one heteroatom and their benzofused derivatives. Five membered heterocyclic compounds with two heteroatoms and their benzofused derivatives are dealt in chapters 7 and 8, respectively. Six membered heterocycles with one heteroatom and their benzofused derivatives, with two heteroatoms and their benzofused derivatives are discussed in chapters 9, 10, 11 and 12, respectively. Chapter 13 focuses on seven membered heterocycles with one heteroatom. Preparation and properties of porphyrins are covered in the final chapter. The mechanisms of important reactions are included at appropriate places in all the chapters. Simple problems with their solutions or hints are given at the end of each chapter for the better understanding of the readers.

Overall we have tried to cover the nomenclature, importance, structure, preparation and reactions of all the important heterocyclic compounds and hope it will give the readers a complete understanding of such compounds. This is a general text book for the undergraduate and post graduate students from all the Universities of India and other countries. This book is also helpful for those appearing for competitive exam and the chemists working in various research and development centers.

Dr. Rakesh Kumar Parashar
Dr. Beena Negi

Acknowledgement

The authors would like to acknowledge Prof. Jitendera M. Khurana, Prof. Ashok K. Prasad, Prof. Diwan S. Rawat and Prof. Sunil K. Sharma for their constructive comments and sound advice. We owe our sincere thanks to our colleagues for their suggestions.

We would like to thank Mrs. Indu Sharma, Dr. Deepak K. Gupta and Dr. Manju K. Saroj for their help and support. Finally, we express our sincere thanks to Mr. Sunil Saxena, Mr. Jai R. Kapoor and the technical staff of Ane Books Pvt. Ltd. associated with this book, for their help and cooperation.

All comments and suggestions will be received with gratitude.

Dr. Rakesh Kumar Parashar
Dr. Beena Negi

Contents

Introduction and Nomenclature

1.1 INTRODUCTION

Heterocyclic compounds are those cyclic compounds which contain at least one heteroatom in their ring. However, organic compounds which contain only carbon atoms in the ring and not any other atom are known as **carbocyclic compounds**. Heterocyclic chemistry is a vast subject of utmost importance as heterocyclic compounds are most common in organic chemistry. More than half of all natural products are heterocycle based compounds. The most commonly present heteroatom in heterocyclic compounds are nitrogen, oxygen and sulfur. Other heteroatoms such as selenium, tellurium, phosphorous, arsenic, silicon, boron, mercury, tin, bismuth, germanium, lead and antimony are also widely present in heterocyclic compounds. Presence of heteroatoms in the ring has their own significance. They are the structural elements of the σ-skeleton and also take part in the π-electron system. The heteroatom may take part in ring formation or cleavage. They represent site for coordination, basicity or hydrogen bonds. They affect polarizibility of the molecule and the reactivity of their neighbouring atoms. The ring maintains the orientation of the heteroatom and its lone pairs. The properties of heterocyclic compounds are influenced by the presence of strain in the ring.

Heterocyclic compounds are more often consist of 3 to 7 membered ring systems. They may be classified into aliphatic and aromatic or saturated and unsaturated compounds. The aromatic heterocyclic compounds are those which obey **Huckel rule** and possess $(4n + 2)$ π-electrons.

Many earlier studies in chemistry discuss about heterocyclic compounds. In **1776**, Scheele isolated a nitrogen based heterocyclic compound from human bladder stone named **uric acid.** Later on, **alloxan** was obtained by oxidation of **uric acid** in the year **1818**, by Brugnatelli. In the year **1834**, **pyrrole** was obtained from the dry distillation of bones. In **1838**, Wohler and Liebig described the derivatives of uric acid (**purines** and **pyrimidines**). **Pyridine** was isolated from bone oil in **1849** by Anderson and its structure was proposed in **1869** by Korner and Dewar. In **1882**, Victor Meyer isolated **thiophene**.

| Uric acid | Alloxan | Pyrrole | Pyridine | Thiophene |

A vast majority of organic compounds used in the pharmaceutical and agrochemical industries are heterocyclic compounds. A large number of heterocyclic compounds have important role in chemistry of life and possess many biological properties. The knowledge of heterocyclic chemistry is useful in biosynthesis and in drug metabolism as well. There are large number of synthetic heterocyclic compounds with additional important applications and many are valuable intermediates in synthesis. Porphyrin, which is the structural unit of heme, chlorophyll and vitamin B12 is also a heterocyclic compound. Three of the naturally occurring amino acids (proline, histidine and tryptophan) contain heterocyclic systems. Nucleic acids (DNA and RNA) which are important in biological processes of heredity and evolution also contain heterocyclic rings. Indigo blue, used to dye jeans is a heterocyclic compound. Tetrahydrofuran, dioxane and pyridine used as solvents as well as crown ethers with the ability to solvate cations, are heterocyclic compounds.

Thus, heterocyclic compounds are of immense significance and it becomes more important to develop a systematic naming for such a huge number of compounds as well as study their synthesis and reactions.

1.2 NOMENCLATURE OF HETEROCYCLIC COMPOUNDS

Some heterocyclic compounds are known by their trivial names. These trivial names are based on the compounds occurrence, preparation or properties. Such names convey little or no information about the structure of the compound. The International Union of Pure and Applied Chemistry (IUPAC) have made certain rules to systematize the nomenclature of heterocyclic compounds. These systematic names for a heterocyclic compound are based on its structure. The IUPAC rules allow two nomenclatures, the **Hantzsch-Widman** and **replacement nomenclature.** The replacement nomenclature is for larger than ten membered rings and based on prior carbocyclic name. The extended **Hantzsch-Widman** system for three to ten membered heterocycles provides a more systematic method of naming heterocyclic compounds that is not dependent on prior carbocyclic names.

The name of a heterocycle is written by combining the "a" prefix(es) for the heteroatom(s) with a stem or suffix. The "a" prefixes for various heteroatoms are shown in Table 1.1 in decreasing order of their priority. These are called "a" prefixes as they end in a. The stem or suffix indicates the ring size and saturation or unsaturation in the ring as given in Table 1.2.

Table 1.1 The "a" prefixes in decreasing order of priority

Element	Symbol (Valence)	Prefix	Element	Symbol (Valence)	Prefix
Oxygen	O (II)	Oxa	Bismuth	Bi (III)	Bisma
Sulfur	S (III)	Thia	Silicon	Si (IV)	Sila
Selenium	Se (II)	Selena	Germanium	Ge (IV)	Germa
Tellurium	Te (II)	Tellura	Tin	Sn (IV)	Stanna
Nitrogen	N (III)	Aza	Lead	Pb (IV)	Plumba
Phosphorous	P (III)	Phospha	Boron	B (III)	Bora
Arsenic	As (III)	Arsa	Mercury	Hg (II)	Mercura
Antimony	Sb (III)	Stiba			

Table 1.2 Suffix indicating ring size and saturation or unsaturation in the ring

Ring Size	Unsaturated	Saturated
3	irene (irine for N)	irane (iridine for N)
4	ete	etane (etidine for N)
5	ole	olane (olidine for N)
6 (If least preferred* heteroatom in the ring is O, S, Se, Te, Bi, Hg)	ine	ane
6 (If least preferred* heteroatom in the ring is N, Si, Ge, Sn, Pb)	ine	inane
6 (If least preferred* heteroatom in the ring is B, P, As, Sb)	inine	inane
7	epin (epine for N)	epane
8	ocin (ocine for N)	ocane
9	onin (onine for N)	onane
10	ecin (ecine for N)	ecane
*Least preferred heteroatom is the one which is part of prefix but directly precedes the stem		

The systematic naming of some heterocyclic compounds according to extended Hantzsch-Widman system is shown in Table 1.3.

Table 1.3 Naming a heterocyclic compound

Structure	Heteroatom	Ring size	Saturated/ unsaturated compound	Name of the compound
	Nitrogen **Az-**	3 **-ir-**	Saturated **-idine**	Aziridine
	Oxygen **Ox-**	3 **-ir-**	Saturated **-ane**	Oxirane
	Sulfur **Thi-**	3 **-ir-**	Saturated **-ane**	Thiirane
	Nitrogen **Az-**	3 **-ir-**	Unsaturated **-ine**	Azirine
	Oxygen **Ox-**	3 **-ir-**	Unsaturated **-ene**	Oxirene
	Sulfur **Thi-**	3 **-ir-**	Unsaturated **-ene**	Thiirene
	Oxygen and nitrogen **Oxaz-**	3 **-ir-**	Saturated **-idine**	Oxaziridine
	Two oxygen atoms **Diox-**	3 **-ir-**	Saturated **-ane**	Dioxirane
	Two nitrogen atoms **Diaz-**	3 **-ir-**	Unsaturated **-ine**	Diazirine
	Nitrogen **Az-**	4 **-et-**	Saturated **-idine**	Azetidine
	Oxygen **Ox-**	4 **-et-**	Saturated **-ane**	Oxetane
	Sulfur **Thi-**	4 **-et-**	Saturated **-ane**	Thietane
	Oxygen **Ox-**	4 **-et-**	Unsaturated **-e**	Oxete
	Sulfur **Thi-**	4 **-et-**	Unsaturated **-e**	Thiete

—Contd...—

Structure	Heteroatom	Ring size	Saturated/ unsaturated compound	Name of the compound
S——S (4-membered ring)	Two sulfur atoms **Dithi-**	4 -et-	Saturated -ane	Dithietane
O——O (4-membered ring)	Two oxygen atoms **Diox-**	4 -et-	Saturated -ane	Dioxetane
HN——NH (4-membered ring)	Two nitrogen atoms **Diaz-**	4 -et-	Saturated -idine	1,2-Diazetidine
NH / HN (4-membered ring)	Two nitrogen atoms **Diaz-**	4 -et-	Saturated -idine	1,3-Diazetidine
S——S (4-membered ring)	Two sulfur atoms **Dithi-**	4 -et-	Unsaturated -e	Dithiete
N–H (5-membered ring)	Nitrogen **Az-**	5 -ol-	Saturated -idine	Azolidine
O (5-membered ring)	Oxygen **Ox-**	5 -ol-	Saturated -ane	Oxolane
S (5-membered ring)	Sulfur **Thi-**	5 -ol-	Saturated -ane	Thiolane
O / O (5-membered ring)	Two oxygen atoms **Diox-**	5 -ol-	Saturated -ane	1,3-Dioxolane
S / S (5-membered ring)	Two sulfur atoms **Dithi-**	5 -ol-	Saturated -ane	1,3-Dithiolane
O / S (5-membered ring)	Oxygen and sulfur **Oxathi-**	5 -ol-	Saturated -ane	Oxathiolane
O / O (6-membered ring)	Two oxygen atoms **Diox-**	6	Saturated -ane	Dioxane
S / S (6-membered ring)	Two sulfur atoms **Dithi-**	6	Saturated -ane	Dithiane

—Contd...—

Structure	Heteroatom	Ring size	Saturated/ unsaturated compound		Name of the compound
	Oxygen and sulfur **Oxathi-**	6	Saturated -ane		1,4-Oxathiane
	Sulfur **Thi-**	7 -ep	Saturated -ane		Thiepane
	Nitrogen **Az-**	7 -ep	Unsaturated -ine		Azepine
	Sulfur **Thi-**	7 -ep	Unsaturated -in		Thiepin
	Oxygen **Ox-**	8 -oc	Saturated -ane		Oxocane
	Nitrogen **Az-**	8 -oc	Unsaturated ine		Azocine
	Oxygen **Ox-**	9 -on	Saturated -ane		Oxonane
	Nitrogen **Az-**	9 -on	Unsaturated -ine		Azonine
	Oxygen **Ox-**	9 -on	Unsaturated -in		Oxonin
	Oxygen **Ox-**	10 -ec	Saturated -ane		Oxecane
	Nitrogen **Az-**	10 -ec	Unsaturated -ine		Azecine

1.2.1 Important points to remember while naming a heterocyclic compound according to extended Hantzsch-Widman system:

1. The types of heteroatom present in a ring are indicated by prefixes mentioned in Table 1.1.

2. When more than one **same heteroatoms** are present in a ring then prefixes di-, tri-, etc are used.

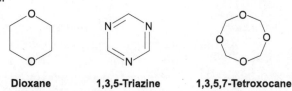

Dioxane	1,3,5-Triazine	1,3,5,7-Tetroxocane

3. If there are more than one **different heteroatoms** present in a ring then the compound is named by combining the prefixes of each heteroatom in order of their preferences as shown in Table 1.1.

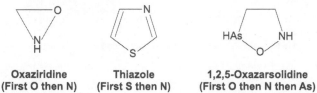

Oxaziridine	Thiazole	1,2,5-Oxazarsolidine
(First O then N)	(First S then N)	(First O then N then As)

4. While numbering of heterocyclic ring the locant 1 is given to the heteroatom with highest priority and then the ring is numbered in such a way so as to give the lowest possible number to the other heteroatoms present in the ring in order of their preferences as shown in Table 1.1. The structure and numbering of some heterocyclic compounds are given below:

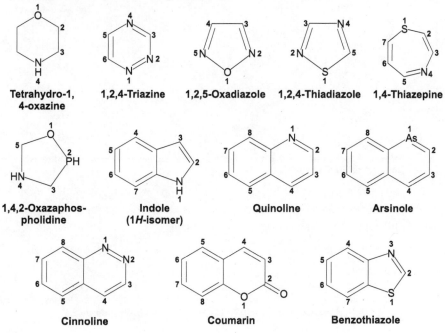

Tetrahydro-1, 4-oxazine 1,2,4-Triazine 1,2,5-Oxadiazole 1,2,4-Thiadiazole 1,4-Thiazepine

1,4,2-Oxazaphos-pholidine Indole (1*H*-isomer) Quinoline Arsinole

Cinnoline Coumarin Benzothiazole

However, in some compounds the numbering is not assigned the way it is explained in point 4. The following heterocycles are exception to the systematic numbering:

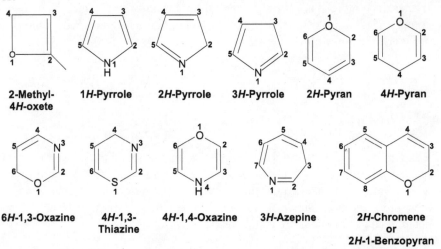

5. For rings containing maximum number of non-cumulative double bonds the position where no multiple bond is attached is specified by presence of an extra hydrogen atom at such position with a numerical locant and by an italicized capital *H*.

6. Numbering starts at a nitrogen atom that carries a substituent rather than at a multiply bonded nitrogen atom.

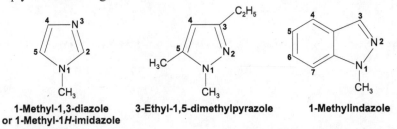

7. When the carbon atom of the ring is involved in the carbonyl group, then the locant specifying the carbonyl carbon is followed by an italicized capital *H* in parenthesis, which indicates the position of extra H in the ring.

2(1*H*)-Quinoxalinone   **2(1*H*)-Phosphininone**

8. Sometimes the position of substituents around the ring is written by Greek letters α, β, γ etc.

9. There are several important ring systems for which trivial and semisystematic names are more commonly used, as shown below.

Pyrrole   **Pyrazole**   **Imidazole**   **Furan**   **Thiophene**

Pyridine   **Indole**   **Purine**   **Quinoline**

10. Systems having a lesser degree of unsaturation require an appropriate prefix, such as "dihydro"or "tetrahydro".

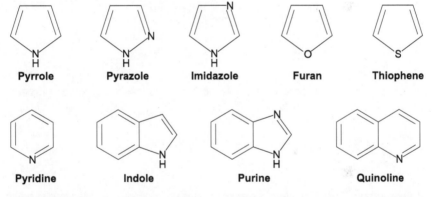

2,3-Dihydropyrrole   **1,2-Dihydropyridine**   **1,4-Dihydropyridine**   **Hexahydropyridine (Piperidine)**

11. The **fused heterocyclic compounds** are considered to be constructed by the combination of two or more cyclic structural units, one of the rings is considered as a base component and others as attached components. The fused heterocyles are named as follows:

(a) The base component should be a heterocyclic system. The prefixes designating an attached component are formed by changing the terminal 'e' of a trivial or extended Hantzsch-Widman name of a component into 'o' (Table 1.4).

Table 1.4: Prefixes designating an attached component

Attached component	Prefix
Pyrazine	Pyrazino
Triazole	Triazolo

However, there are some exceptions to this rule. For them the prefixes used are shown in Table 1.5.

Table 1.5: Prefixes designating an attached component exception to rule mentioned in step 11a

Attached component	Prefix
Thiophene	Thieno
Furan	Furo
Imidazole	Imidazo
Benzene	Benzo
Pyridine	Pyrido
Pyrimidine	Pyrimido
Quinoline	Quino
Isoquinoline	Isoquino

(b) The location of a fused ring may be indicated by a lowercase letter (a, b, c, d, etc.) which designates the edge of the heterocyclic ring (base component) involved in the fusion.

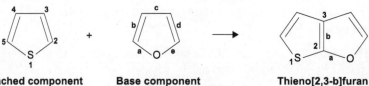

Benzo[b]pyridine	Attached component	Base component	Benzo[c]pyridine

Benzo[b]furan	Attached component	Base component	Benzo[c]furan

(c) If the two heterocyclic rings of same size contain different heteroatoms, the component which contains heteroatom appearing highest in the Table 1.1 is selected as base component. The numerals separated by comma in square bracket indicate the site of fusion of the attached component.

Attached component   Base component   Thieno[2,3-b]furan

(d) In heterocyclic system containing rings of unequal size, the component with the largest size of the ring is considered as a base component.

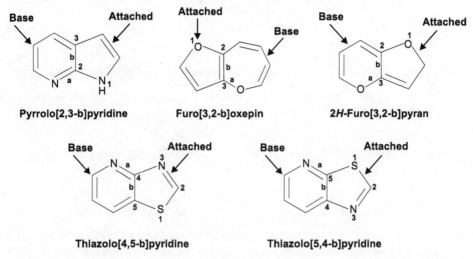

Pyrrolo[2,3-b]pyridine   Furo[3,2-b]oxepin   2*H*-Furo[3,2-b]pyran

Thiazolo[4,5-b]pyridine   Thiazolo[5,4-b]pyridine

(e) Fused heterocycles containing, rings of same sizes with different number and kind of heteroatoms are named with base component as the ring containing more number of heteroatoms as shown below:

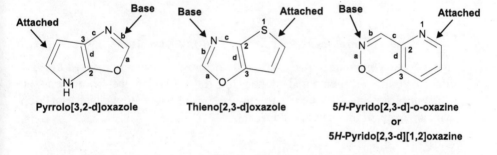

Pyrrolo[3,2-d]oxazole   Thieno[2,3-d]oxazole   5*H*-Pyrido[2,3-d]-o-oxazine

or

5*H*-Pyrido[2,3-d][1,2]oxazine

(f) Fused heterocycles containing rings of same size with equal number of different heteroatoms are named with base component containing the greatest variety of hetero atoms.

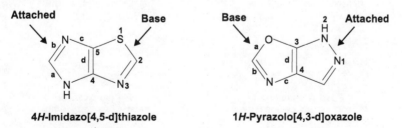

4*H*-Imidazo[4,5-d]thiazole   1*H*-Pyrazolo[4,3-d]oxazole

(g) In fused heterocycles containing rings of same sizes with same numbers and same kinds of heteroatoms, the base component is that ring which contains more number of carbon atoms adjacent to the fusion.

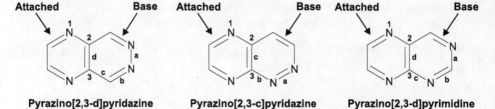

Pyrazino[2,3-d]pyridazine   **Pyrazino[2,3-c]pyridazine**   **Pyrazino[2,3-d]pyrimidine**

(h) If a heteroatom is present at the position of fusion then, the names of both the component rings fused should contain the heteroatom.

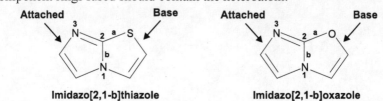

Imidazo[2,1-b]thiazole   **Imidazo[2,1-b]oxazole**

(i) If a benzene ring is fused to the heterocyclic ring, then the compound is named by placing number(s) indicating position(s) of the heteroatom(s) before the prefix benzo.

3-Benzooxepin   **4H-1,4-Benzothiazine**

(j) Heterocyclic compounds, consisting of two benzene rings ortho-fused to a 1,4-diheteroatomic six membered monocyclic ring in which the heteroatoms are different, are named by adding the prefix 'pheno-' to the extended Hantzsch-Widman name of the heteromonocycle.

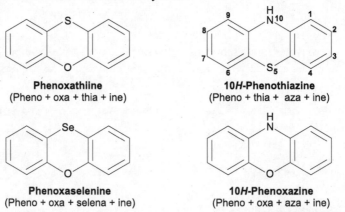

Phenoxathiine
(Pheno + oxa + thia + ine)

10H-Phenothiazine
(Pheno + thia + aza + ine)

Phenoxaselenine
(Pheno + oxa + selena + ine)

10H-Phenoxazine
(Pheno + oxa + aza + ine)

(k) Heterocyclic compounds, consisting of two benzene rings ortho-fused to a 1,4-diheteroatomic six-membered monocyclic ring in which the heteroatoms are the same are named by adding the prefix for the heteroatom to the term anthrene, without 'a'.

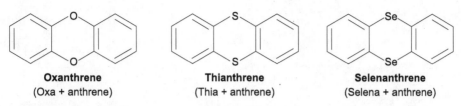

Oxanthrene	**Thianthrene**	**Selenanthrene**
(Oxa + anthrene)	(Thia + anthrene)	(Selena + anthrene)

1.2.2 The names of some common heterocyclic compounds are shown below:

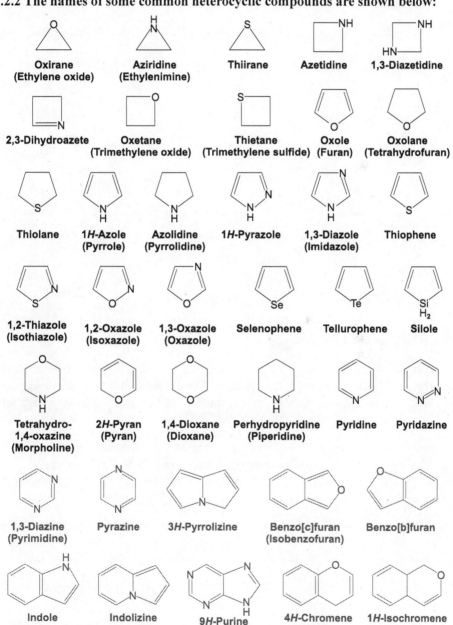

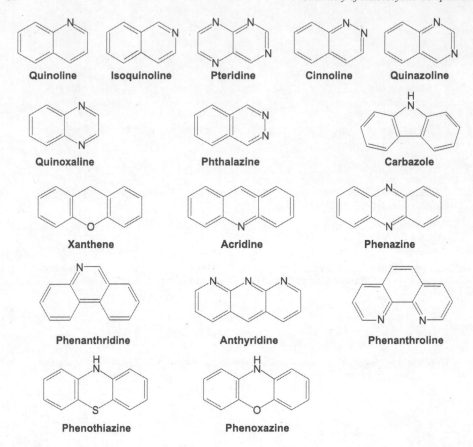

Quinoline   **Isoquinoline**   **Pteridine**   **Cinnoline**   **Quinazoline**

Quinoxaline   **Phthalazine**   **Carbazole**

Xanthene   **Acridine**   **Phenazine**

Phenanthridine   **Anthyridine**   **Phenanthroline**

Phenothiazine   **Phenoxazine**

1.2.3. Replacement nomenclature

In replacement nomenclature the name of the carbocyclic ring without the heteratom is used for naming the heterocyclic compounds. The carbon ring is named as according to IUPAC rules and the heteroatom present is indicated by a prefix. This nomenclature can be used for naming fused-, spiro- and bridged- heterocycles.

Replacement nomenclature is mainly used for greater than ten membered monocyclic rings. For lower sized monocyclic compounds the Hantzsch-Widman system or trivial names are most preffered though they can also be named according to replacement nomenclature.

The replacement nomenclature for some lower and higher ring compounds is shown below:

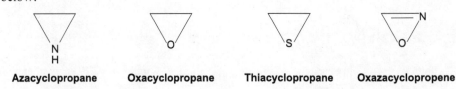

Azacyclopropane   **Oxacyclopropane**   **Thiacyclopropane**   **Oxazacyclopropene**

Azacyclopentane   **Oxacyclopentane**   **Thiacyclopentane**   **1-Oxa-3-azacyclopenta-2,4-diene**

Azacyclohexane   **Oxacyclohexane**   **Thiacyclohexane**   **Azabenzene**

Phosphabenzene   **1,4-Diazabenzene**   **2-Azanapthalene**   **5-Azaphenanthrene**

Thiacyclododecane   **1,5-Dithiacyclodecane**   **1-Thia-5-selenacyclododecane**

1.3 SPIROHETEROCYCLIC COMPOUNDS

The compounds in which two rings are fused at a common point are known as spiro compounds. The common atom is known as a "**spiro atom**". Some naturally occuring spiroheterocycles such as, **horsfiline** (analgesic), **spirotryprostatin A** (anti-mitotic) and **coerulescine** (cytotoxic) exhibit pharmacological properties.

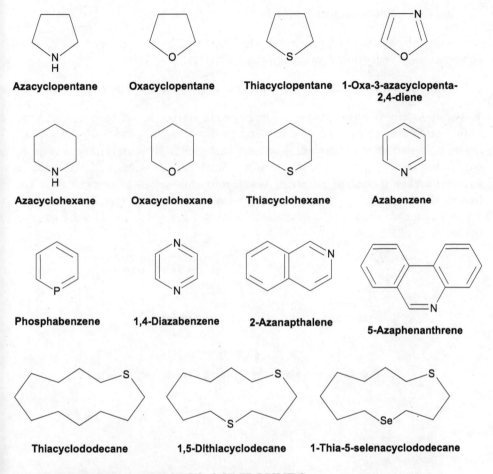

Horsfiline   **Spirotryprostatin A**   **Coerulescine**

Naming a spiroheterocyclic compound

The nomenclature and name **spirane** were proposed by **von Baeyer** for bicyclic compounds with only one atom common to both rings.

Important points to remember while naming a spiroheterocyclic compound:

1. A spiro heterocycle is named by adding a prefix **spiro.** The total number of atoms of the spiroheterocycle are counted first and a suffix with the same number of carbon atoms as in **parent hydrocarbon** is written after prefix. The atoms of each ring are written between the prefix 'spiro' and suffix 'hydrocarbon name' enclosed within a **square bracket** in ascending order in **Arabic numerals** separated by a **full stop**. The **hetero atoms** present are indicated by the replacement prefixes (aza, oxa, thia etc.). The position and prefix for the heteroatom is written before the word spiro as shown below.

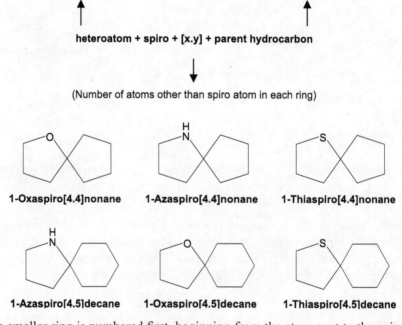

1-Oxaspiro[4.4]nonane   1-Azaspiro[4.4]nonane   1-Thiaspiro[4.4]nonane

1-Azaspiro[4.5]decane   1-Oxaspiro[4.5]decane   1-Thiaspiro[4.5]decane

2. The smaller ring is numbered first, beginning from the atom next to the spiro atom (common atom) then further proceeding to the larger ring. The heteroatom is assigned the lowest number locant.

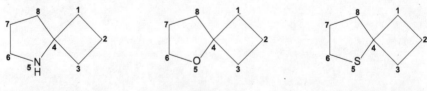

5-Azaspiro[3.4]octane   5-Oxaspiro[3.4]octane   5-Thiaspiro[3.4]octane

6-Azaspiro[4.5]decane

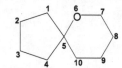

6-Oxaspiro[4.5]decane

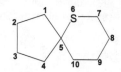

6-Thiaspiro[4.5]decane

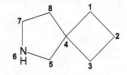

6-Azaspiro[3.4]octane

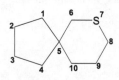

7-Thiaspiro[4.5]decane

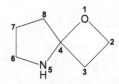

1-Oxa-5-azaspiro[3.4]octane

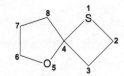

5-Oxa-1-thiaspiro[3.4]octane

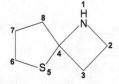

5-Thia-1-azaspiro[3.4]octane

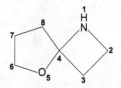

5-Oxa-1-azaspiro[3.4]octane

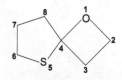

1-Oxa-5-thiaspiro[3.4]octane

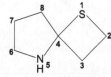

1-Thia-5-azaspiro[3.4]octane

3. If more than one similar heteroatoms are present they are indicated by di, tri, tetra etc. before the prefix along with their position.

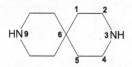

3,9-Diazaspiro[5.5]undecane   **2,4-Dithiaspiro[5.5]undecane**

4. If two different heteroatoms are present they are numbered according to their priority as shown in table 1.1 (O > S > N).

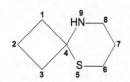

5-Thia-9-azaspiro[3.5]nonane

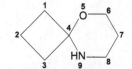

5-Oxa-9-azaspiro[3.5]nonane

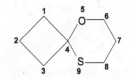

5-Oxa-9-thiaspiro[3.5]nonane

5. If the rings with and without heteroatom are of equal size, then the heterocyclic ring is preferred over the carbocyclic ring.

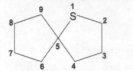

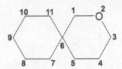

1-Thiaspiro[4.4]nonane   2-Oxaspiro[5.5]undecane

6. If both the rings contain heteroatom, the heteroatom appearing first in the preference Table 1.1 is preferred.

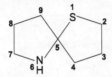

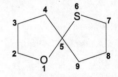

1-Thia-6-azaspiro[4.4]nonane   1-Oxa-6-thiaspiro[4.4]nonane

7. If any of the rings contain unsaturation then numbering is carried out as usual but heteroatom is given more preference than multiple bond. The unsaturation is indicated by the endings ene, diene, etc.

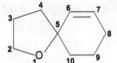

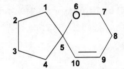

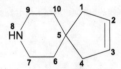

1-Oxaspiro[4.5]dec-6-ene   6-Oxaspiro[4.5]dec-9-ene   8-Azaspiro[4.5]dec-2-ene

Some more examples of nomenclature of spiroheterocycles using these rules are as follows:

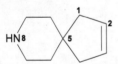

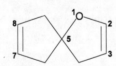

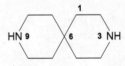

8-Azaspiro[4.5]dec-2-ene   1-Oxaspiro[4.4]nona-2,7-diene   3,9-Diazaspiro[5.5]undecane

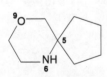

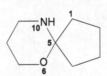

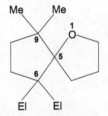

9-Oxa-6-azaspiro[4.5]decane   6-Oxa-10-azaspiro[4.5]decane   6,6-Diethyl-9,9-dimethyl-
1-oxaspiro[4.4]nonane

EXERCISE

Q.1 What are heterocyclic compounds? Why is it essential to have their systematic name?

Q.2 Write the name of the following heterocyclic compounds:

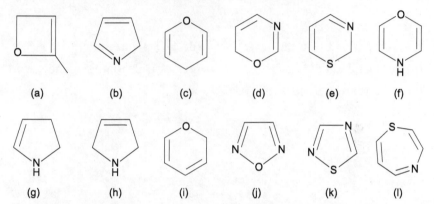

Ans. (a) 2-Methyl-4*H*-oxete   (g) 2,3-Dihydro-1*H*-pyrrole

     (b) 2*H*-Pyrrole   (h) 2,5-Dihydro-1*H*-pyrrole

     (c) 4*H*-Pyran   (i) 2*H*-Pyran

     (d) 6*H*-1,3-Oxazine   (j) 1,2,5-Oxadiazole

     (e) 2*H*-1,3-Thiazine   (k) 1,2,4-Thiadiazole

     (f) 4*H*-1,4-Oxazine   (l) 1,4-Thiazepine

Q.3 Give the nomenclature of the following fused heterocyclic compounds:

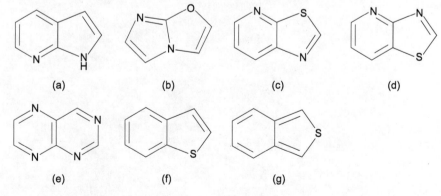

Ans. (a) 1*H*-Pyrrolo[2,3-b]pyridine   (e) Pyrazino[2,3-d]pyrimidine

(b) Imidazo[2,1-b]oxazole   (f) Benzo[b]thiophene

(c) Thiazolo[5,4-b]pyridine   (g) Benzo[c]thiophene

(d) Thiazolo[4,5-b]pyridine

Q.4 Write the name of the following spiroheterocycles:

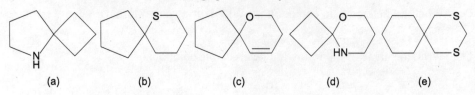

(a)   (b)   (c)   (d)   (e)

Ans. (a) 5-Azaspiro[3.4]octane   (d) 5-Oxa-9-azaspiro[3.5]nonane

(b) 6-Thiaspiro[4.5]decane   (e) 2,4-Dithiaspiro[5.5]undecane

(c) 6-Oxaspiro[4.5]dec-9-ene

2

Importance of Heterocyclic Compounds

Heterocyclic compounds are present in many natural products such as carbohydrates, amino acids, nucleic acids, vitamins, alkaloids etc. The ability of heterocyclic rings to incorporate different substituent around them generates a large number of compounds. Many of the organic compounds used in biology, pharmacology, electronics, optics, material sciences etc. contain heterocyclic moieties. A large number of pharmaceuticals, agrochemicals, additives, cosmetics, dyes and plastics are heterocyclic in nature. Thus heterocyclic compounds and their derivatives are of vital interest to the pharmaceutical and agrochemical industries.

2.1 HETEROCYCLIC COMPOUNDS AS THERAPEUTICS

Heterocyclic rings act as bioisostere for a variety of functional groups in various drugs. Therefore, in medicine many heterocycle containing chemical compounds with diverse structures are used. These drugs are used for the treatment or prevention of various diseases. They interact with biological molecules causing a physiological effect. Many natural products such as **atropine, morphine, codeine, papaverine, theobromine, theophylline, emetine** and **taxol** contain heterocyclic rings. Some of the heterocyclic compounds such as morphine and quinine were present as active ingredients in many natural remedies even before the development of modern chemistry.

Atropine

R = H, Morphine
R = CH₃, Codeine

Papaverine

Theobromine

Theophylline

Emetine

Taxol

Heterocylic nucleus is present in various drug classes such as antimicrobial, antiviral, antimalarial, antihistaminic, antitubercular, anticancer, anti-inflammatory, analgesic, antipsychotic, antiepileptic, antineoplastic, antihypertensive, local anaesthetic, antianxiety, antidepressant, antioxidant, anti-Parkinson's, antidiabetic, antiplatelet, antiobesity, antischizophrenic etc. Some drugs containing heterocyclic ring system are **cilostazol** (antiplatelet), **nitazoxanide** (antidiarrhoeal), **captopril** (antihypertensive), **atorvastatin** (reduce cholesterol), **amlodipine** (antihypertensive), **omeprazole** (proton pump inhibitor), **quetiapine** (antischizophrenic), **clopidogrel** (antiplatelet), **risperidone** (antischizophrenic), **olanzapine** (antischizophrenic), **chlorpromazine** (antipsychotic), **diazepam** (antianxiety) and **pioglitazone** (antidiabetic).

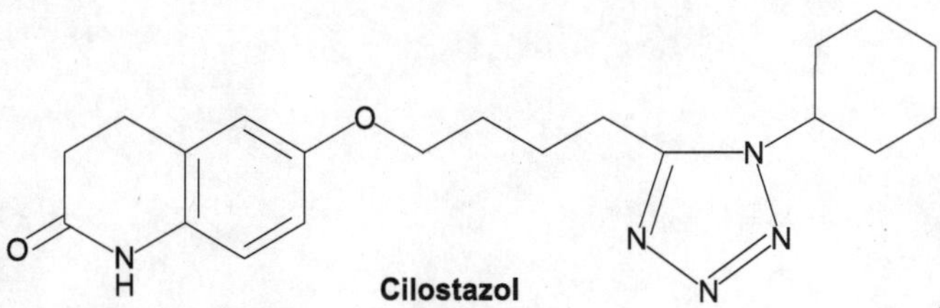

Cilostazol

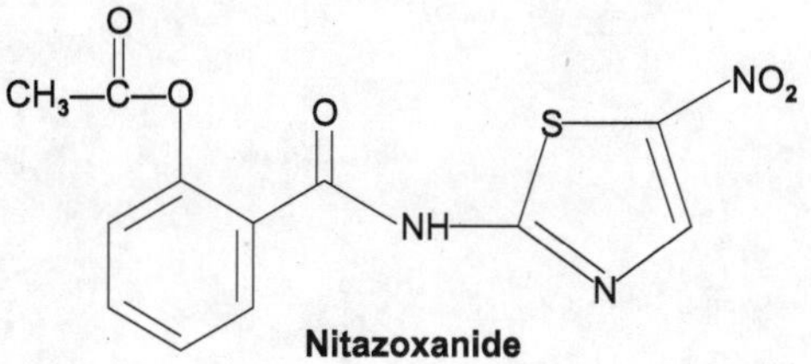

Nitazoxanide

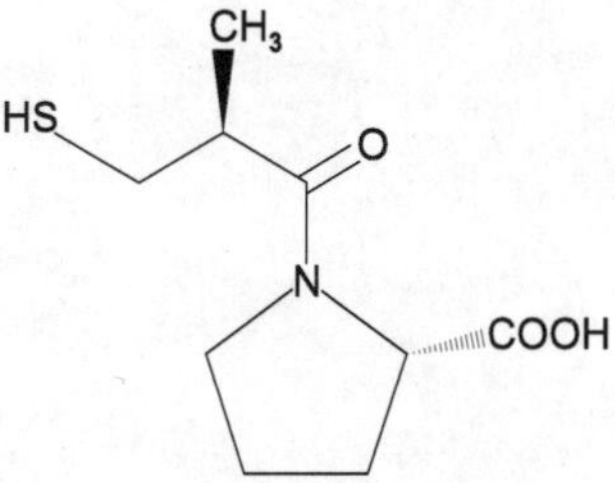

Captopril

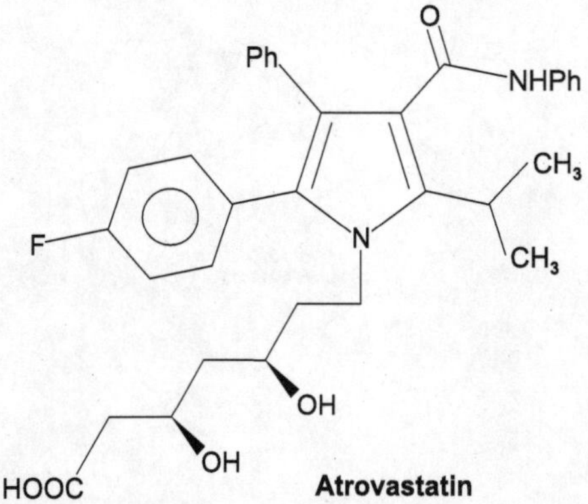

Atrovastatin

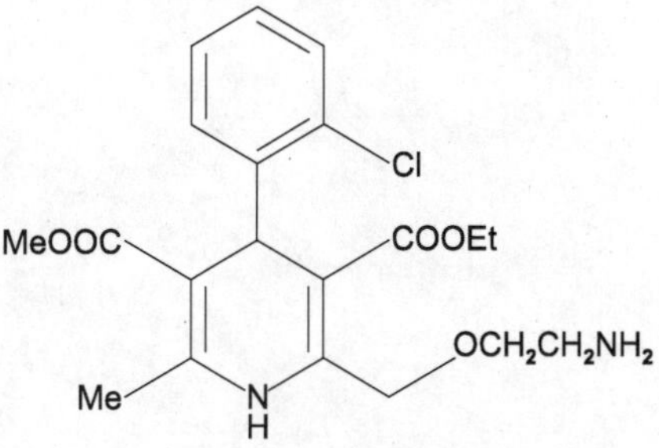

Amlodipine

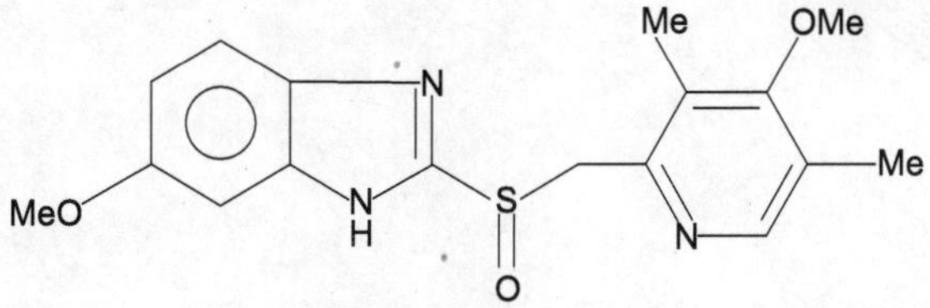

Omeprazole

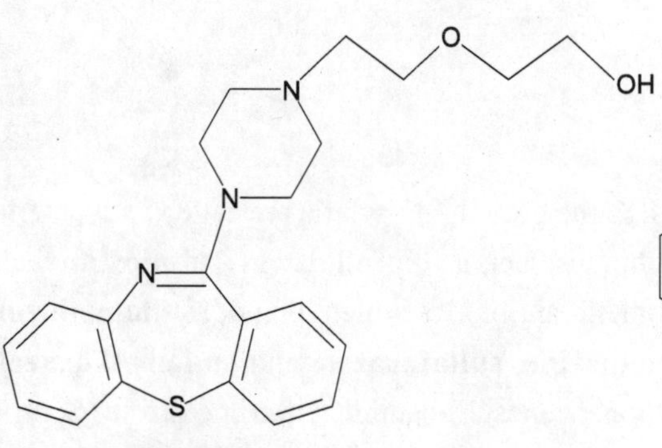

Quetiapine

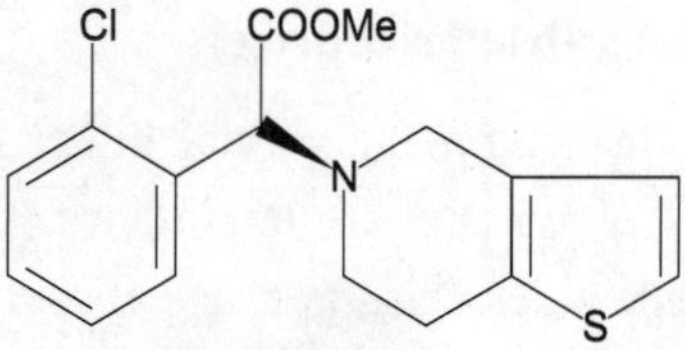

Clopidogrel

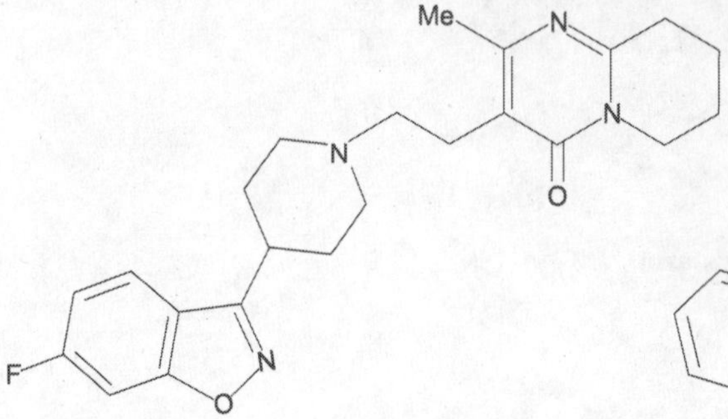

Risperidone

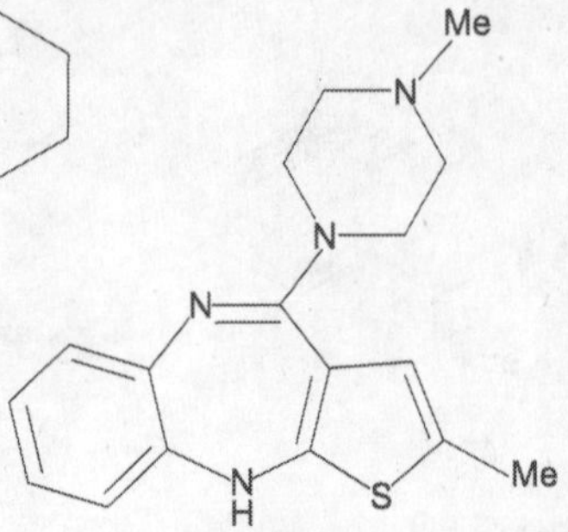

Olanzapine

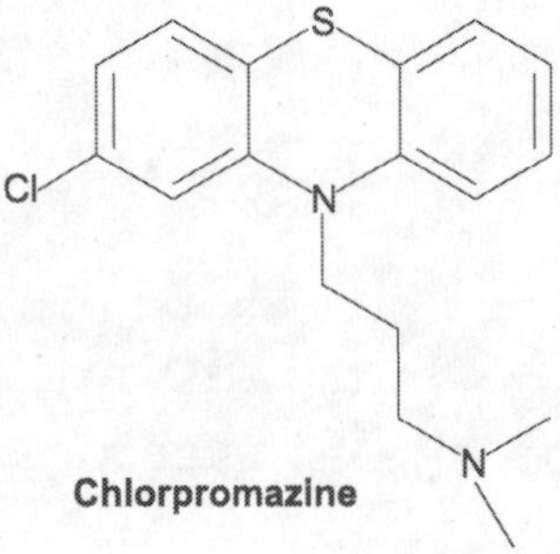

Chlorpromazine

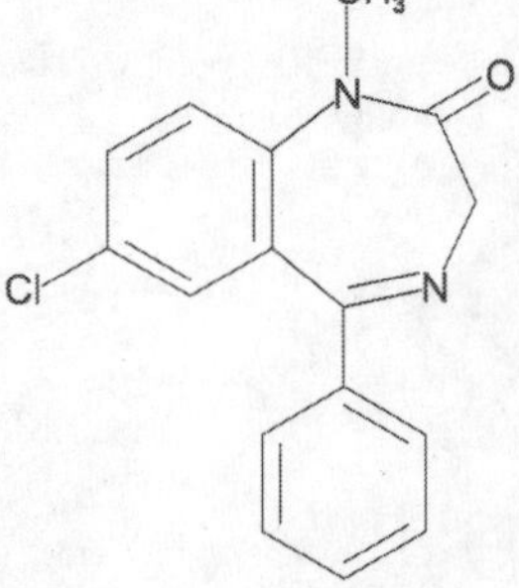

Diazepam

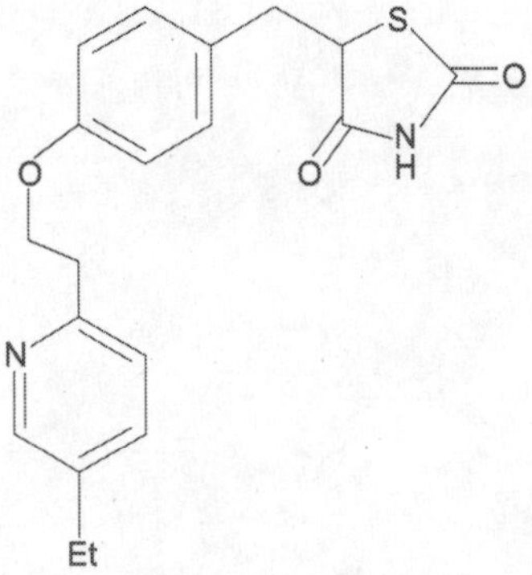

Pioglitazone

2.1.1 Antibacterial drugs

Chemical compounds which kill or stop the growth of bacteria are called antibacterial. Many quinolones are important antibiotics such as **ciprofloxacin** and **moxifloxacin.** Sulfonamides or sulfa drugs are synthetic antibiotics which include, **sulfamethizole, sulfadiazine, sulfathiazole, sulfamethazine, sulfafurazole** and **sulfamethoxazole.** They all are derivatives of para-aminobenzenesulfonamide. Co-trimoxazole (Septrin) is a combination of **sulfamethoxazole** with **trimethoprim**. Co-trimoxazole is used for the treatment of pneumonia, urinary tract infection, bronchitis etc. The β-lactam antibiotics such as **penicillins** and **cephalosporins** contain saturated heterocyclic rings. **Isoniazid** is a pyridine based heterocyclic compound which is used for the treatment of tuberculosis in combination with other drugs under the DOTS (directly observed treatment short course) programme.

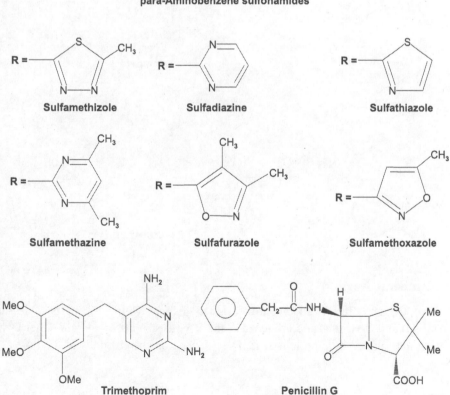

Ciprofloxacin

Moxifloxacin

para-Aminobenzene sulfonamides

Sulfamethizole

Sulfadiazine

Sulfathiazole

Sulfamethazine

Sulfafurazole

Sulfamethoxazole

Trimethoprim

Penicillin G

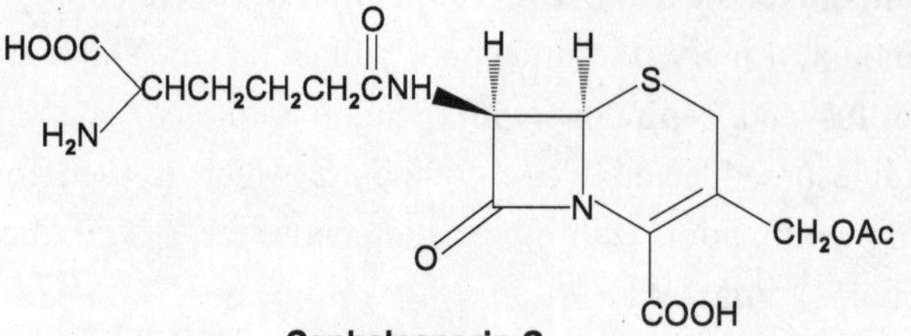

Cephalosporin C

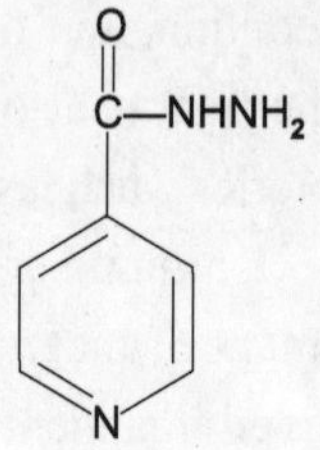

Isoniazid

2.1.2 Antifungal drugs

An antifungal drug is a medication used to treat fungal infections such as athlete's foot, candidiasis, ringworm, cryptococcal meningitis. The most common heterocyclic antifungal agents are triazoles such as **fluconazole, isavuconazole, voriconazole** and **ravuconazole**.

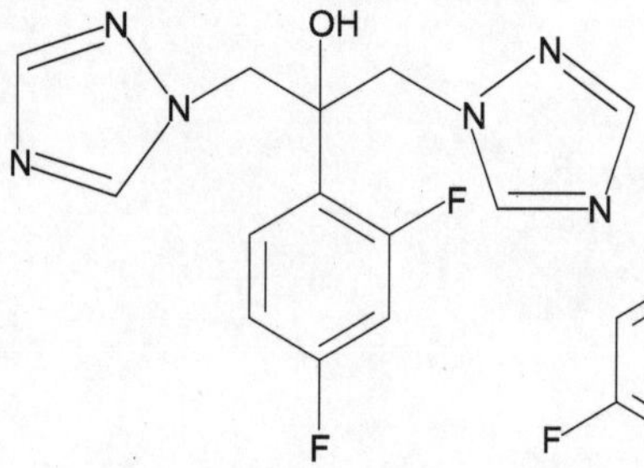

Fluconazole

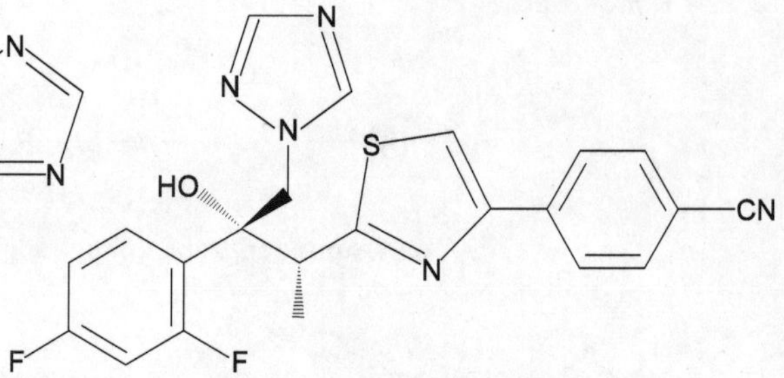

Isavuconazole

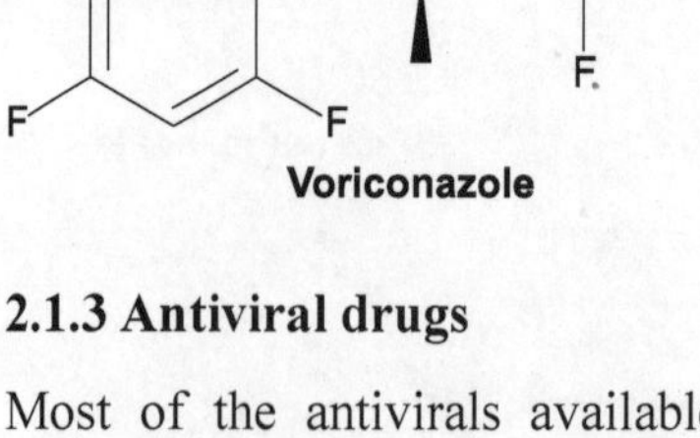

Voriconazole

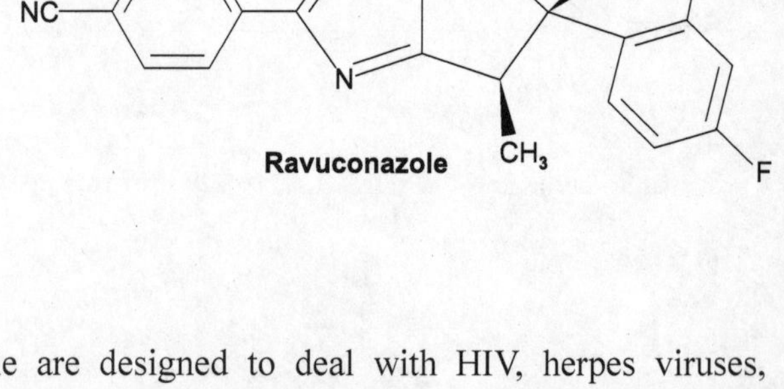

Ravuconazole

2.1.3 Antiviral drugs

Most of the antivirals available are designed to deal with HIV, herpes viruses, hepatitis B and C viruses, and influenza A and B viruses. Viruses contain either DNA or RNA surrounded by a protein coat. They depend on the enzymes of the host cell to complete their life cycle. Most of the antiviral drugs are modified nucleosides. These drugs interfere with DNA or RNA and thus prevent replication of the virus. In 1983, HIV was discovered to be the cause of AIDS. Since then it has claimed the life of 20 millions people worldwide. The first drug licensed for the treatment of HIV infection was **AZT (Azidothymidine or Zidovudine)**, a heterocyclic compound first synthesized at the Detroit Institute of Cancer Research in 1964 as a potential anticancer agent. To date deoxynucleosides, such as **AZT** (3'-azido-3'-deoxythymidine), **ddC** (2',3'-dideoxycytidine) and **ddI** (2',3'-dideoxyinosine) are the drugs for the treatment of acquired immune deficiency syndrome (AIDS). **Abacavir**, the guanosine analogue was also approved in 1998 as Nucleoside Reverse Transcriptase Inhibitor to be used in combination with other drugs for the treatment of HIV and AIDS. **Lamivudine** is an effective anti-AIDS drug when used in combination with zidovudine. The oral formulation of **acyclovir** is effective against both first and second-degree recurrence-genital herpes with minimal side effects. The antiherpes drug **iodoxuridine** and antiviral compound **ribavirin** generally work by disrupting the synthesis of viral DNA.

3'-Azido-3'-deoxythymidine
or Zidovudine (AZT)

2',3'-Dideoxycytidine (ddc)
or Zalcitabine

2',3'-Dideoxyinosine (ddI)
or Danidosine

Abacavir

Lamivudine (3-TC)

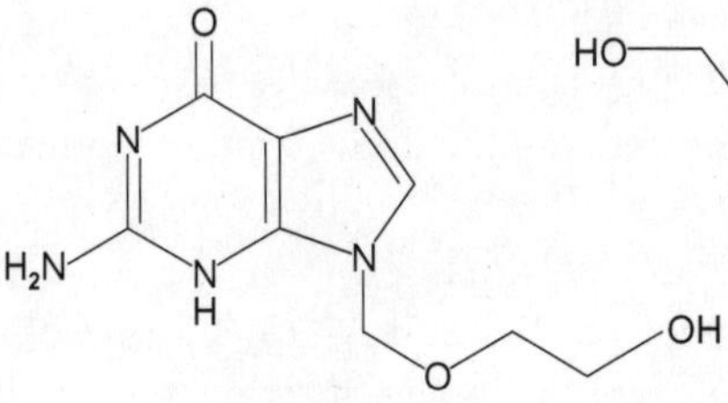

Acyclovir 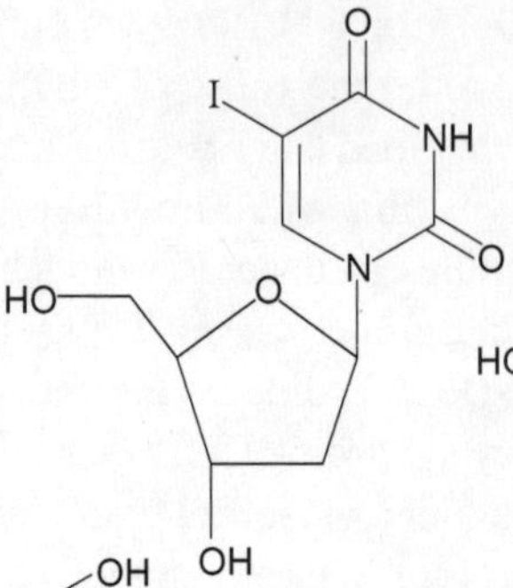   **Idoxuridine** 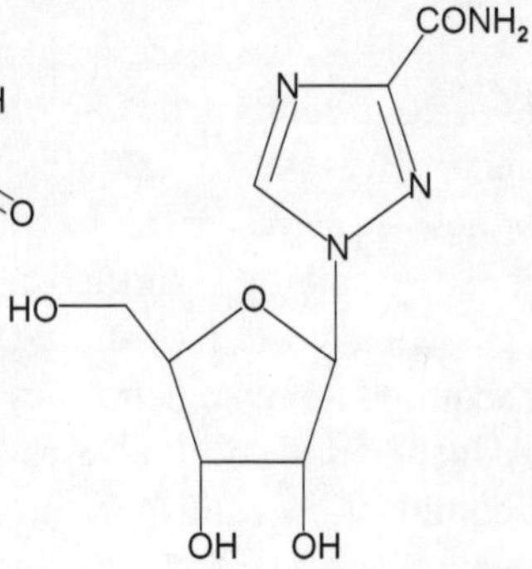   **Ribavirin**

There are some other heterocyclic compounds which are non-nucleoside based anti-HIV agents such as **delavirdine** a reverse transcriptase inhibitor and **saquinavir** a protease inhibitor which act by binding close to the active site of an enzyme, altering its conformation and thereby deactivating it.

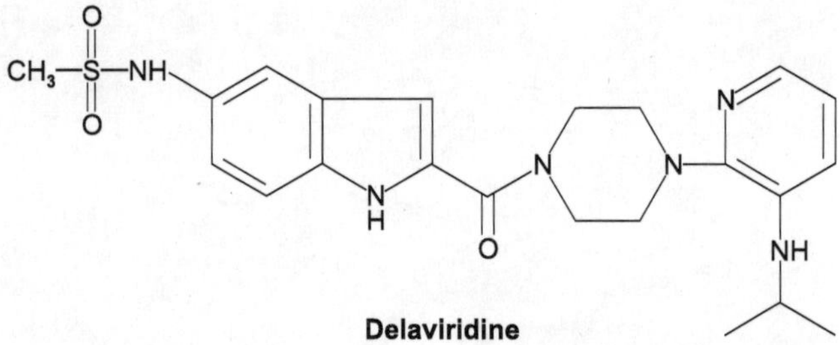

Delaviridine

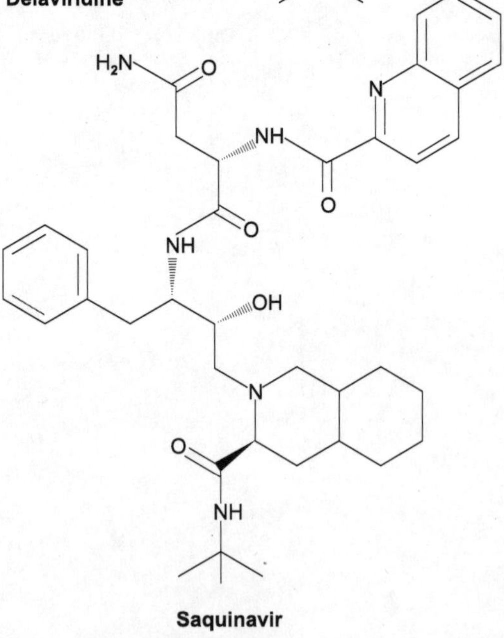

Saquinavir

2.1.4 Antiparasitic drugs

Malaria is one of the most devastating parasitic diseases in the developing countries, caused by protozoan parasite of the genus *Plasmodium*. **Quinine**, a natural product isolated from cinchona bark was used as an antimalarial agent for a long period of time. The most important antimalarial drugs are modified quinoline derivatives. **Chloroquine**, the main drug among the 4-aminoquinoline class, is one of the most successful antimalarial agents ever produced. **Primaquine** is also a drug of the 8-aminoquinoline class with free amino group that is connected to amino acids by forming peptide bond.

Quinine

Chloroquine

Primaquine

Metronidazole (Flagyl) having imidazole ring is used for the treatment of anti-infectious diseases caused by protozoa such as *Trichomonas vaginalis*, *Entamoeba histolytica*, *Giardia intestinalis*, and infections caused by Gram-negative anaerobes such as *bacteroides* and Gram-positive anaerobes such as *clostridia*.

Metronidazole

Benzimidazole, pyrazine, isoquinoline, tetrahydropyrimidine, tetrahydroquinolone, piperidine, piperazine, triazoles, indole, isoxazole derivatives are the different types of heterocyclic compounds used as anthelmintics drugs. **Albendazole** and **mebendazole** (Pripsen) are the most active benzamidazole drugs used for the treatment of a variety of parasitic worm infestations.

Albendazole

Mebendazole (Pripsen)

2.1.5 Anticancer drugs

Cancer may be caused due to environmental toxins, genetic problems, radiation, viruses, obesity, excessive exposure to sunlight, intake of excess alcohol and exposure to toxic chemicals. All forms of cancers together account for approximately 13% of all deaths each year, the most common being: lung, stomach, colorectal, liver and breast cancer. Cancer is characterized by uncontrolled growth of cells. Anticancer drugs generally act by disrupting the growth of cells and therefore oppose the excessive and abnormal growth. **Temozolomide** is an anticancer agent which resembles nucleic acid base, it degrades *in vivo* into diazomethane (an alkylating agent). It is used for glioma, a type of tumor that starts in the brain or spine and melanoma, a form of skin cancer. **6-Mercaptopurine** and **5-azacytidine** are antileukemics. **5-Fluorouracil** (5-FU) is a cytotoxic agent used to treat solid tumors, such as those of the breast and colon. It is also very useful for the treatment of certain skin cancers. **Gemcitabine** is used against breast and lung cancer. **Methotrexate** is used to treat certain types of cancer of the skin, head, breast, neck or lung. It is also used to treat severe psoriasis and rheumatoid arthritis.

Temozolomide **6-Mercaptopurine** **5-Azacytidine** **5-Fluorouracil**

Gemcitabine **Methotrexate**

Over 62% of the anticancer drugs approved during 1983 to 1994 belong to natural product or natural product analogues. Some examples include **vincristine** (*Catharanthus roseus*), **camptothecin** (*Camptotheca acuminata*), **paclitaxel** or **taxol** (*Taxus baccata, T. brevifolia, T. canadensis*). Vincristine is a naturally occurring complex indole alkaloid which binds to tubulin, a protein essential to cell division. Vincristine contains azonane, indole, piperidine and pyrrolidine heterocycles in its core structure.

Vincristine

Camptothecin

2.1.6 Neurotransmitters

The conduction of signals along nerves is dependent on the release of neurotransmitters at synapses. The neurotransmitters such as **5-hydroxytryptamine** and **histamine** are heterocyclic compounds. It has been discovered that these H1-histamines are actually inverse agonists at the histamine H1-receptor and are used to treat urticaria, anaphylaxis, asthma and allergic rhinitis.

**5-Hydroxytryptamine
(Serotonin)**

Histamine

The heterocyclic compounds most used as histamine antagonist are phenothiazine derivatives such as **promethazine, methdilazine** and **mequitazine**. Promethazine is used as a sedative and for the treatment of motion sickness.

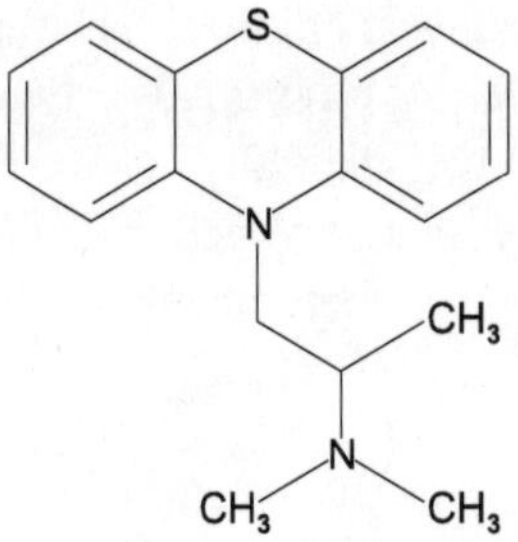

Promethazine 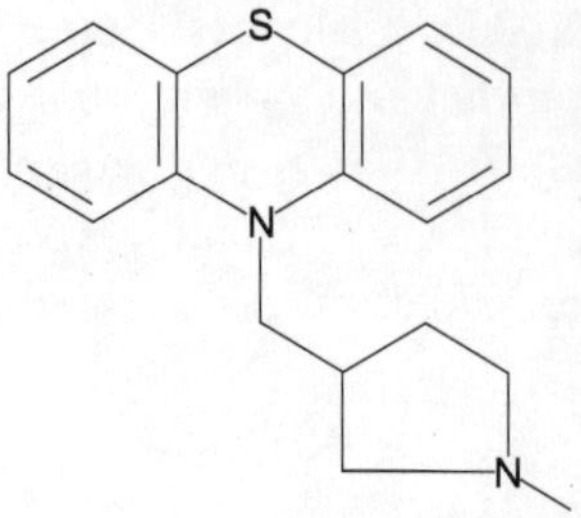   **Methdilazine** 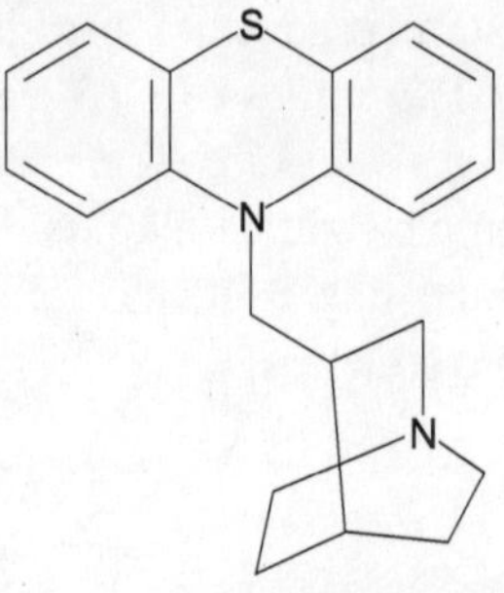   **Mequitazine**

The H2 receptor is involved in release of acid into the stomach. Although it is essential for digestion however, excessive amounts of acid can lead to indigestion or in more serious cases leads to peptic ulcers. **Ranitidine**, **cimetidine** and **omeprazole** are H2 antagonist that inhibit the release of gastric acid and is very effective in treating ulcers. Omeprazole launched by AstraZenca was the first compound to act as a proton pump inhibitor. Ranitidine contains furan ring, cimetidine has imidazole and omeprazole has benzimidazole and pyridine rings in their core structure.

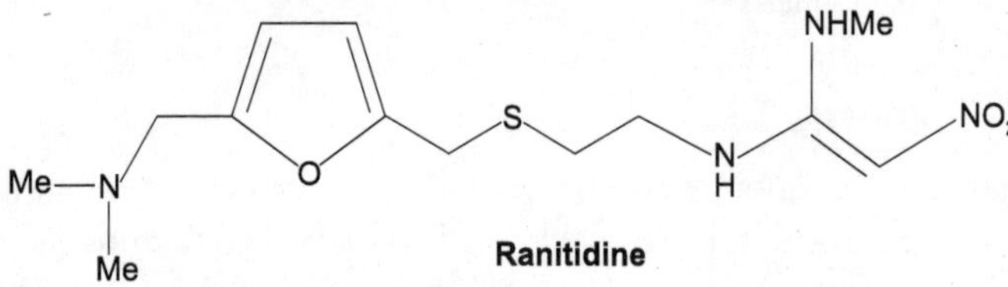

Ranitidine

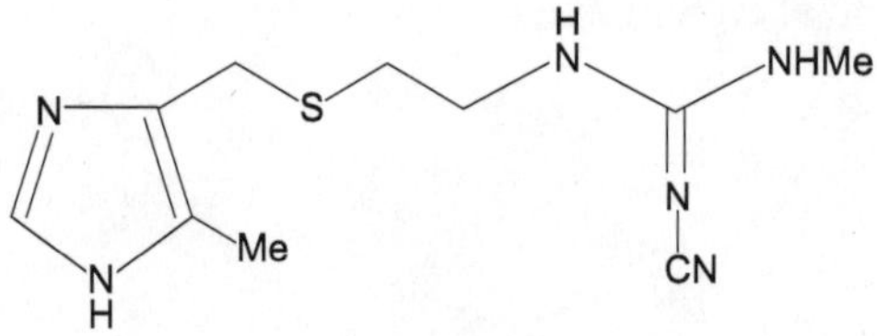

Cimetidine

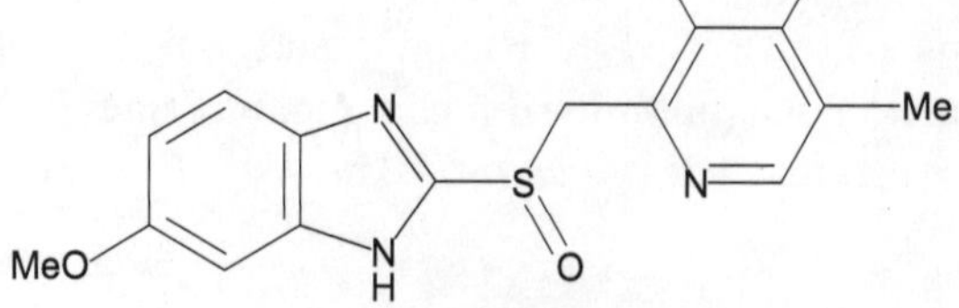

Omeprazole

2.1.7 Antipyretics and non-steroidal antiinflammatory drugs

Non-steroidal antiinflammatory drugs are the drugs with analgesic, antipyretic and antiinflammatory effects. Antipyrine (phenazone) was the first synthetic heterocyclic drug synthesized by Ludwig Knorr in 1883. It contains a pyrazolone ring and shows analgesic and antipyretic effect. The pyrazolone derivatives, **antipyrine, metamizole, aminophenazone** and **phenylbutazone** are non-steroidal antiinflammatory drugs (NSAIDs). **Celecoxib, etoricoxib** and **valdecoxib** are used in osteoarthritis, rheumatoid arthritis, acute pain and menstrual symptoms.

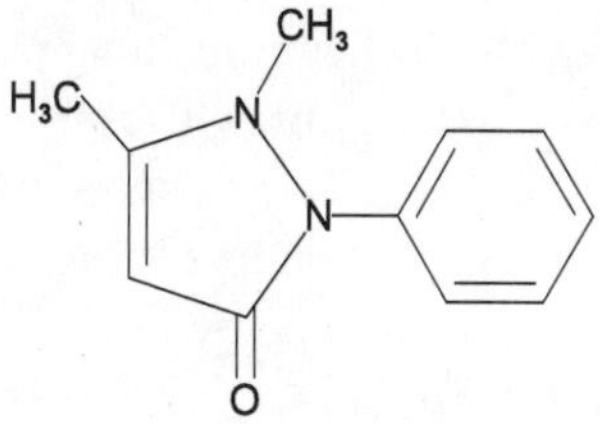

Antipyrine

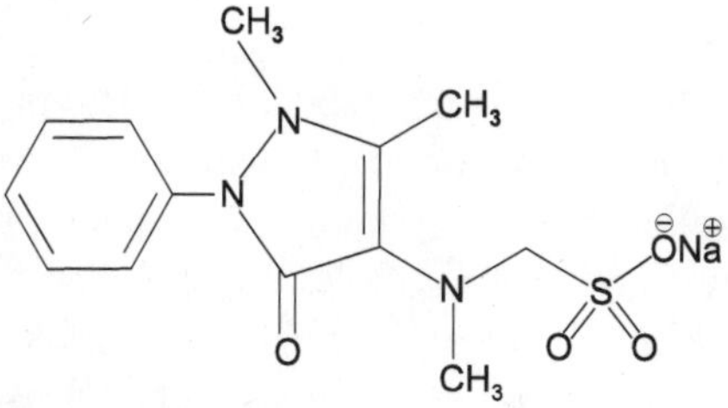

Metamizole

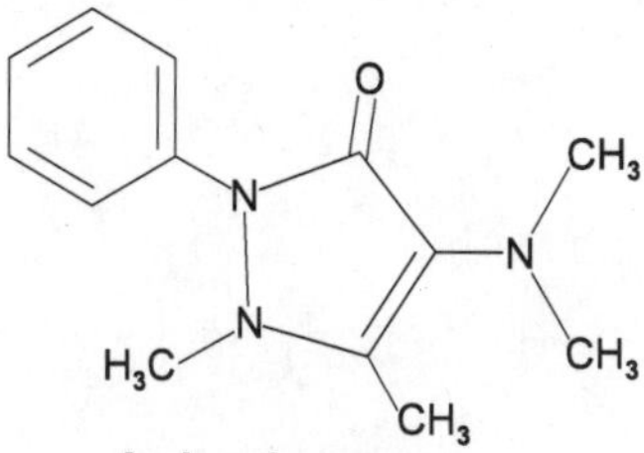

Aminophenazone

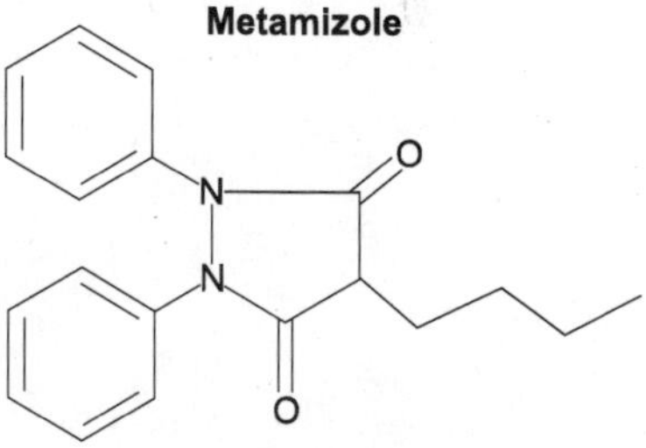

Phenylbutazone

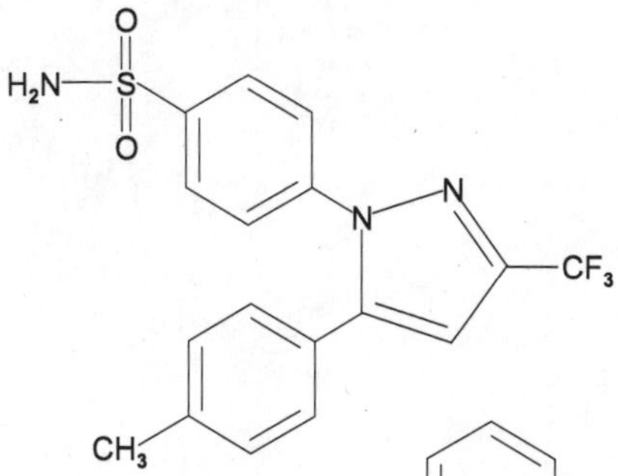

Celecoxib

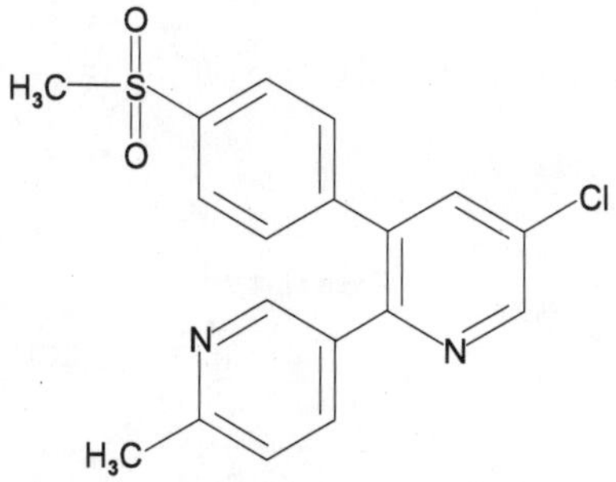

Etoricoxib

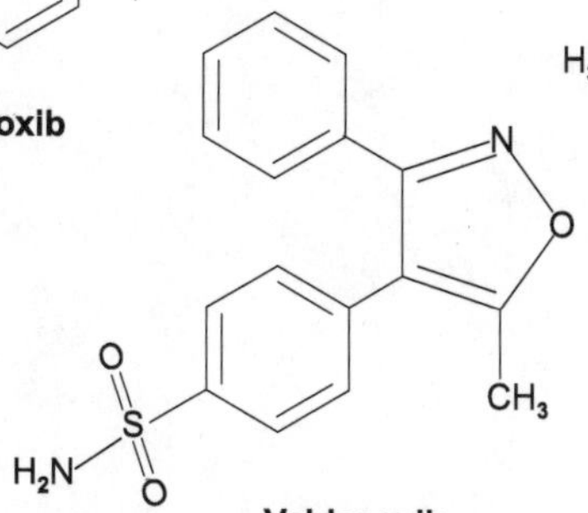

Valdecoxib

2.2 HETEROCYCLIC COMPOUNDS AS AGROCHEMICALS

Those chemical compounds which are used in agriculture for increasing crop production and crop protection are known as agrochemicals. A large number of agrochemicals are heterocyclic compounds. The agrochemicals used are pesticides, fertilizers and plant growth hormones. Pesticides are chemicals, which are used to kill or control pests such as insects, slugs, snails, mice, rats and weeds which can damage approximately 30% of the crop. Many of the agrochemicals used contain either a five membered or a six membered heterocyclic moiety.

2.2.1 Fungicides

Fungi can damage crops leading to loss of yield and quality. Therefore, fungicides are used to inhibit or kill fungal spores. **Iprodione, benomyl, fuberidazole and vinclozolin** are the common fungicides containing five membered heterocycles. Benomyl is a benzimidazole fungicide and it is also selectively toxic to earthworms. Iprodione and vinclozolin are dicarboximide fungicide. Vinclozolin is used to control blights, rots and molds. Triazoles such as **propiconazole, triadimefon, cyproconazole** and **paclobutrazol** are used as fungicides. Paclobutrazol is also plant growth retardant, agonist of the plant hormone gibberllin.

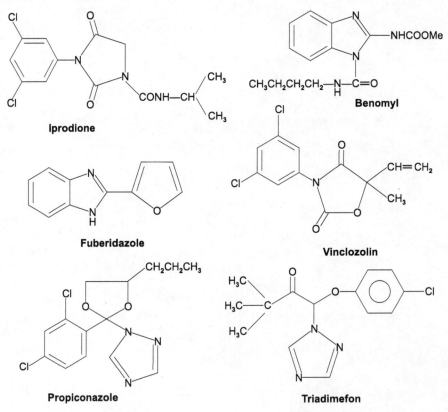

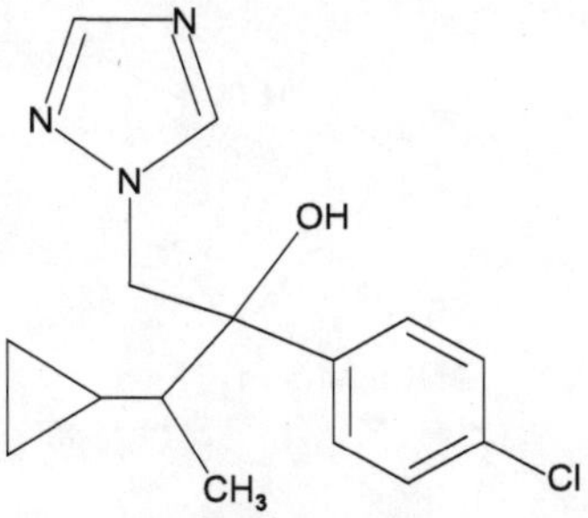

Cyproconazole

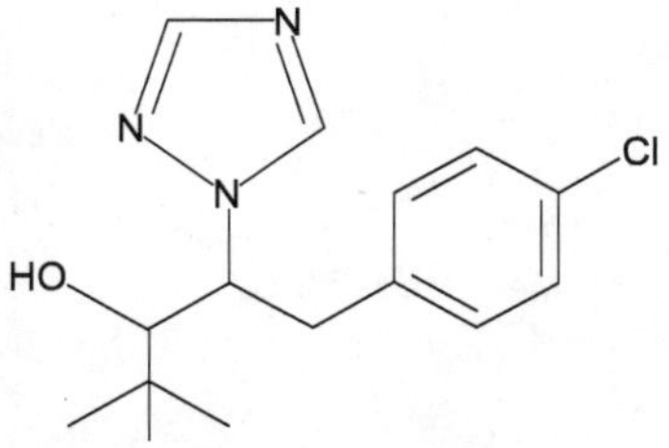

Paclobutrazol

Dimethirimol and **ethirimol** are pyrimidine containing heterocyclic fungicides, which act through interference with the metabolism of adenine in the plant.

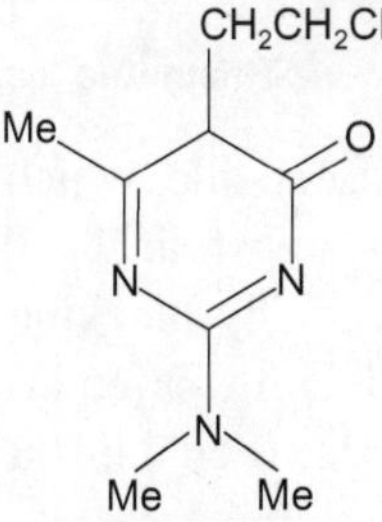

Dimethirimol

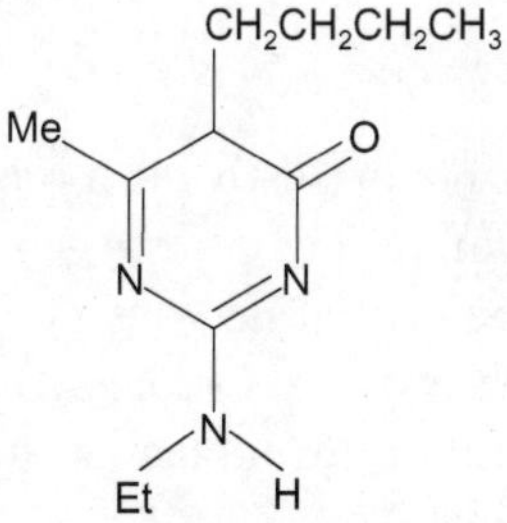

Ethirimol

2.2.2 Insecticides

Chemicals used to kill or control insects are called insecticides. **Isolan, pyrazophos, karphos, phosalone, methabenzthiazuron, methidathion** and **triazophos** are commonly used insectides. Methidathion is an organophosphate insecticide. **Thiamethoxam** is a thiazole ring containing insecticide which acts as agonists of the nicotinic acetylcholine receptor. It is especially used in the protection of tomato crops. Thiamethoxam is effective against aphids, thrips, beetles, centipedes, millipedes, sawflies, leaf miners, stem borers and termites.

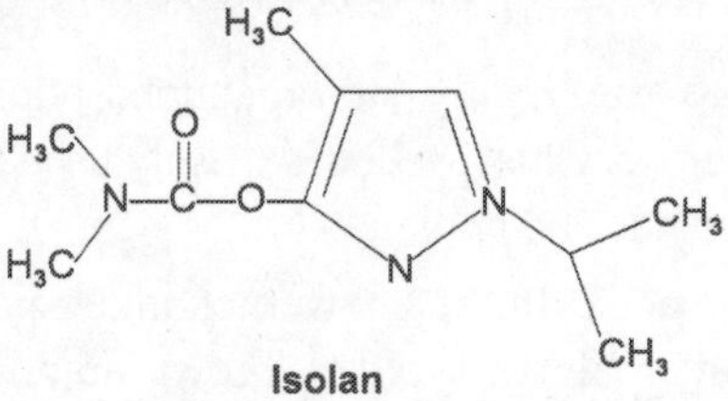

Isolan

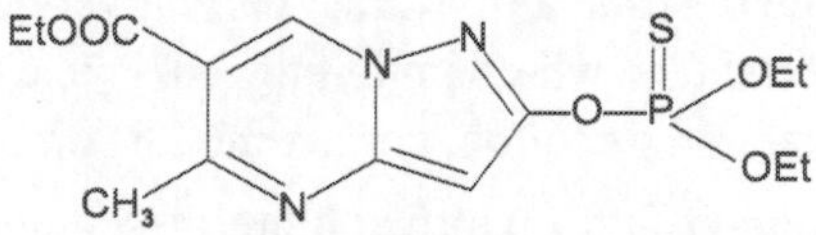

Pyrazophos

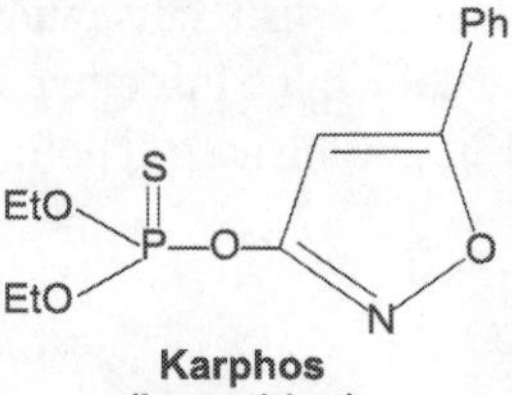

Karphos (Isoxathion)

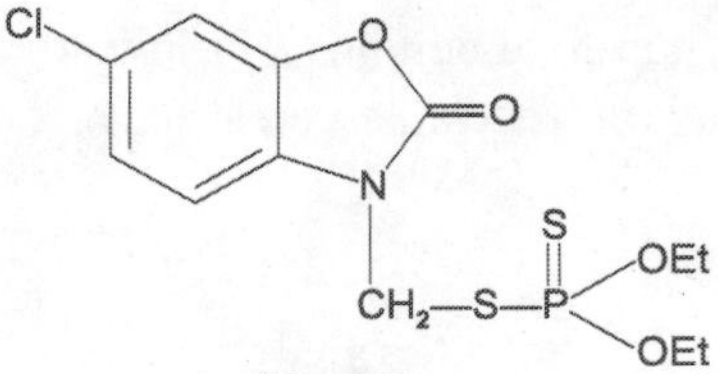

Phosalone

Methabenzthiazuron

Methidathion

Triazophos

Thiamethoxam

Fipronil a pyrazole derivative is a broad spectrum insecticide, which is effective for the elimination of various pests, such as wasps, bees, cockroaches, fleas, etc. It acts by disrupting the central nervous system of the insects, by blocking the passage of chloride ion in the system. **Chlorantraniliprole** is a diamide based insecticides which is commercially used for protection from various pests. It contains a pyrazole and a pyridine nucleus.

Fipronil

Chlorantraniliprole

Chlorpyrifos a pyridine derivative belongs to the family of organophosphorus insecticides, which inhibit the enzyme cholinesterase. Chlorpyrifos is mainly used for protection of cotton, corn, almonds, oranges, bananas and apples.

Diazinon and **pirimicarb** are insecticides that react with the enzyme cholinesterase that is involved in the nervous system. Diazinon a colorless to dark brown liquid, is a thiophosphoric acid ester used to control cockroaches, silverfish, ants, and fleas in residential, non-food buildings. Pirimicarb is a selective carbamate insecticide used to control aphids on vegetable, cereal and orchard crops by inhibiting acetylcholinesterase activity.

Chlorpyrifos   Pirimicarb   Diazinon

2.2.3 Herbicides

Herbicides are agrochemicals used to kill or control weeds or unwanted plants. **Oxadiazon** and **tebuthiuron** are the commonly used herbicides. Tebuthiuron is a nonselective broad spectrum herbicide of the urea class used to control weeds, woody and herbaceous plants and sugar cane. The herbicidal heterocycle **pyraclonil** with its two pyrazole rings is useful in the control of broadleaf weeds and grass in rice fields.

Tebuthiuron

Oxadiazon

Pyraclonil

The pyridine derivatives **picloram** is a herbicide used to remove broadleaf weeds. Picloram has growth promoting properties and thus leads to fast growth of plant using all the nutrients and thereby kills the weed. During the Vietnam War, a mixture of picloram and 2,4-dichlorophenoxyacetic acid (2,4-D) was used to make Agent White (Tordon 101) by U. S. forces to clear inland forests.

The bipyridyl salts **paraquat** and **diquat** are nonselective herbicides. They can interfere with the photosynthetic electron transport system of all green plants. Excessive exposure to paraquat may cause Parkinson's disease. The name is derived from the para positions of the quaternary nitrogens. **Fluridone** is a herbicide that is a gamma pyridone derivative.

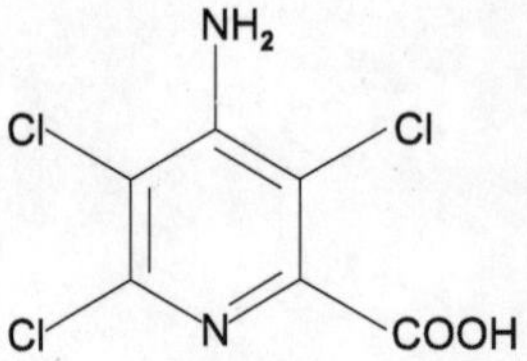

Picloram

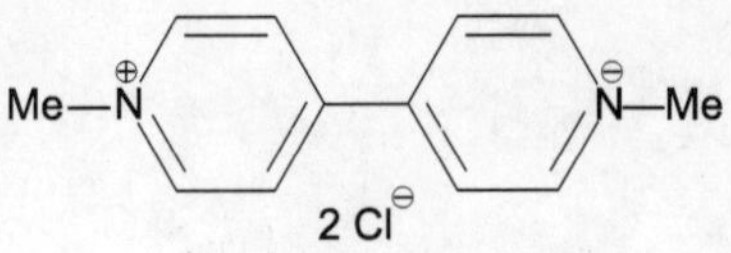

Paraquat

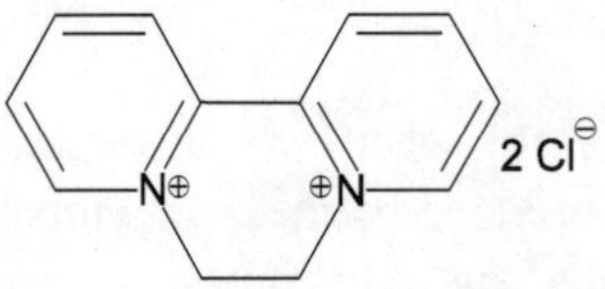

Diquat

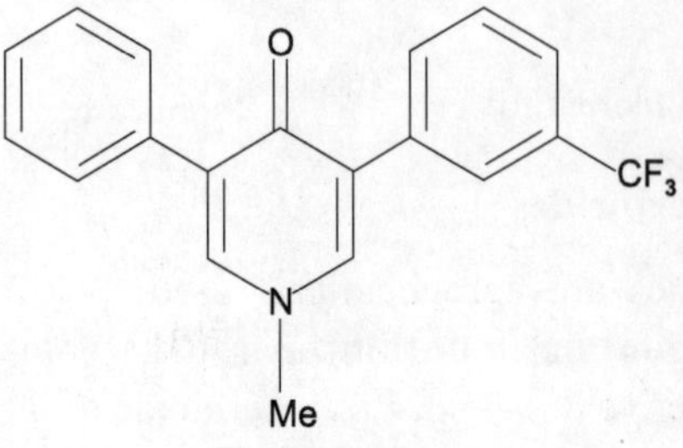

Fluridone

Bromacil and **lenacil** containing pyrimidine nucleus are used as total herbicides. Bromacil is a substituted uracil which is known as broad spectrum herbicide and used for nonselective weed and brush control on non-croplands. Bromacil is also found to be excellent at controlling perennial grasses.

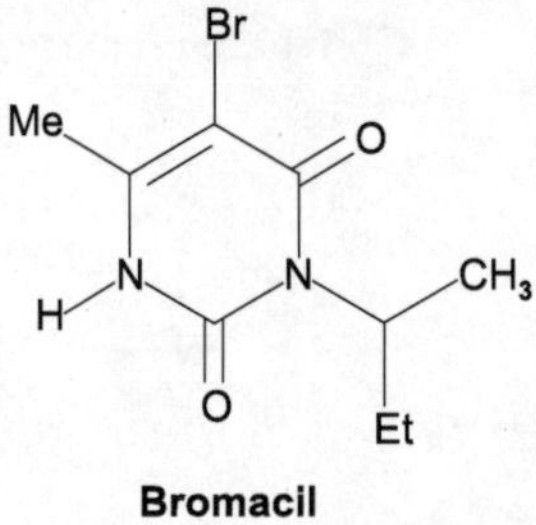

Bromacil

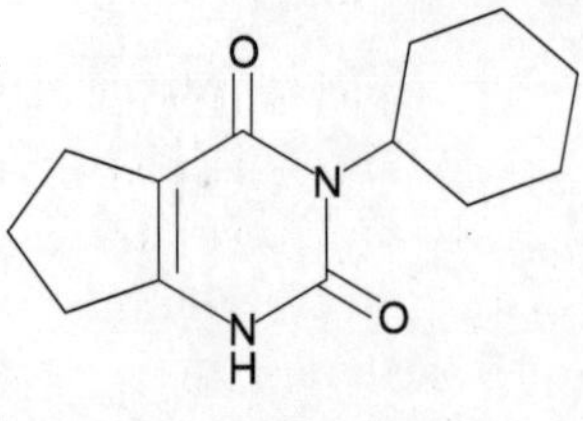

Lenacil

Pyridate and **maleic hydrazide** are herbicides containing pyridazine heterocyclic moiety.

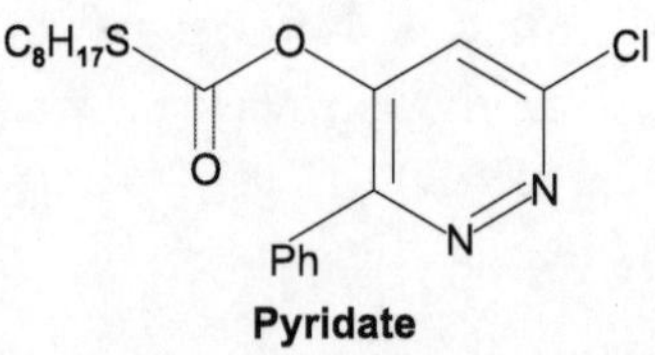

Pyridate

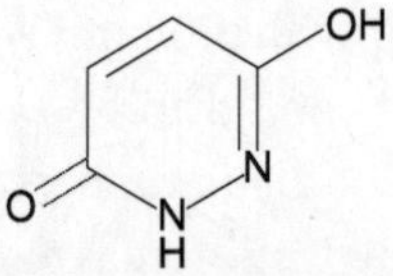

Maleic hydrazide

Atrazine, simazine, terbumeton and **hexazinone** contain 1,3,5-triazine nucleus and are used as herbicides. Atrazine is used to stop pre- and post-emergence broadleaf and grassy weeds in crops (maize and sugarcane). Simazine is used to control broad leaved weeds and annual grasses.

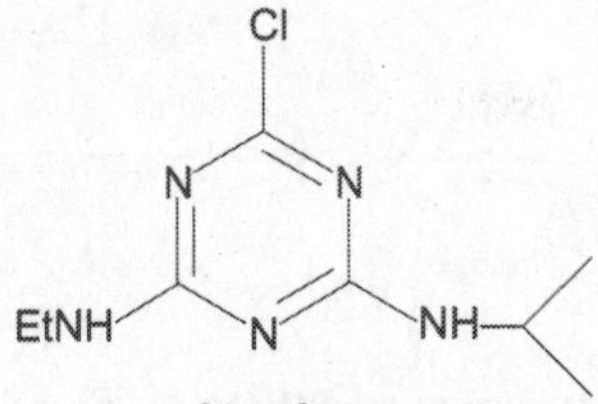

Atrazine

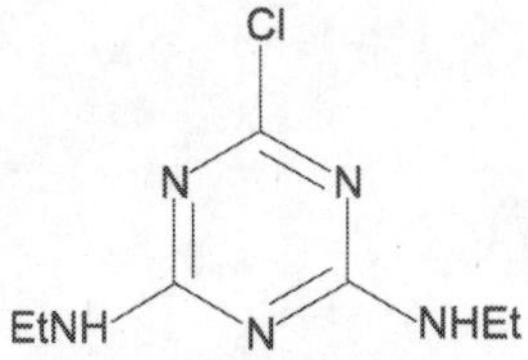

Simazine

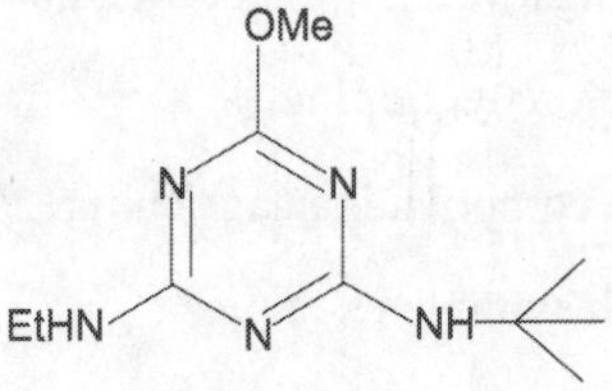

Terbumeton

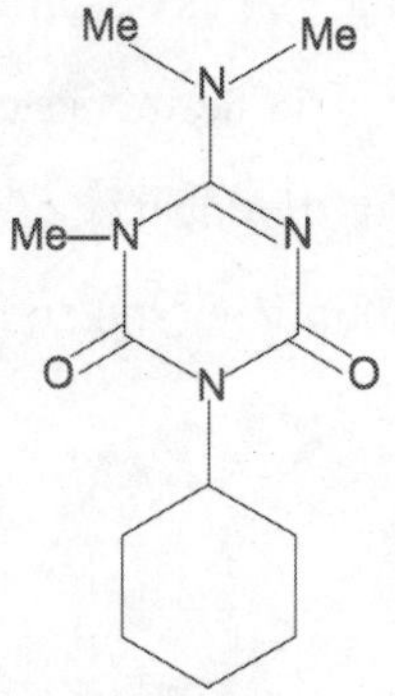

Hexazinone

2.2.4 Plant growth hormones

Chemical compounds which effect the plant growth are called plant growth hormones. Auxins and cytokinins class of plant growth hormones are heterocyclic compounds. Auxins such as **indole-3-acetic acid** stimulate root growth, affects cell elongation and fruit drop or retention. **Cytokinins** promote cell division and aging of leaves. Cytokinins are purine derivatives.

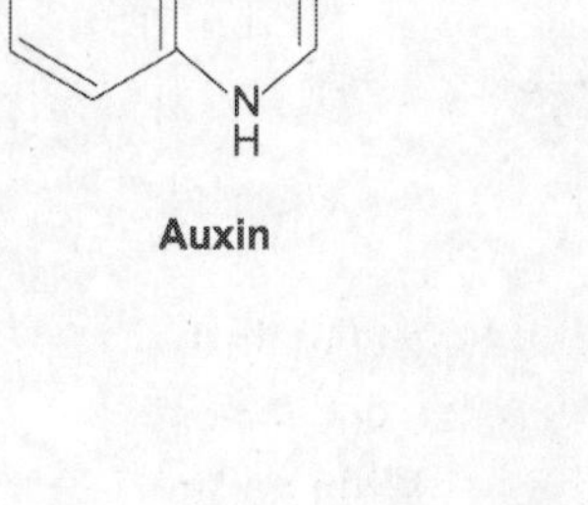

Auxin

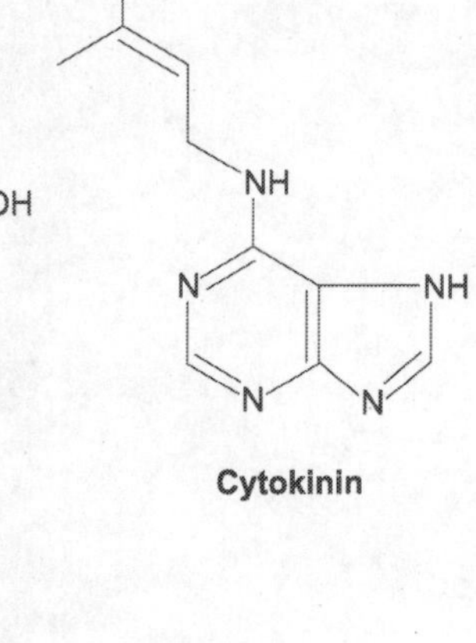

Cytokinin

EXERCISE

Q.1   Explain the significance of heterocyclic compounds as antibacterial agents.

Q.2   Mention the names of some antiviral drugs which are modified nucleosides.

Q.3   Name the plant hormones containing heterocyclic ring.

Q.4   What is the significance of heterocyclic compounds in agriculture.?

Q.5   Explain antibiotics and antipyretics with suitable examples.

Three Membered Heterocyclic Compounds with One Heteroatom

Aziridine, Oxirane, Thiirane and 1-Azirine (2*H*-Azirine)

3.1 AZIRIDINE, OXIRANE AND THIIRANE

3.1.1 Introduction

The three membered heterocyclic compounds with one heteroatom are formally derived from cyclopropane by replacing one of the methylene groups with the corresponding heteroatom. The saturated three membered heterocycle with one nitrogen is **aziridine**, with one oxygen atom is **oxirane** and with one sulfur atom is **thiirane**. Aziridine is a water soluble colorless liquid (bp 56°C). It is toxic and causes irritation of eyes, skin and internal inflammation. Oxiranes (ethylene oxide), are also known as epoxides. Many of the oxiranes are carcinogenic. Oxirane itself is a colourless water soluble gas (bp 10.5°C). It is extremely poisonous. Oxirane is important as an intermediate in the petrochemical industry. Thiiranes are also known as episulfides. Thiirane itself is a volatile pale yellow liquid (bp 56°C).

Aziridine   **Oxirane**   **Thiirane**

3.1.2 Biological importance

Monocyclic aziridines and 2*H*-azirines have found broad application in the synthesis of complex natural products. The ease with which these small ring nitrogen-containing compounds can be converted to important pharmaceutical products under mild conditions and with wide functional group compatibility makes these molecules quite useful.

Aziridine and its derivatives are of importance in industry and biology. Certain antibiotics and anticancer agents possess aziridine ring. The **mitomycins A, B** and **C** are natural products isolated from *Streptomyces caespitosus* or *Streptomyces lavendulae* containing aziridine ring. **DAPP** or diethyl 1-(2-hydroxymethyl)aziridin-1-yl)-1-oxopropan-2-yl phosphonate has antibacterial activity against *Escherichia*

coli, Ataphylococus aureus, Pseudodomonas aeruginsosa and *Enterococcus faecalis.*
Azadirachtin contains tetra-substituted oxirane and it is present in neem seeds.

	Mitomycin A	Mitomycin B	Mitomycin C
R^1	OCH_3	OCH_3	NH_2
R^2	CH_3	H	CH_3
R^3	H	CH_3	H

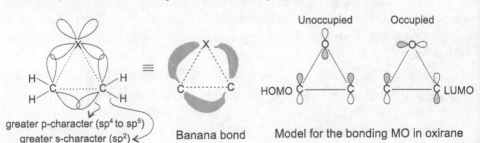

DAPP

Azadirachtin

3.1.3 Structure

The bonding in aziridine, oxirane and thiirane is similar to that in cyclopropane. Aziridine ring is planar and rigid. The three bonds within the ring are formed by the oblique overlapping of sp^3 hybrid orbitals resulting in the formation of bent or **banana bonds**. In a bent bond the hybridisation states of the orbitals involved in bond formation has either more or less s-character. Bent or banana bonds are found in cyclopropane, aziridine, oxirane and thiirane. The C-H bonds and the lone pair on heteroatom have more s-character than normal sp^3-hybridized atoms. The inner C-C or C-X bond orbitals have more p-character, than sp^3 hybrid orbitals.

greater p-character (sp^4 to sp^5)
greater s-character (sp^2)

Banana bond   Model for the bonding MO in oxirane

In oxirane HOMO of ethene interact with the unoccupied atomic orbital of oxygen atom, and the LUMO of ethene interacts with occupied atomic orbital of oxygen, to form the bonding molecular orbital.

The properties of three membered heterocycles are due to high bond angle strain (Baeyer strain). The ring strain makes these compounds highly reactive. The order of ring strain in saturated three membered ring compounds is cyclopropane > aziridine ~ oxirane > thiirane. Thiirane has the lowest ring strain and also lowest electron density at the heteroatom. The strain energies of cyclopropane, aziridine, oxirane, and thiirane amounts to 27.3, 26.2, 24.6 and 18.9 kcal/mol, respectively.

Microwave spectra as well as electron diffraction studies show that the oxirane ring is close to an equilateral triangle. However in thiirane due to greater atomic radius of the S-atom, the three atoms form an acute angled triangle. In spite of the smaller strain enthalpy, thiirane is thermally less stable than oxirane. It is evident from the fact that, even at room temperature, linear macromolecules are formed because of polymerization of ring-opened products. Substituted thiiranes are thermally more stable. The existence of ring strain in aziridine has been further confirmed by IR spectroscopy, this is reflected in increase in C-H vibrational frequency from 1465 cm^{-1} (for secondary amines) to 1475 cm^{-1} and decrease in N-H vibrational frequency to 1441 cm^{-1} which is lower than observed for secondary amines (1460 cm^{-1}).

Bond lengths and bond angles of aziridine are the same as those in oxirane.

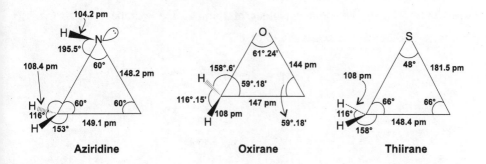

Aziridine　　　　**Oxirane**　　　　**Thiirane**

The plane in which the N-atom, its nonbonding electron pair and the N-H bond are situated is perpendicular to the plane of the aziridine ring. Therefore, 2-methylaziridine display diastereoisomerism. The trivalent N-atoms are, however, liable to pyramidal inversion. The activation enthalpy of this process is $\Delta G^{\ddagger} = 70$ kJ/mol (or 16.7 kcal/mol) and inversion occurs rapidly at room temperature. Therefore the diastereoisomers of 2-methylaziridine are not separable. However, in the case of 1-chloro-2-methylaziridine, where $\Delta G^{\ddagger} = 112$ kJ/mol (or 26.8 kcal/mol) the mixture of stereoisomers can be separated.

2-Methylaziridine

$\Delta G^{\ddagger}$ = 16.7 kcal/mol (inversion at 25°C)

trans   *cis*

$\Delta G^{\ddagger}$ = 26.8 kcal/mol (diastereomers isolable at 25°C)

(R and S) *cis*-1-Chloro-2-methylaziridine   (R and S) *trans*-1-Chloro-2-methylaziridine

Diastereoisomerism in substituted aziridines

3.1.4 Dipole moment

The dipole moment values for aziridine, oxirane and thiirane are 1.89, 1.88 and 1.84 D, respectively. The dipole moment values of thiirane and oxirane indicate that the polarity of C-O bond is larger than C-S bond.

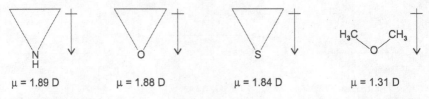

μ = 1.89 D μ = 1.88 D μ = 1.84 D μ = 1.31 D

3.1.5 Spectral data

The ^{1}H and ^{13}C NMR (given in brackets) chemical shift values for aziridine, oxirane and thiirane are shown below:

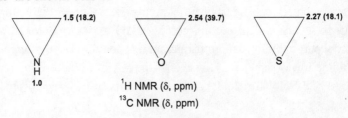

^{1}H NMR (δ, ppm)
^{13}C NMR (δ, ppm)

3.1.6 Basicity

Aziridine (pK$_a$ value of the aziridinium ion is 7.98) is a weakly basic compound than the open chain compound dimethylamine (pK$_a$ of conjugate acid is 10.87), but stronger than aniline (pK$_a$ of conjugate acid is 4.62). The reason for weak basicity for aziridine and its derivatives is the strain in the three membered ring. In case of aziridine the internal bond angle is smaller than typical secondary amines. The external bond angles are greater. Thus it has greater s-character than normal sp^3 hybrid orbital and consequently more electronegative, and less basic. As the lone pair is present in an orbital with greater s-character than normal sp^3 hybrid orbital.

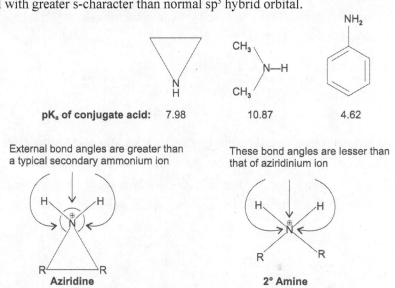

pK$_a$ of conjugate acid: 7.98   10.87   4.62

External bond angles are greater than a typical secondary ammonium ion

These bond angles are lesser than that of aziridinium ion

Aziridine   2° Amine

3.1.7 Methods of preparation of saturated three membered heterocycles

3.1.7.1 Methods of preparation of aziridine

(*i*) **Intramolecular cyclizations of 1,2-aminohalides:** The **Gabriel-Marckwald** synthesis of aziridines involves reaction of a 1,2-aminohalide with a base. In this reaction, 1,2-aminohalide undergoes an intramolecular cyclization through nucleophilic displacement of halogen by the nitrogen of the amino group. This reaction is stereospecific. For example, cyclization of (-)-erythro-1,2-diphenyl-2-bromoethylamine gives *trans*-2,3-diphenylaziridine.

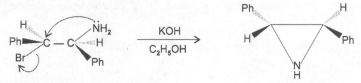

Erythro-1,2-diphenyl-2-bromoethylamine   ***trans*-2,3-Diphenylaziridine**

(*ii*) **Intramolecular cyclizations of 1,2-aminoalcohols:** 1,2-Aminoalcohols, which are conveniently made from oxiranes on reaction with ammonia or amines, react with thionyl chloride to give 1,2-chloroamines, which can be cyclized to aziridines by alkali hydroxide. The mechanism of the reaction involves S_N2 process and the reaction is stereospecific.

2-Amino ethanol

Sulfate esters, obtained from 1,2-aminoalcohols and sulfuric acid, when treated with alkali also form aziridines. This is known as **Wencker synthesis**.

*Wenker, H., J. Am. Chem. Soc. **1935**, 57, 2328.*

(*iii*) **Thermal or photochemical reaction of azides with alkenes:** Phenyl azide reacts with alkenes to give 4,5-dihydro-1,2,3-triazoles *via* 1,3-dipolar cycloaddition. Further photochemical decomposition or thermal decomposition results in the formation of aziridines through loss of nitrogen. This reaction is called **Scheiner aziridine** synthesis.

(*iv*) **Addition of nitrene to alkenes:** Addition of nitrene to alkenes leads to the formation of aziridines. For example, ethyl azidoformate gives ethoxycarbonyl nitrene thermally or photochemically, which undergoes [2+1] cycloaddition reaction with alkene to form corresponding aziridine.

Ethyl azidoformate   **Ethoxycarbonyl nitrene**

[2+1] Cycloaddition

(*v*) Addition of carbene to imines: This method involves the addition of a carbene to an imine, in which new C-N and C-C bonds are simultaneously formed.

(*vi*) From 1,2-iodoazides: Iodine azide (IN_3), obtained *in situ* by the reaction of iodine monochloride (ICl) and sodium azide, adds to alkenes to give 1,2-iodoazides. The addition occurs in anti-fashion. Reduction of 1,2-iodoazides with $LiAlH_4$ produces 1,2-aminoiodide, which displaces the neighbouring iodide through intramolecular S_N2 process to give aziridine. This reaction is known as **Hassner aziridine synthesis**.

Erythro-iodoazide   **Erythro**   *trans* (95%)

For example, 2,3-dimethylbutene forms 2,2,3,3-tetramethylaziridine by Hassner aziridine synthesis.

Iodoazide

$LiAlH_4$
Ether, 10h
$-N_2$

2,2,3,3-Tetramethylaziridine

(*vii*) From iodine isocyanate: Addition of iodine isocyanate (I-N=C=O) to alkene yields 1,2-iodoisocyanate of the alkene. This on treatment with CH_3OH yields carbamate, which cyclizes in the presence of a base to give aziridine derivative.

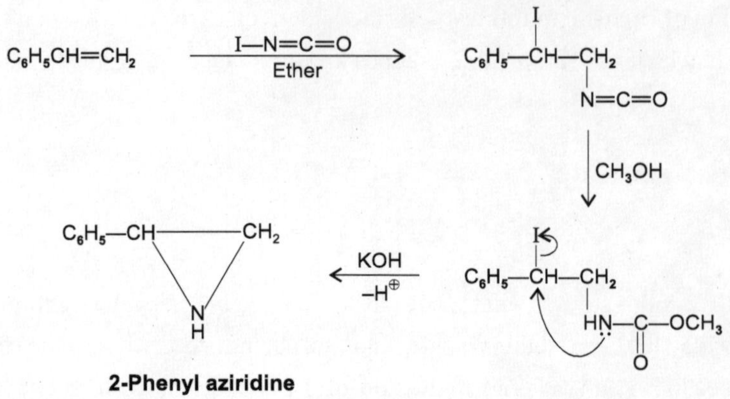

$$C_6H_5CH=CH_2 \xrightarrow[\text{Ether}]{I-N=C=O} C_6H_5-\overset{\overset{\displaystyle I}{|}}{C}H-CH_2$$

2-Phenyl aziridine

(*viii*) **From ketoxime:** Aziridine can be synthesized form ketoxime by reaction with Grignard reagent followed by hydrolysis. This is known as **Hoch-Campbell** Synthesis.

Hoch, J., Compt. Rend. **1934**, *198, 1865; Campbell, K.N.; Campbell, B.K.; McKenna, J.F.; Chaput, E.P., J.Org. Chem.* **1943**, *8, 103.*

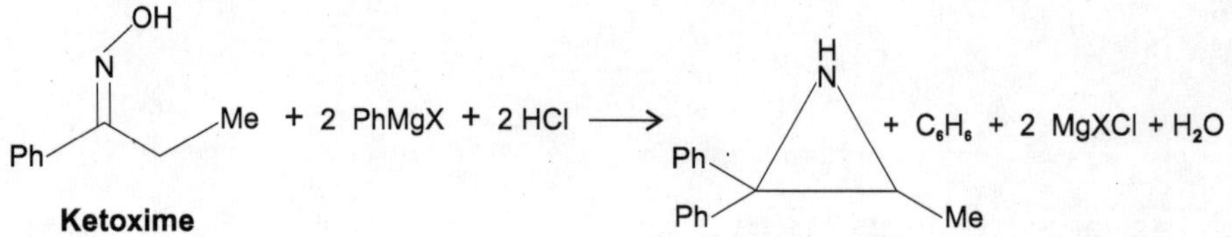

Ketoxime

Mechanism:

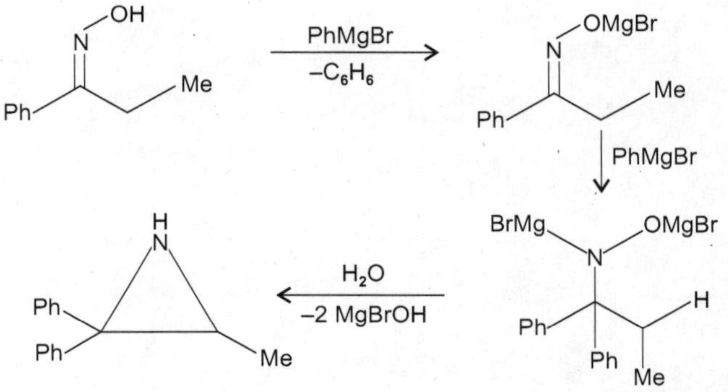

(*ix*) **From α-chloroimines:** Reaction of α-chloroimines with nucleophiles such as hydride, cyanide, or Grignard reagents leads to the formation of aziridines. This is known as **De Kimpe** azirdine synthesis.

De Kimpe, N.; Verhe, R.; De Buyck, L.; Schamp, N., Recl. Trav. Chim. Pays-Bas, **1977**, *96, 242-246.*

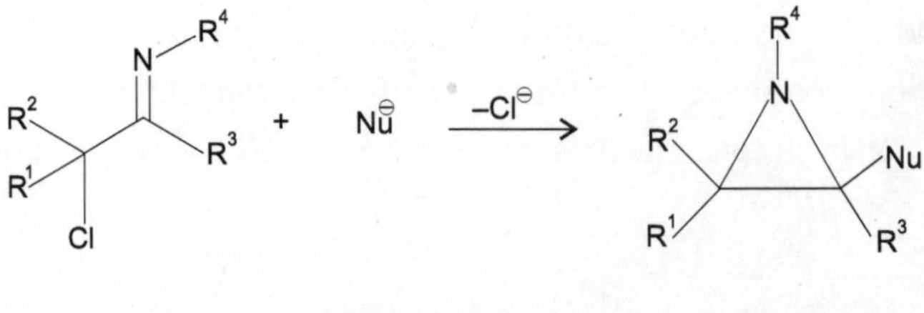

Mechanism: This reaction involves the nucleophilic addition to the imino carbon of α-chloroimines and subsequent intramolecular nucleophilic substitution. Both ketimines and aldimine can be used in this reaction. Normally, the intermediate anion cannot be isolated. However, if a group that stabilizes the anion such as tert-butanesulfinyl is present then aziridine is formed by applying the more forcing conditions, for example, with additional base under reflux conditions.

3.1.7.2 Methods of preparation of oxirane

(*i*) **Cyclodehydrohalogenation of β-halo alcohols:** The first synthetic method for the preparation of oxirane was given by **Wurtz** in the year 1859. The 2-chloroethanol on reaction with sodium hydroxide gives oxirane.

Oxirane

trans-Chlorohydrin prepared by the addition of hypochlorous acid to an alkene undergoes S_N2 intramolecular substitution in the presence of a base to give the oxirane ring. Base deprotonates β-halo alcohols to give the intermediate alkoxide anion. This is followed by an intramolecular displacement of the halogen atom in the rate-determining step. In spite of the ring strain in the product and the considerable activation enthalpy, the reaction occurs rapidly at room temperature owing to favorable entropy.

***trans*-Chlorohydrin**

(*ii*) **Prilezhaev reaction:** Oxiranes are often prepared by the direct oxidation of alkene with peroxy acids such as peroxybenzoic acid, meta-chloroperoxybenzoic acid or monoperoxyphthalic acid. This reaction is known as **Prilezhaev (Prileschajew) reaction.** For example, styrene oxide is prepared by peroxidation of styrene.

Styrene → **Styrene oxide**

Mechanism: In weakly polar solvents, the reaction occurs in a concerted manner, stereospecifically. Thus, *cis*-alkenes yield *cis*-oxiranes while *trans*-alkenes yield *trans*-oxiranes.

For example, *trans*-stilbene on reaction with meta-chloroperbenzoic acid (m-CPBA) gives *trans*-stilbene oxide.

trans-Stilbene   **trans-Stilbene oxide**

(*iii*) **Darzens reaction (glycidic ester synthesis): Darzen** in 1904 gave a method for the synthesis of 2-(ethoxycarbonyl)oxiranes also known as glycidic esters by reaction of α-halo esters with carbonyl compounds in the presence of sodium ethoxide. First the α-halo ester is deprotonated by a base to the corresponding carbanion. This acts as a nucleophile and adds to the carbonyl compound in a rate-determining step. In the final step halogen atom is substituted intramolecularly.

(*iv*) **Corey synthesis:** S-ylide nucleophile (dimethylsulfoxonium methylide) obtained from trimethylsulfoxonium iodide reacts with carbonyl compounds to form oxiranes. This method was given by **Corey** in 1962.

Dimethylsulfoxide (DMSO)

Trimethylsulfoxonium iodide

NaH
$-$NaI, $-H_2$

S-ylide (Dimethylsulfoxonium methylide)

(*v*) **Oxidation of ethylene:** A very important industrial method for the preparation of oxirane is oxidation of ethylene by air over a silver catalyst at high temperature. The disadvantage of this method is that only 50% ethylene oxide is obtained. The propylene and isobutylene give only CO_2 and H_2O.

$$CH_2 = CH_2 \xrightarrow[\text{Ag}]{[O]}$$

Ethylene

(50%)

3.1.7.3 Methods of preparation of thiirane

(*i*) **Cyclization of 2-substituted thiols:** Just as oxirane are obtained by ring closure methods, thiirane can also be prepared by cyclization of 2-halomercaptans with alkali.

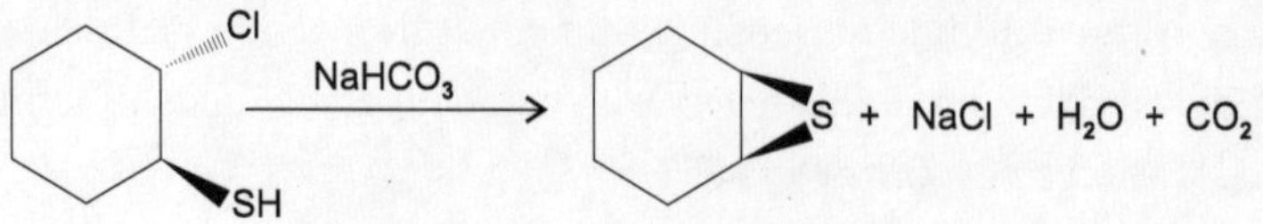

2-Sulfanylethanol on reaction with phosgene (COCl$_2$) in the presence of base such as pyridine gives l,3-oxathiolan-2-one, which on heating at 200°C undergoes decarboxylation to give thiirane.

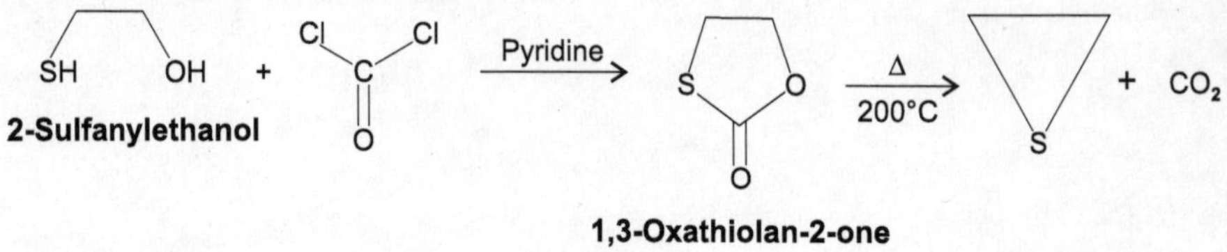

1,3-Oxathiolan-2-one

(*ii*) **Ring transformation of oxiranes:** Thiirane can be synthesized from oxirane. This conversion of one heterocycle into another is known as ring transformation. Oxiranes react with aqueous ethanolic potassium thiocynate (KSCN) to give thiiranes.

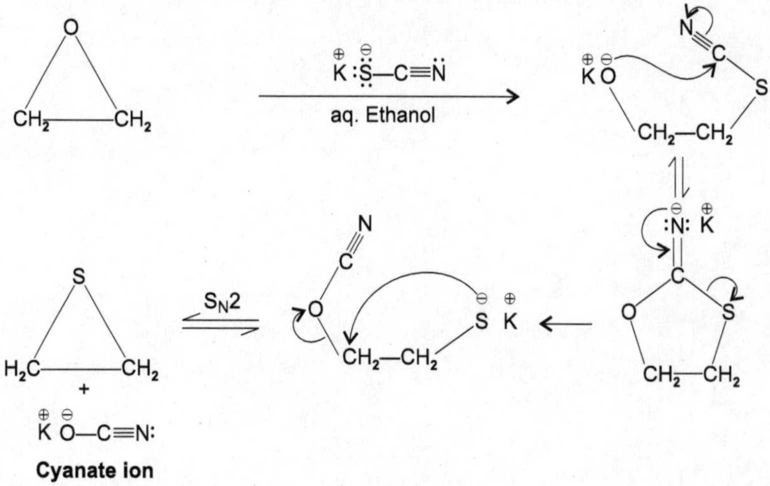

Cyanate ion

(*iii*) **From thiadiazolines:** Thiadiazolines on heating gives thiocarbonyl ylides which cylizes to give thiiranes.

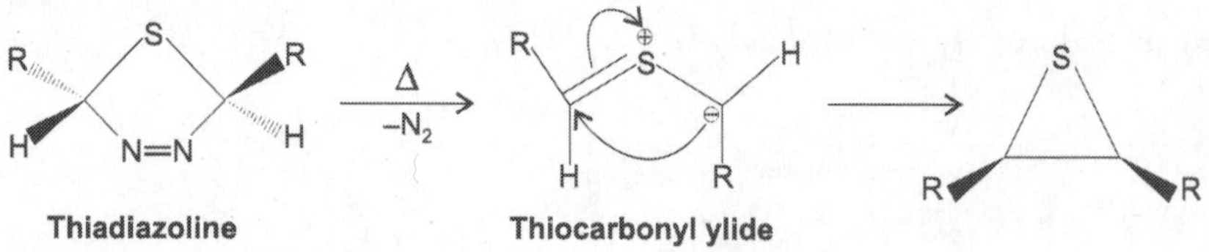

Thiadiazoline   **Thiocarbonyl ylide**

3.1.8 Reactions of saturated three membered heterocycles

3.1.8.1 Reactions of aziridine

Acid base reactions

Unsubstituted aziridines act like secondary amines. They react with acids to form aziridinium salts.

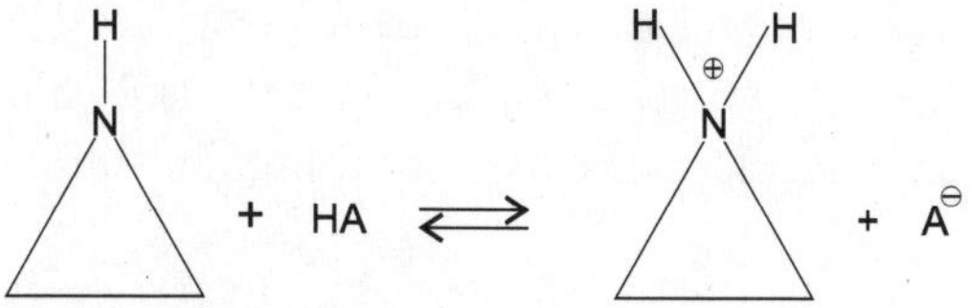

Ring-opening by nucleophiles

Salt formation destabilizes the aziridine ring and in such cases ring-opening by nucleophiles is favored. Aziridine itself reacts with acids explosively to give polymeric products. Aziridines are protonated and acylated by treatment with acid and acid chloride, respectively. The unstable aziridinium ion thus formed undergoes ring opening reaction. These rings with positively charged nitrogen open up by attack of an even relatively poor nucleophile such as chloride ion.

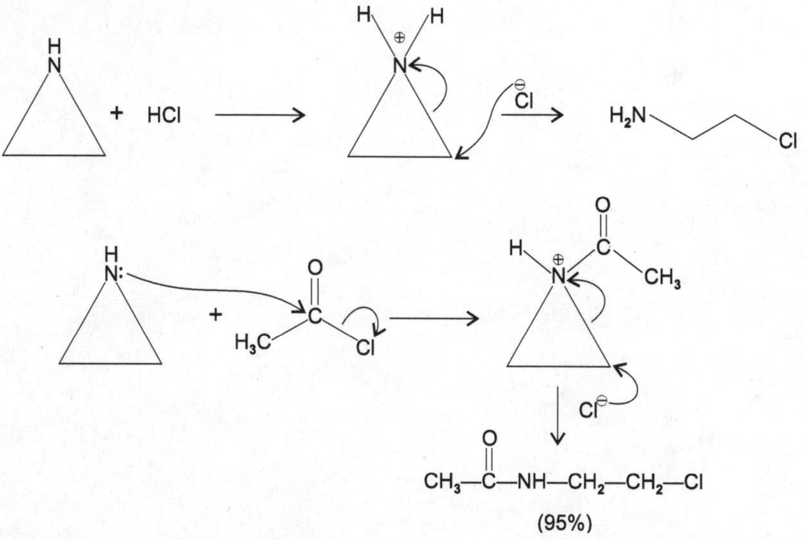

(95%)

The ring-opening of the aziridines is catalyzed especially effectively by acids. The acid-catalysed hydrolysis of aziridine gives amino alcohol.

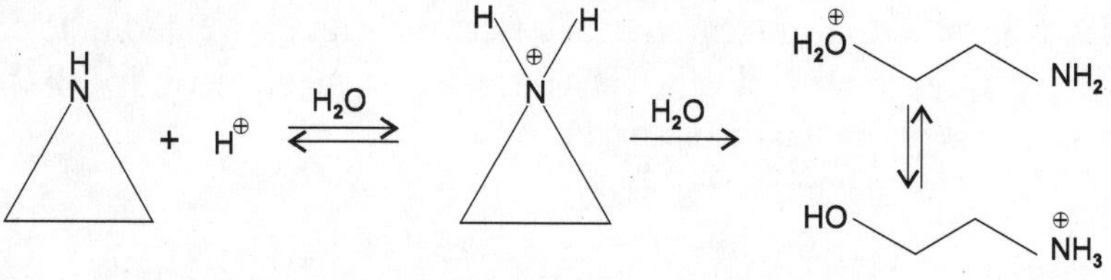

On reaction with ammonia and primary amines the azirdines yield 1,2-diamines.

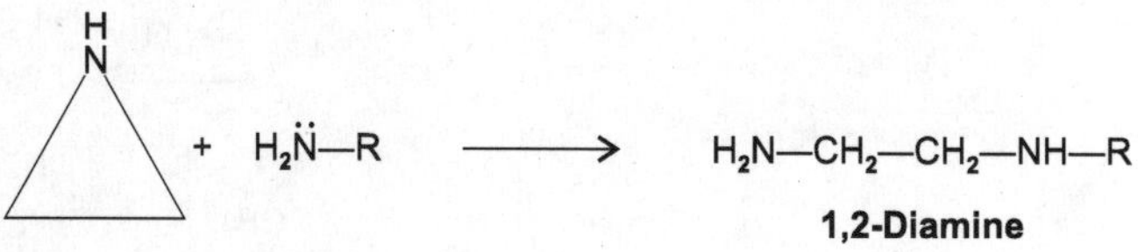

$H_2N-CH_2-CH_2-NH-R$

1,2-Diamine

These type of reactions cause alkylation of the nucleophile. That is why aziridines and bis(2-chloroethyl)amine act as cytotoxic agents. Bis(2-chloroethyl)amine and 1-(2-chloroethyl)aziridinium chloride exist in equillibrium with each other.

bis(2-Chloroethyl)amine

Reactions with electrophilic reagents

Aziridines are nucleophilic and react with various electrophiles to give N-substituted aziridines.

Deamination

Aziridines with an unsubstituted N-atom reacts with nitrosyl chloride to form the corresponding N-nitrosoderivative which undergoes deamination to give alkene as the final product. This reaction is stereospecific.

3.1.8.2 Reactions of oxirane

Oxygen heterocycles are cyclic ethers which are least reactive of all common functional groups. However, epoxides are particularly reactive because ring opening release ring strain and drives the reaction to forward direction. Apart from the ring strain, a significant property of oxiranes is their Brönsted and Lewis basicity, because of the non-bonding electron pairs on the O-atom. Consequently, they react with acids.

Acid-catalysed hydrolysis

Epoxides undergo both acid-catalyzed and base-catalyzed ring opening reactions. Epoxides are protonated by acid and the protonated epoxides can undergo attack by solvent or number of nucleophilic reagents. In this reaction, an acid-base equilibrium precedes the nucleophilic ring-opening of the oxirane ring.

1,2-Ethanediol

Acid catalyzed epoxide ring opening reaction is a stereospecific reaction involving bimolecular mechanism. Thus, (±)-butane-2,3-diol is formed from *cis*-2,3-dimethyloxirane and *meso*-butane-2,3-diol from *trans*-2,3-dimethyloxirane.

cis-**2,3-Dimethyloxirane**   (±)-**Butane-2,3-diol**

trans-**2,3-Dimethyloxirane**   *meso*-**Butane-2,3-diol**

Cyclohexene epoxide yields *trans*-cyclohexane-1,2-diol on acid catalysed ring opening.

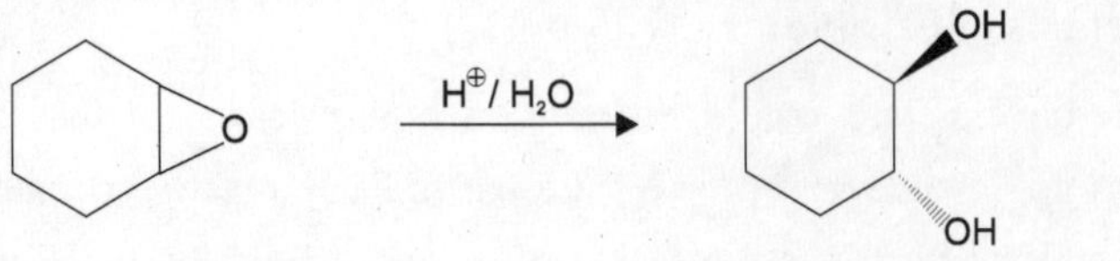

Cyclohexene epoxide   ***trans*-Cyclohexane-1,2-diol**

Epoxides can also be cleaved under basic conditions. Reaction of ethylene oxide with sodium salt of phenol gives the 2-phenoxyethanol.

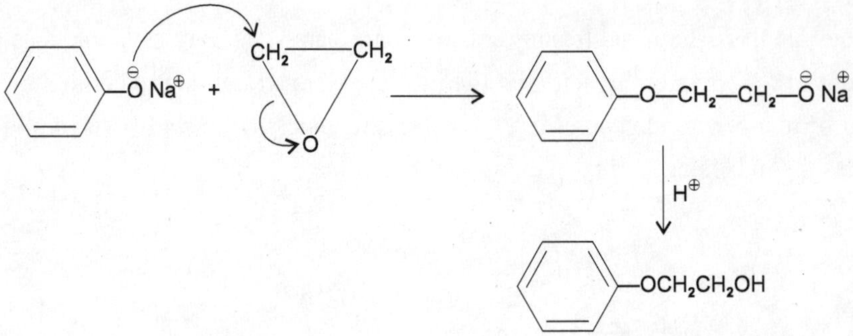

2-Phenoxyethanol

In an unsymmetrical epoxide like isobutylene oxide the two carbons are not equivalent and the product obtained depends upon which one is preferentially attacked by nucleophile or base. The regioselectivity reverses with the reaction conditions. In general the nucleophile attacks the more substituted carbon atom of the epoxide in an acid catalyzed cleavage. Whereas, in a base catalyzed cleavage of epoxide the nucleophile attacks the least substituted carbon. This can be explained by looking at the transition state formed in acid catalyzed ring opening reaction.

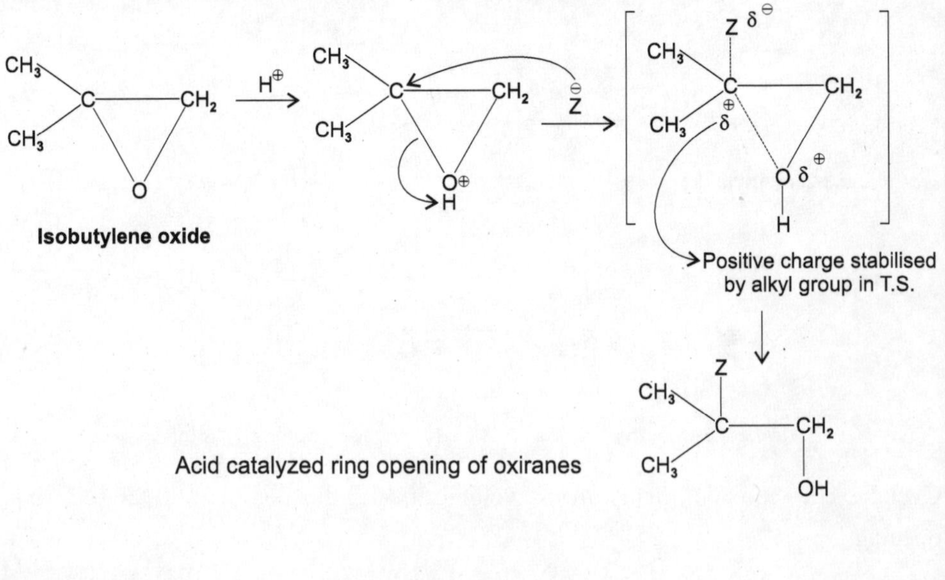

Positive charge stabilised by alkyl group in T.S.

Isobutylene oxide

Acid catalyzed ring opening of oxiranes

In the transition state of acid catalyzed cleavage, bond breaking occurs faster than bond making, and the most substituted carbon has acquired a considerable positive charge. The substituents such as alkyl groups stabilize the positive charge at the carbon atom. Thus the weak nucleophile or solvent attacks here.

In case of base catalyzed ring opening reaction the base attacks at the least substituted carbon. In base catalyzed reaction there is no protonation and no-build up of positive charge in transition state. The strong nucleophile attacks only the least substituted carbon of the epoxide. Nucleophile approaches from the less hindered side. Therefore steric hindrance becomes the controlling factor here.

Base catalyzed ring opening in oxiranes

Ring-opening by nucleophiles

Nucleophiles such as ammonia or amines opens the oxirane ring to yield amino alcohols.

It is a concerted reaction which proceeds *via* an S_N2 mechanism. A nucleophilic substitution takes place at the saturated C-atom. For example, *cis*-2,3-dimethyloxirane, yields (±)-threo-3-aminobutan-2-ol and *trans*-2,3-dimethyloxirane, yields (±)-erythro-diastereomer This reaction is thus sterespecific.

cis-2,3-Dimethyloxirane   (±) *Threo*-3-Aminobutan-2-ol

trans-2,3-Dimethyloxirane   **(±) Erythro-3-Aminobutan-2-ol**

Oxirane is converted to β-haloalcohols on treatment with hologen and triphenylphosphine.

Reduction

Oxiranes are reduced to alcohols using sodium borohydride. In this reaction the ring-opening occurs *via* attack of nucleophilic hydride ion. Oxirane is reduced to ethanol. In case of unsymmetrically substituted oxiranes mixture of alcohols is generated.

Isomerization

Oxiranes isomerize to carbonyl compounds on reaction with catalytic amounts of Lewis acids, such as boron trifluoride, magnesium iodide, or nickel complexes. Oxirane is converted to acetaldehyde. Substituted oxiranes generate mixture of products.

Deoxygenation

Oxirans are deoxygenated to olefins. For example, a *trans*-oxirane yields a (Z)-olefin (or *cis*-olefin) with triphenylphosphine at 200°C. This reaction proceeds *via* formation of betaine intermediate, which is formed after the C-C bond rotation. Therefore, *trans*-

2,3-dimethyloxirane yields *cis*-2-butene on reaction with PPh_3.

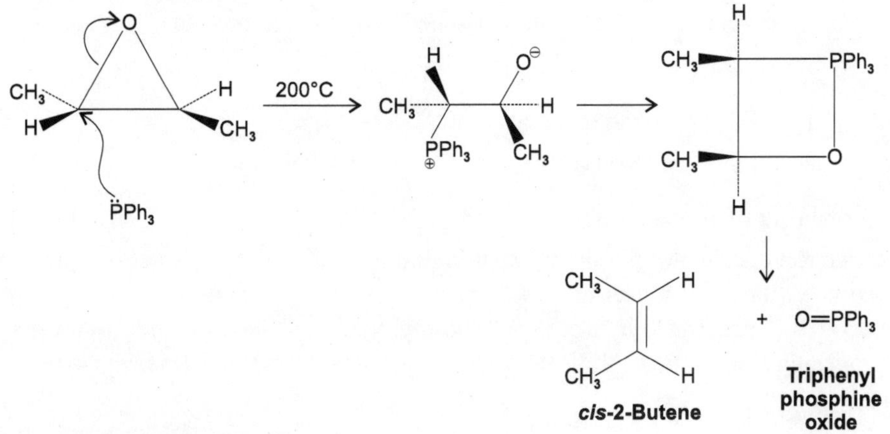
200°C
cis-2-Butene
Triphenyl phosphine oxide

3.1.8.3 Reactions of thiirane

Thiiranes are generally less reactive than oxiranes.

Polymerization

Thiirane undergoes polymerization in the presence of acid and base to give polythiirane $(C_2H_4S)_n$.

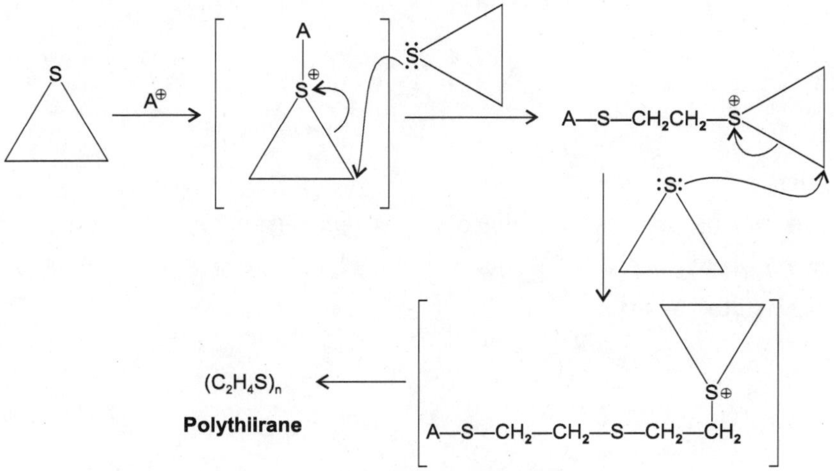
Polythiirane

Acid catalysed polymerisation

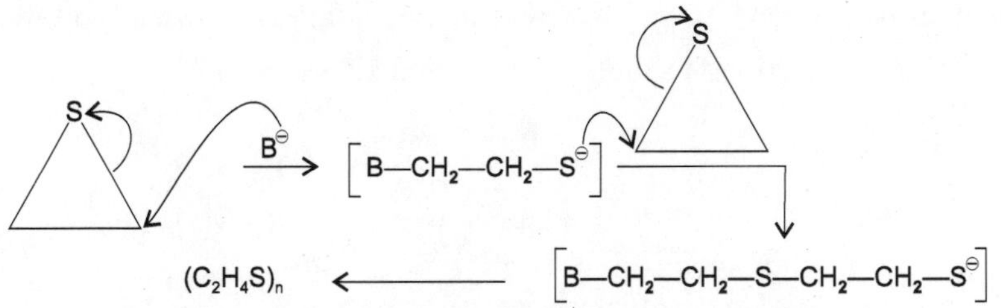

Base catalysed polymerisation

Polythiirane is an amorphous white insoluble substance of the composition $(C_2H_4S)_n$. It can also be obtained by reaction of ethylene chloride or bromide with sodium or potassium sulfide.

$$n(CH_2=CH—X) \xrightarrow[\text{Na}_2\text{S}]{\text{K}_2\text{S or}} (C_2H_4S)_n$$

X = Cl or Br   **Polythiirane**

Ring-opening by nucleophiles

The electron density at the sulphur atom in thiirane is lower than that at oxygen atom in oxiranes. Thus, thiiranes are less reactive towards electrophilic reagents. In the presence of concentrated hydrochloric acid protonation of the S-atom of thiirane and ring-opening by the nucleophilic chloride ion gives β-chlorothiols (mercaptans).

$$\xrightarrow{\text{HCl}} R—\underset{\underset{Cl}{|}}{CH_2}—CH_2SH$$

β-Chlorothiol

Ammonia, or primary or secondary amines, react with thiiranes to give β-amino thiols. The yields are lower, due to competing polymerization.

$$R—\overset{..}{N}H_2 \ + \ \triangle \longrightarrow R—NH—\overset{\beta}{C}H_2—\overset{\alpha}{C}H_2—SH$$

β-Aminothiol

Oxidation

Thiiranes can be oxidized to thiirane oxides (ethylene sulphoxide) using sodium periodate or peroxy acids. At high temperature the thiirane oxide decompose to yield ethene and sulfur monoxide.

$$\xrightarrow[\substack{\text{aq EtOH} \\ 20-25°C \\ -\text{NaIO}_3}]{\text{NaIO}_4}$$

$$\xrightarrow{\Delta} CH_2{=}CH_2 \ + \ SO$$

Thiirane does not forms dioxide, however thiirane dioxides known as episulfone can be obtained by the reaction of sulfenes with diazoalkanes.

Sulfenes   **Diazoalkanes**   **Ethylenesulfone**

The formation of episulfone as an intermediate is also detected in **Rambery-Backlund rearragement** reaction of α-halo sulfones in presence of base.

(82%)

Desulfurization to alkenes

A variety of reagents can be employed to remove sulfur atom of thiirane. Triphenylphosphine and trialkyl phosphites, desulfurizes thiirane to give alkene. The reaction is stereospecific as *cis*-thiiranes yield *cis*-olefins and *trans*-thiiranes yield *trans*-olefins.

Metallic reagents, e.g. n-butyllithium, also bring about a stereospecific desulfurization of thiiranes.

Thiirane on pyrolysis or in the presence of ultraviolet light also undergoes desulfurization to yield alkenes.

Ph—CH=CH—Ph + S

$$\xrightarrow{\Delta}$$

CH$_2$=CH$_2$ + S

$$\xrightarrow{hv}$$

Dimerization

Thiirane gives dithiirane in the presence of alumina at 220°C or in presence of Pt, PdCl$_2$.

$$2 \xrightarrow[\text{Pt, PdCl}_2\text{/CO}_2\text{, 60 atm}]{\text{Al}_2\text{O}_3\text{, 220°C}}$$

3.2 1-AZIRINE (OR 2*H*-AZIRINE)

3.2.1 Introduction

The smallest unsaturated nitrogen heterocycle is azirine, which exists in two forms the **2*H*-azirine** (1-azirine or imine) and **1*H*-azirine** (2-azirine or enamine).

1-Azirine
(2*H*-Azirine)

2-Azirine
(1*H*-Azirine)

Azirines with phosphorus substituents regulate many important biological functions. Azirines are of interest due to their biological potential and as intermediates in synthesis to produce amino acid derivatives and heterocycles. 1*H*-Azirines occur as reactive intermediates. 2*H*-Azirines are naturally occurring antibiotics. 2*H*-Azirine is thermally unstable and has to be stored at very low temperature. Substituted 2*H*-azirines are more stable and are liquids or low melting solids. The basicity of substituted 2*H*-azirines is lower than that of comparable aliphatic compounds. This is clear from the insolubility of 2-methyl-3-phenyl-2*H*-azirine in hydrochloric acid.

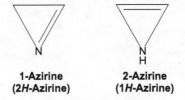

$$\xrightarrow{\text{HCl}} \text{Insoluble}$$

The ring strain in 2*H*-azirines is greater than that of the saturated three membered heterocycles. The 2*H*-azirine ring is stable due to bond shortening, angle compression, and presence of the electron-rich nitrogen atom. The strain energy is due to deformation of the normal bond angles between the atoms of the ring. The total ring strain energy of 2*H*-azirine is 48 kcal/mol.

3.2.2 Spectral data

Infrared spectra of 2*H*-azirine has abnormal C=N stretching frequency at 1800 cm^{-1}, while the normal stretching frequency of C=N bond is at 1650 cm^{-1}. ^{13}C NMR also shows a higher chemical shift for carbon at 176 ppm.

^{13}C NMR (δ, ppm)

3.2.3 Methods of preparation of 1-azirine (or 2*H*-Azirine)

(*i*) **From alkene:** Alkenes on reaction with bromine followed by treatment with sodium azide and then NaOH give vinyl azides. The loss of nitrogen from vinyl azide on thermolysis or photolysis gives 2*H*-azirine *via* vinyl nitrene intermediate.

(*ii*) **From N,N-disubstituted acid amides:** 3-(Dialkylamino)-2*H*-azirines can be prepared from N,N-disubstituted acid amides by reaction with phosgene (COCl$_2$) followed by reaction with sodium azide.

(*iii*) **Neber rearrangement of oxime sulfonates:** Another method for the preparation of 2*H*-azirine system is the Neber rearrangement of oxime sulfonates. The presence of strong electron withdrawing groups in the α-position to the oxime increases the acidity of those protons, and leads to cycloelimination in basic medium. The reaction occurs either through an internal concerted nucleophilic displacement or *via* a vinyl nitrene intermediate.

Vinyl Nitrene

Similarly, methyl iodides of dimethyl hydrazones (or N,N,N-trimethylhydrazonium iodide) in the presence of sodium propoxide as a strong base gives 2*H*-azirines.

3.2.4 Reactions of 1-azirine (or 2*H*-azirine)

The ring strain, presence of reactive π-bond and ability to undergo regioselective ring cleavage leads to high chemical reactivity of 2*H*-azirines. The 2*H*-azirine ring can be cleaved by thermal and photochemical excitation. The lone pair of electrons present on the nitrogen atom can interact with electrophiles. The π-bond as well as the ring strain makes 2*H*-azirines good dienophiles and dipolarophiles. The trigonal carbon atom in 2*H*-azirines is an electrophilic center.

Reaction with nucleophiles

2*H*-Azirines are more susceptible to nucleophilic addition than imines, due to their ring strain. The primary addition compounds are aziridines, which might be isolated or undergo ring cleavage generating diverse types of products.

Reaction with methanol: Electrophilic reagents react with the N-atom while, nucleophilic reagents attack C-atom. Reaction of methanol with 2*H*-azirines is such a reaction where methanol in the presence of a catalytic amount of sodium methoxide produces 2-methoxyaziridines.

Reaction with Grignard reagent: Grignard reagents add to the C=N bond of 2*H*-azirines leading to isolable aziridines. The nucleophilic attack occurs selectively on the least hindered face of the ring. If a carboxylic ester group is present, in that case the attack is still selective, but opposite to the previous, taking place by the more hindered face of the azirine. This effect is described as a consequence of Grignard reagents pre-chelation with the carboxyl ester groups.

R = Me (80%)
R = Ph (100%)

Reaction with carboxylic acid: Addition of carboxylic acids to 2*H*-azirines leads to ring-opened products. In acidic conditions azirines protonate at the nitrogen first and then the carboxylate anion attack takes place. Benzoic acid and chloro-, bromo-, iodo- and cyanoacetic acids are reported to react with 3-phenyl-2*H*-azirine to yield corresponding amides.

3-Phenyl-2*H*-azirine

Amide

Reaction with amines: C-Nitrogen aziridine adducts are rare, but azirine reacts with pyrrolidine and piperidine giving isolable aziridine compounds.

Cycloaddition reaction

Azirines with an ester group on the imine carbon undergoes cycloaddition. The imine unit acts as dienophile on reaction with cyclopentadiene (a diene).

Diene   Dienophile

Reduction

The C=N bond of 2*H*-azirine is reduced in the presence of lithium aluminium hydride or sodium borohydride to give aziridine. Reactions is highly selective giving *cis*-aziridines in good yields.

cis-**Aziridine**

Several azirines have also been efficiently reduced with $NaBH_4$ in ethanol leading to the respective *cis*-aziridines in good yields together with small amounts of ethanol-adducts.

Thermal reaction of 2*H*-azirines

The major thermal reaction of 2*H*-azirines involves C(2)–N bond cleavage to form vinyl nitrene intermediates. Which may undergo typical reactions of nitrenes. For example, the thermolysis of aryl-substituted 2*H*-azirines results in the formation of indoles by intramolecular electrocyclization of the intermediate vinyl nitrene with the aromatic ring.

Vinyl nitrene   **Indole derivative**

Photochemical reactions of 2*H*-azirines

2*H*-Azirines are photochemically highly active substances. Upon irradiation the strained three membered azirine ring opens selectively at the C–C bond in a heterocyclic fashion leading to the formation of a nitrile ylide intermediate. The nitrile ylide can be trapped by reactive dipolarophiles (A=B) to give five membered ring. In alcohols as solvents, the nitrile ylides are protonated to yield azallenium cations which are then trapped by the alcohol to furnish alkoxyimines. The protonation rate of the ylide in alcohol increases with the acidity of the alcohol.

Nitrile ylide

Azallenium ion   **Alkoxyimines**

EXERCISE

Q.1   How is aziridine synthesized by Wencker synthesis and discuss its mechanism?

[Hint. Treatment of sulfate esters with alkali to form aziridines]

Q.2   What is Hassner aziridine synthesis and discuss its mechanism?

[Hint. Reaction of iodine azide with alkenes followed by reduction and S_N2 displacement]

Q.3   Why *cis*-alkenes yield *cis*-oxiranes while *trans*-alkenes yield *trans*-oxiranes, by the direct oxidation of alkene with peroxy acids?

[Hint. reaction occurs in a concerted manner, stereospecifically]

Q.4   What is Prilezhaev (Prileschajew) reaction.

Ans. Synthesis of oxiranes by the oxidation of alkene with peroxy acids.

Q.5   Draw the diasteromeric forms of 2-methylaziridine. (Refer sec. 3.1.3)

Q.6   What is polythiirane? How is it obtained.

Ans. Polythiirane, $(C_2H_4S)_n$ is an amorphous white insoluble substance, obtained by reaction of ethylene chloride with sodium sulfide.

Q.7   Discuss Neber rearrangement of oxime sulfonates.

Q.8   How oxiranes are prepared by Corey synthesis?

[Hint. Reaction of S-ylide with carbonyl compounds]

Q.9   What products are obtained in the acid- or base- catalysed ring opening of an unsymmetrical oxirane? Explain with suitable example.

[Hint. In an acid catalyzed cleavage the nucleophile attacks more substituted carbon atom of the epoxide whereas, in a base catalyzed cleavage of epoxide the nucleophile attacks the least substituted carbon]

Q.10 How can be an oxirane converted to episulfide (or thiirane) with the help of thiourea or thiocyanates? (Refer sec. 3.1.7.3)

Q.11 Explain why aziridine (pK$_a$ value of the aziridinium ion is 7.98) is weakly basic compound than the open chain compound dimethylamine (pK$_a$ of conjugate acid is 10.87). (Refer sec. 3.1.6)

Q.12 What products will be obtained when aziridine is treated with

(a)   Nitrosyl chloride

(b)   HCl

(c)   Ethylamine

Ans. (a)   N-nitrosoaziridine which converts to ethene

(b)   2-Chloroethanamine

(c) $NH_2CH_2CH_2NHCH_2CH_3$

Q.13 Complete the following reactions:

    (a) Oxirane   + BF_3 $\longrightarrow$

    (b) Styrene   +   m-CPBA $\longrightarrow$

    (c) 2-Aminoethanol + $SOCl_2$ $\longrightarrow$

    (d) Ethene   +   Ag   +   air $\longrightarrow$

    (e) Aziridine + benzoyl chloride + NaOH $\longrightarrow$

    (f) Oxirane   + PPh_3 + I_2 $\longrightarrow$

Ans. (a) CH_3CHO

    (b) Styrene oxide

    (c) Aziridine

    (d) Oxirane

    (e)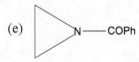

    (f) ICH_2CH_2OH

Q.14 Out of selenirane and tetrahydro-2*H*-selenopyran which is more basic?

Ans. Tetrahydro-2*H*-selenopyran is more basic than selenirane. In the case of tetrahydro-2*H*-selenopyran the selenium atom is sp³ hybridized, the lone pair near a sp³ hybridized orbital makes it less stable, hence protonation will lead to stability in this compound. Whereas, in selenirane the lone pair on selenium are in sp² hybridized orbital making it more stable and less basic.

In other words greater the s-character less will be the basicity. Thus Se in selenirane has sp² hybridisation and more s-character than sp³ hybrid Se of tetrahydro-2*H*-selenopyran.

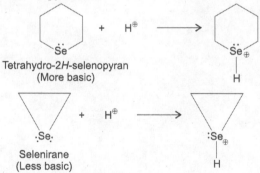

Q.15 Mention the missing reagents/conditions:

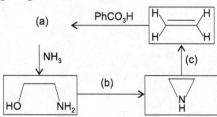

Ans. (a)

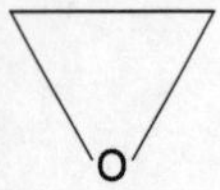

(b) (i) H_2SO_4,   (ii) KOH

(c) HNO_2

Q.16 Write the reaction sequence involved in the ring opening of the following oxirane
with

(a) Methanol in presence of acid

(b) Sodium methoxide

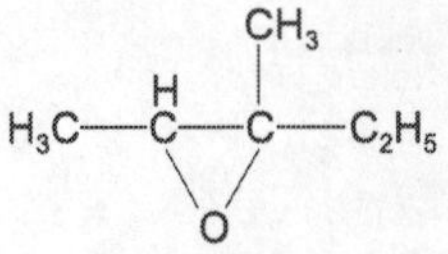

Hint. (a)

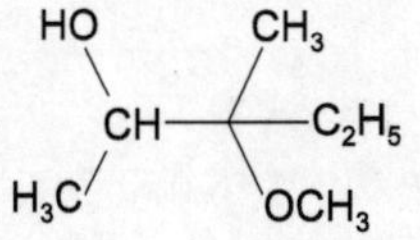

(b)

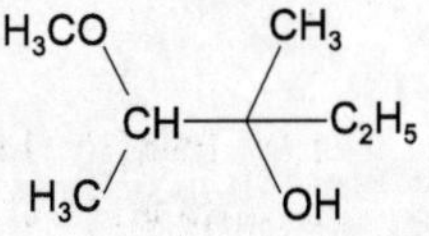

Four Membered Heterocyclic Compounds with One Heteroatom

Azetidine, Oxetane, Thietane and 2-Azetidinone

4.1 AZETIDINE, OXETANE AND THIETANE

4.1.1 Introduction

The saturated four membered heterocyclic compounds containing nitrogen, oxygen and sulfur are **azetidine**, **oxetane** and **thietane**, respectively.

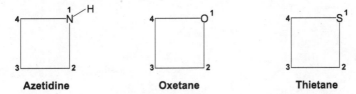

The four membered unsaturated heterocycles containing one nitrogen atom are azete, 1-azetine and 2-azetine.

The four membered unsaturated compounds with one oxygen atom and sulfur atom are known as 2-oxetene (or oxete) and 2-thietene (or thiete), respectively.

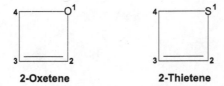

Earlier azetidine was called trimethyleneimine. Azetidine is a water miscible, colorless liquid (bp 61.5°C). It smells like ammonia and fumes in air. Oxetane is a colorless, water miscible liquid (bp 48°C). Thietane is a colorless, water insoluble liquid (bp 94°C), which polymerizes slowly at room temperature and faster on exposure to light.

4.1.2 Biological importance

The oxetane ring is present in **paclitaxel** (taxol) which is used in cancer chemotherapy and isolated from the bark of the Pacific yew tree (*Taxus brevifolia*). Paclitaxel is now used to treat patients with lung, ovarian, breast, head and neck cancer, as well as the advanced forms of Kaposi's sarcoma. (2R,3S)-3-Amino-2-oxetane carboxylic acid (**oxetin**), the first natural product possessing an oxetane ring is an antibiotic.

Shimada et al isolated **oxetanocin A** from a strain of bacteria *Bacilllus megaterium* in 1986 which is potent against viruses such as HSV-1, HSV-2, HCMV, MCMV, VZV, HIV-1 and 2. Oxetanones, are widely applied in polymer manufacturing and in agriculture as herbicides, fungicides and bactericides.

    Mono and **dialkyl thietanes** are present in anal gland secretions of small mammals such as polecats, minks, stoats, weasels, ferrets, kiores and voles. Pfizer has developed an artificial sweetener **alitame** containing thietane ring which is 2000 times more efficient than sucrose. As alitame does not contain phenylalanine it can therefore be used by people with phenylketonuria. Thietanes are used in the production of polymers, as bactericides and fungicides in paint, and as iron corrosion inhibitors.

Paclitaxel (Taxol)

Oxetin

Oxetanocin A

R^1 = Me/Et/Pr/*i*-pr, R^2= R^3 = H
R^1 = Me, R^2= H, R^3= Et
R^1 = Me, R^2= Me, R^3= H

Alkyl thietanes

Alitame

4.1.3 Structure

The bonds of four membered heterocyclic compounds are formed by the oblique overlapping of sp^3 hybrid orbitals. The oblique overlapping and ring strain is similar to cyclobutane but less than the three membered heterocyclic compounds.

The nitrogen containing compound azetidine has a puckered ring. Azetidine exhibits conformational isomerism, have N-H bond at either equitorial or axial positions. The activation energy for conversion into these two forms is 5.5 kJ/mole.

(equitorial N—H)
low energy

axial N-H

However, the oxetane ring is regarded planar. The eclipsing of atoms is leser in oxetane due to lone pair on oxygen, as compared to lone pair on nitrogen of azetidine. Due to the ring strain C-C-C and C-O-C bond angles are slightly deviated from normal tetrahedral angle. The bond angle at oxygen is 91°44' and oxetane ring is slightly distorted square.

Ring puckering : 7.9°

The thietane ring is planar and rigid, however substituents present in the ring leads to puckering.

4.1.4 Basicity

Azetidine is a stronger base (pK$_a$ = 11.29) than aziridine (pK$_a$ = 7.98) and dimethylamine (pK$_a$ = 10.87). The pK$_a$ values given are for the corresponding protonated compounds (or for the conjugated acids).

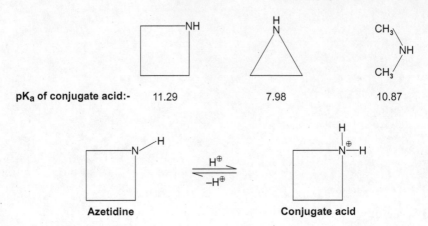

pKa of conjugate acid:-

pK$_a$ of conjugate acid:-   11.29   7.98   10.87

Azetidine   **Conjugate acid**

4.1.5 Dipole moment

The calculated dipole moment values of oxetane (2.01 D) and aliphatic ethers (dimethyl ether 1.31 D) indicate more electron density at the oxygen atom in oxetane than in acyclic aliphatic ethers.

μ = 2.01 D μ = 1.31 D

4.1.6 Spectral data

The ^{1}H and ^{13}C NMR (given in brackets) chemical shift values for azetidine are shown below.

H 3.95

(19) 3.4   3.7 (45)

^{1}H NMR (δ, ppm)
^{13}C NMR (δ, ppm)

4.1.7 Methods of preparation of azetidine, oxetane and thietane

Because of the inherent ring strain in four membered heterocyclic compounds, the synthesis of these compounds is not always straightforward.

4.1.7.1 Methods of preparation of azetidine

(*i*) **Cyclization of γ-substituted amines:** Azetidine can be prepared by dehydrohalogenation of γ-halogen substituted amines in presence of base.

CH₃ ... Base, Δ / −Br⊖ ... −H⊕

1,3,3-Trimethylazetidine

(*ii*) **From 1,3-dihaloalkanes:** Azetidine can also be prepared from 1,3-dihaloalkanes by the action of para-toluenesulfonamide in the presence of base, followed by the reductive removal of tosyl group from the 1-tosylazetidine.

H_2N-SO_2 — ⟨⟩ — CH_3 ; ⊖OH ; Na/n-pentanol

In an alternate manner 2,4-dimethylazetidine can also be prepared by nucleophilic displacement of tosyl group.

NaOEt ; −OTs⊖ ; Na / n-pentanol

The N-phenylazetidine can be synthesized by cyclization of aniline with 1,3-dichloropropane using microwave (80-100 W).

4.1.7.2 Methods of preparation of oxetane

(*i*) **Cyclization of γ-haloesters:** Oxetane may be prepared by ring closure reaction of 3-chloropropyl acetate in hot potassium hydroxide.

Mechanism: The mechanism of this reaction involves the hydrolysis of ester to alkoxide, which then undergoes an intramolecular substitution reaction to give oxetane.

(*ii*) **From 1,3-diols:** Oxetanes can also be prepared from 1,3-diols or γ-hydroxysulfonates.

(*iii*) **Paterno-Buchi reaction:** In this reaction a photochemical [2+2] cycloaddition of carbonyl compounds with alkenes takes place to form oxetanes.

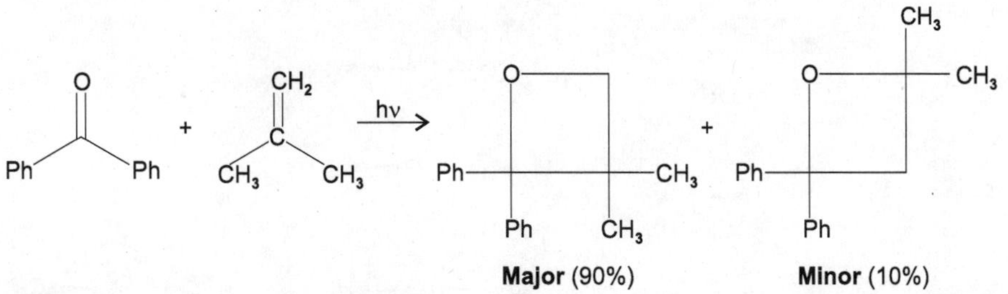

Major (90%)   **Minor** (10%)

Mechanism: The Paterno-Buchi reaction involves absorption of quantum of light by the carbonyl compound and it gets converted to an electronically excited singlet state. This singlet state is then converted to a low energy triplet state. The excited triplet carbonyl compound then reacts with alkene, to form an intermediate triplet diradical. Unsymmetrical alkenes may give rise to two types of triplet diradical. But, the more stable diradical leads to the formation of major product.

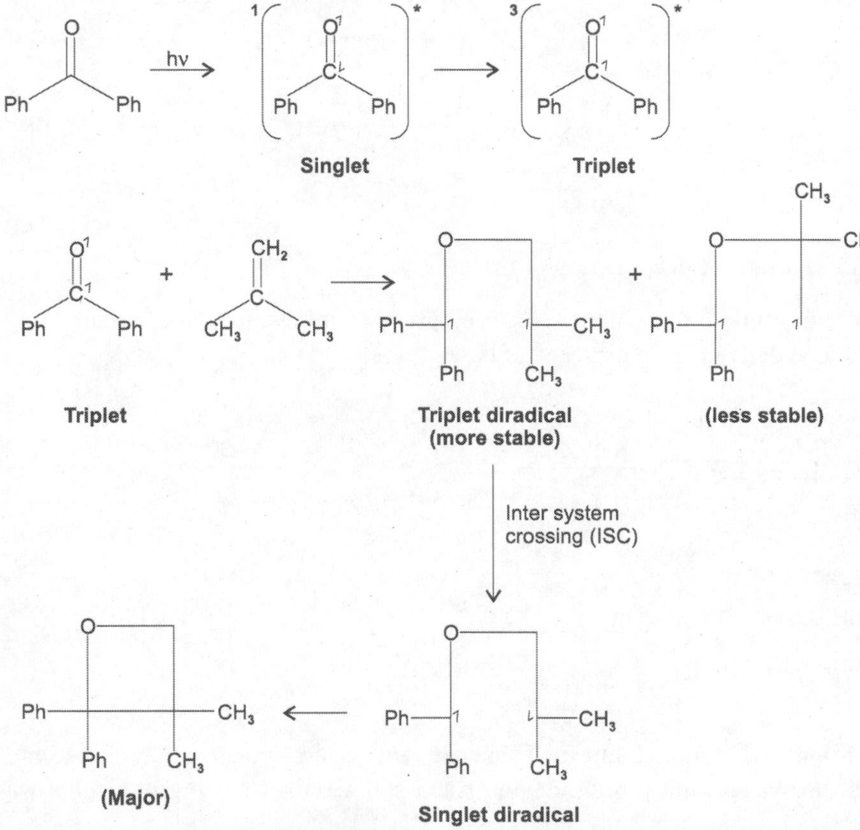

Singlet   **Triplet**

Triplet   **Triplet diradical**   **(less stable)**
                (more stable)

Inter system crossing (ISC)

(Major)   **Singlet diradical**

According to the Woodward-Hoffmann rules, addition of the alkenes possessing electron-withdrawing groups (-CN etc.) occurs in a concerted and stereospecific manner. Thus, *cis*-alkene will give *cis*-substituted oxetanes. But alkenes with donor substituents (alkyl group) react through biradical intermediates as shown below, and give mixture of *cis* and *trans* products.

4.1.7.3 Methods of preparation of thietane

(*i*) **From γ-halothiols:** Thietanes can be prepared by cyclization of γ-halothiols or their acetyl derivatives in presence of base.

(*ii*) **From 1,3-dihaloalkanes:** Thietanes are easily obtained by reaction of 1,3-dihaloalkanes with sodium or potassium sulfide either heating in alcohol-water solution or by using phase transfer catalyst (PTC).

(iii) **From (3-chloropropyl)isothiuronium bromide**: Thietanes can also be obtained from (3-chloropropyl)isothiuronium bromide which is prepared from 1-bromo-3-chloropropane on reaction with thiourea.

4.1.8 Reactions of four membered heterocyclic compounds

4.1.8.1 Reactions of azetidine

Azetidines are thermally stable and less reactive than aziridines. They behave in their reactions almost like secondary alkylamines.

A positive charge on the N-atom of azetidine destabilizes the ring, as is the case with the aziridines. Ring-opening by nucleophiles proceeds with acid catalysis. For example, azetidine on reaction with hydrogen chloride yields γ-chloroamines.

Azetidines unsubstituted on the N atom react with alkyl chloride to give 1-alkylazetidines which can react further to give quaternary azetidinium salts. These 1,1-dialkylazetidinium chlorides formed isomerize on heating to give tertiary γ-chloroamines.

With acyl halides, azetidines produce N-acylazetidines.

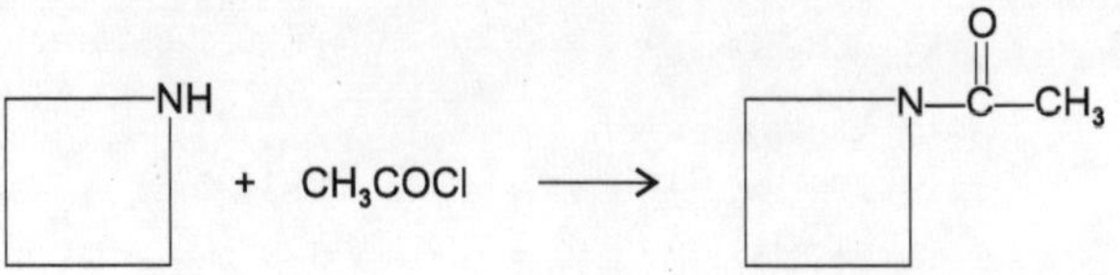

N-Acetylazetidine

Azetidines with nitrous acid give N-nitrosoazetidines.

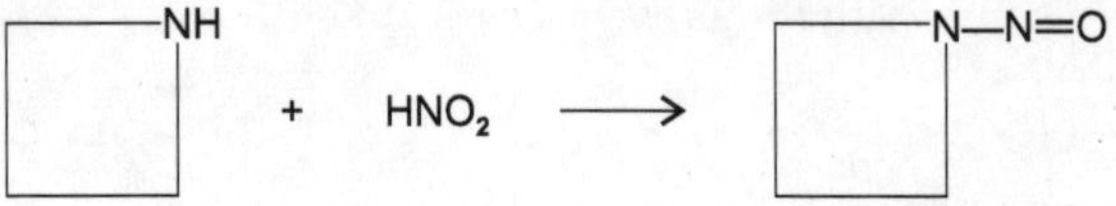

N-Nitrosoazetidine

In the presence of hydrogen peroxide ring cleavage takes place. Thus azetidine gives acrylaldehyde on reaction with hydrogen peroxide.

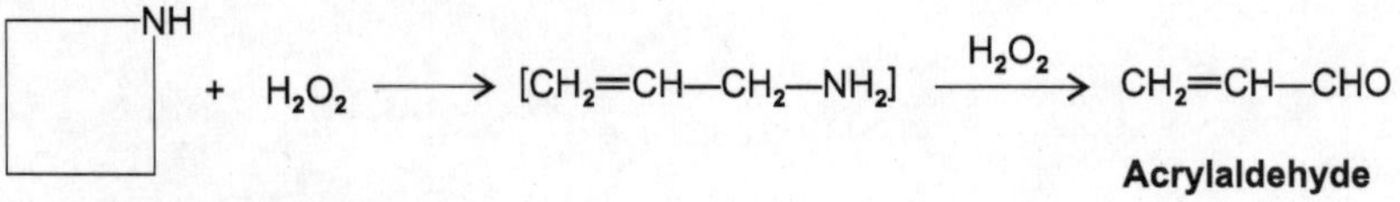

Acrylaldehyde

Azetidine reacts with carbon sulfide to give a salt.

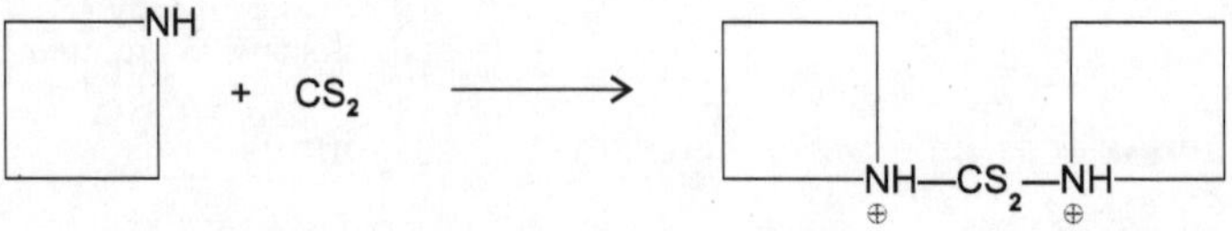

Azetidine also reacts with formaldehyde to give N-hydroxymethyl azetidine.

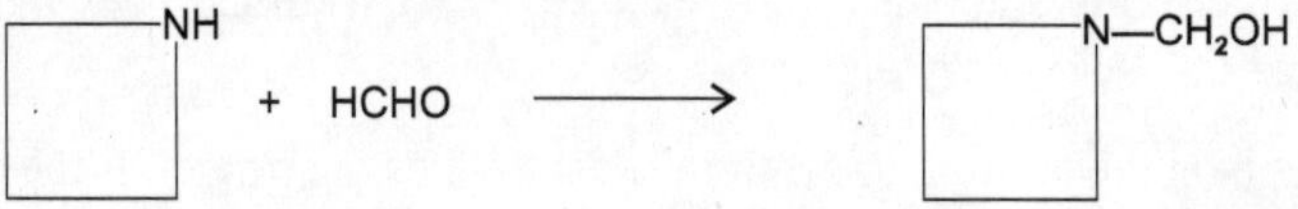

N-Hydroxymethyl azetidine

4.1.8.2 Reactions of oxetane

Oxetanes are cyclic ethers and are the least reactive of all the common functional groups. Oxetanes react like oxiranes with ring-opening at a slower rate and under forcing conditions. To make oxetane more reactive, it must be treated with strong Lewis acid such as $BF_3.OEt_2$. The complex thus formed will react with nucleophiles such as BuLi to give n-heptanol in a good yield.

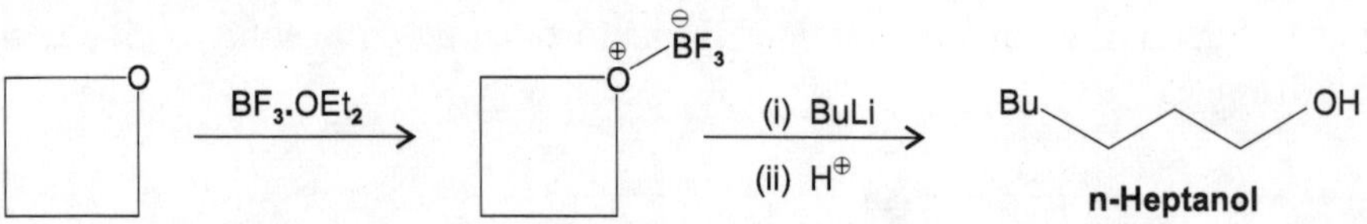

n-Heptanol

Acid-catalysed ring-opening by nucleophiles

Hydrogen halides (HX) react with oxetanes to give 3-haloalcohols with the formation of an intermediate oxonium ion. The acid-catalyzed hydrolysis leads to the formation of 1,3-diols.

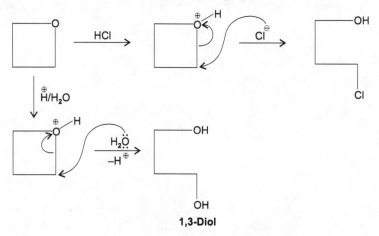

1,3-Diol

Oxetane ring also undergoes base catalysed ring opening reactions.

Cyclooligomerization and polymerization

In the presence of Lewis acids, such as boron trifluoride, in dichloromethane as solvent, oxetane undergoes cyclooligomerization. However, in the presence of water, linear polymers are formed.

1,5,9-Trioxacyclododecane

Oxetane can also react with carbon dioxide to form trimethylene carbonate in the presence of tetraphenyl antimony iodide as catalyst.

4.1.8.3 Reactions of thietane

Thietanes are less reactive towards nucleophiles as compared to thiiranes. Thietane does not react at all with ammonia or amines at room temperature. Electrophilic attack at the S atom can cause ring-opening. Thus, addition of acids results into polymerization.

Thietane undergoes ring opening by haloalkanes. This demonstrates that a positive charge on the heteroatom destabilizes the ring.

Reaction of thietane with one mole of hydrogen peroxide, nitrous acid or chromic acid in glacial acetic acid yields cyclic 1-sulfoxides. The sulfoxides on treatment with excess of hydrogen peroxide or potassium permanganate or peracetic acid yields cyclic sulfones

On heating thietane gives ethylene and thioformaldehyde.

When reacted with ammonia at high temperature thietane gives 3-aminopropanethiol (or 3-aminopropyl mercaptan).

3-Aminopropyl mercaptan

Thietane reacts with chlorine, allyl bromide and acetyl chloride under different conditions to yield corresponding open chain compounds.

At 250°C thietane decomposes over aluminum oxide to give hydrogen sulfide, ethene, and hydrogen. In an atmosphere of hydrogen sulfide the pyrolysis of thietane on Al_2O_3 at 250°C yields small quantities of 1,2-dithiolane. 2-Methylthietane isomerizes with ring enlargement rearrangement to thiophene at temperatures between 350 and 400°C.

Thietanes on treatment with Raney nickel in boiling benzene undergoes reductive desulfurization with ring splitting.

Thietanes undergo ring enlargement to 1,2-dithiolanes by heating them with elemental sulfur. For example, 5-(2-thiethanyl)valeric acid yields α-lipoic acid on heating with sulfur.

5-(2-Thiethanyl)valeric acid   **α-Lipoic acid**

When 3,3-pentamethylene thietane is treated with iodine a crystalline stable iodine adduct is obtained.

3,3-Pentamethylene thietane

Thietane when added dropwise with intense cooling to a solution of chlorine chloroform (1:1 molar ratio), yields 3-chloropropanesulfenyl chloride-1. This further reacts with another mole of thietane to form γ-chloropropyl disulfide. Also, if 0.5 mole of chlorine or bromine is added to 1 mole of thietane, the γ-chloropropyl disulfides is obtained as the main product.

4.2 β-LACTAM OR 2-AZETIDINONE

4.2.1 Introduction

The four membered cyclic amide **2-azetidinone**, is the smallest cyclic system capable of accommodating the amide group. The 2-carbonyl derivative of azetidine known as 2-azetidinones (or β-lactams), are part of the **penicillins**, **cephalosporins**, **carbapenems** and **monobactams**, series of antibiotics. The β-lactams serve as synthons for many biologically important classes of organic compounds.

Penicillins   **Cephalosporins**   **Carbapenems**   **Monobactams**

Penicillin   **Cephalosporin**

Carbapenem

The β-lactam ring containing compounds show various biological activities such as antibacterial, antifungal, antitubercular, antitumor, cholesterol absorption inhibition and enzyme inhibition. **Penicillin** contains a bicyclic system consisting of a four membered β-lactam ring fused to a five membered thiazolidine ring. In 1928, **Sir Alexander Fleming** noted that a bacterial culture which had been left for several weeks open to the air had become infected by a fungal colony. Bacterial colonies around the fungal colony died as the fungal colony produced an antibacterial agent (species of *Penicillium*). In 1928, penicillin was discovered and by 1940 effective means of its isolation was developed by Florey and Chain. Penicillin is a toxic chemical produced by a fungus to kill bacteria. **Fleming, Florey** and **Chain** shared the 1945 Nobel Prize in medicine or physiology for their contributions.

4.2.2 Methods of preparation of 2-azetidinone

(*i*) **From isocyanates:** The [2+2] cycloaddition of two components of the ring, each containing a double bond may be utilized for the synthesis of β-lactams in one step. For example, addition of isocyanate to an olefin occurs through a concerted process.

(*ii*) **From ketenes:** [2+2] Cycloaddition of a ketene and an imine gives β-lactam. The first β-lactam was prepared by **Staudinger** in 1907 by the reaction of diphenylketene with benzylidene aniline.

Mechanism: Both ketene and imine can act as either nucleophiles or electrophiles. The imine adds to the ketene as a nucleophile which is followed by cycloaddition to form β-lactam. Since, the first β-lactam synthesis, many synthetic methods have evolved which seem to the variation of the **Staudinger** method.

(*iii*) **Cyclization of β-amino acid derivatives:** Unsubstituted β-lactam is prepared by the treatment of a 3-aminopropane ester with a Grignard reagent, in very low yield.

Similarly, treatment of β-amino acid chloride with triethyl amine, N,N-dimethylaniline or diazomethane gives the corresponding β-lactams.

(*iv*) **Cyclization of acrylamides:** Acrylamides (R^1, R^2 and R^3 are electron withdrawing groups) undergo intramolecular cyclization upon the influence of a suitable base to give β-lactams.

(*v*) **Cyclization of N-substituted β-bromopropano amides:** The cyclization of N-substituted β-bromopropano amides in presence of a strong base leads to alkyl or aryl substituted β-lactams. The base (alkali metal amide or lithium or sodium carbonate in paraffin oil) converts the amide into its conjugate base, which on internal displacement of halide gives the β-lactam.

Ethyl 3-bromo-2-oxopropanoate on treatment with SF_4 yields ethyl 3-bromo-2,2-difluoropropanoate which on reaction with $ClSO_3H$ gives 3-bromo-2,2-difluoropropanoyl chloride. This on reaction with aromatic amine gives a β-bromopropionamide derivative. The β-bromopropionamide derivative undergoes Wasserman cyclization using sodium hydride to give the N-(3-carboxy-6-methylphenyl)-3,3-difluoro-2-azetidinone.

β-Bromopropionamide derivative

4.2.3 Reactions of 2-azetidinone

2-Azetidinones or β-lactams are attacked by nucleophile on carbonyl group which results in the opening up of the four membered ring. Opening of four membered ring relieves the strain.

An amine opens the ring at high temperature.

The β-lactams may undergo the reactions of carbonyl group without opening of the ring. For example, β-lactam derivative reacts with Wittig reagent to give the corresponding alkene.

However, the lithium dibutylcuprate reacts with the carbonyl group leading to opening of the β-lactam ring.

If electron donating groups (such as amino group or alkoxy group) present at 4-position, N1-C4 cleavage occurs.

The high reactivity of β-lactam is the key to activity of antibiotics such as penicillins and cephalosporins. For example, penicillin reacts with the key enzyme involved in the synthesis of the bacterial cell wall protein. Thus, penicillin interferes with the bacterial cell wall construction. Bacteria develops resistance to penicillin by secreting **penicillinase**, an enzyme that destroys penicillin by hydrolyzing its β-lactam ring before the drug can interfere with bacterial cell wall synthesis.

Certain molecules used in conjunction with a β-lactam antibiotic extend their activity, by inhibiting β-lactamases.

One such penicillinase inhibitor is a sulfone, which is easily prepared from penicillin by oxidizing the sulfur atom with a peroxyacid.

EXERCISE

Q.1   How will you synthesize oxetanes *via* Paterno-Buchi reaction?

[Hint. Photochemical [2+2] cycloaddition of carbonyl compounds to alkenes]

Q.2   How will you prepare 2-azetidinones from a ketene and an imine by Staundinger method?

Q.3   Explain:

(a) Azetidine exhibit conformational isomerism.

(b) Oxetane undergo cyclooligomerization in presence of BF_3.

Q.4   How does bacteria develops resistance to penicillin?

Ans. Bacteria develops resistance to penicillin by secreting Penicillinase, an enzyme that destroys penicillin by hydrolyzing its β-Lactam ring before the drug can interfere with bacterial cell wall synthesis.

Q.5   Why do the product formed in a Paterno-Buchi reaction depend on the nature of substituent present in the alkene? Explain with suitable examples.

[Hint. Addition of the alkenes possessing EWG occurs in a concerted and stereospecific manner, but alkenes with EDG react through biradical intermediates.]

Q.6   What will happen on heating thietane?

Ans. On heating thietane gives, ethylene and thioformaldehyde.

Q.7   Complete the following reactions:

(a) Azetidine + HCl $\longrightarrow$ (b) Azetidine + HNO_2 $\longrightarrow$

(c) Azetidine + HCHO $\longrightarrow$ (d) Oxetane + $NaOCH_3$ + CH_3OH $\longrightarrow$

(e) Thietane + NH_3 $\xrightarrow{200°C}$

Ans. (a) $ClCH_2CH_2CH_2NH_2.HCl$ (b) N-Nitrosoazetidine

(c) N-Hydroxymethyl azetidine   (d) $HOCH_2CH_2CH_2OCH_3$

(e) $HSCH_2CH_2CH_2NH_2$

Q.8   The following photochemical transformation proceeds through which process.

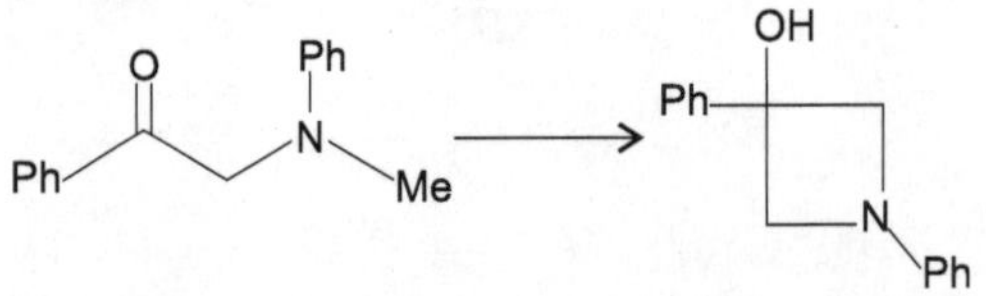

Ans. Norrish type II reaction

Five Membered Heterocyclic Compounds with One Heteroatom

Pyrrole, Furan and Thiophene
Furan Derivatives: Furfural and 2-Furoic Acid

The five membered heterocyclic compounds, both saturated and unsaturated, are widespread in the living world. The saturated five membered rings with one heteroatom are pyrrolidine, tetrahydrofuran, thiolane, borolane, phospholanes, arsolane, stibolanes, bismolanes, silolanes and stannolanes.

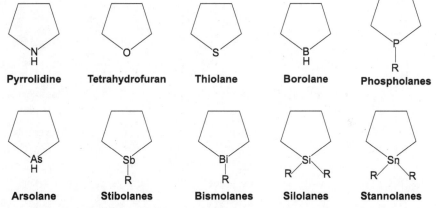

Pyrrolidine	Tetrahydrofuran	Thiolane	Borolane	Phospholanes

Arsolane	Stibolanes	Bismolanes	Silolanes	Stannolanes

While five membered unsaturated ring compounds, with one heteroatom are pyrrole, furan, thiophene, borole, phosphole, arsole, stibole, bismole, silole and stannole.

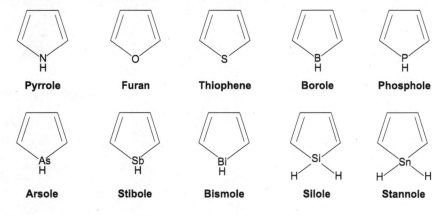

Pyrrole	Furan	Thiophene	Borole	Phosphole

Arsole	Stibole	Bismole	Silole	Stannole

Examples of saturated five membered ring compounds containing two heteroatoms are imidazolidine, pyrazolidine, oxazolidine, isoxazolidine, thiazolidine, isothiazolidine, 1,3-dioxolane, 1,2-dioxolane, 1,3-dithiolane and 1,2-dithiolane.

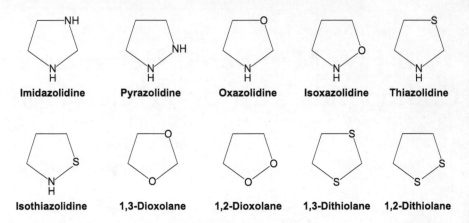

The unsaturated five membered ring compounds with two heteroatoms are imidazole, pyrazole, oxazole, isoxazole, thiazole and isothiazole.

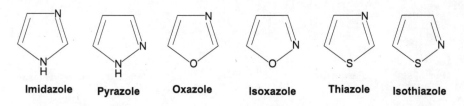

Examples of five membered ring compounds with three heteroatoms are triazole, thiadiazole and dithiazoles. Although, five membered ring compounds with four heteroatoms exists e.g., tetrazole, but five membered ring compounds with five heteroatoms compound is inorganic not a heterocyclic compound e.g. pentazole.

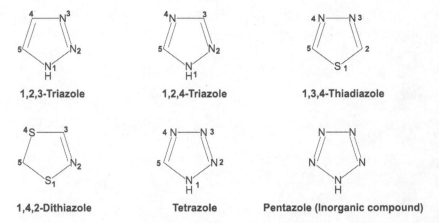

5.1 PYRROLE, FURAN AND THIOPHENE

5.1.1 Introduction

The three main five membered aromatic heterocyclic compounds with one heteroatom are **pyrrole, furan** and **thiophene**.

| Pyrrole | Furan | Thiophene |

Pyrrole is a five membered heterocyclic aromatic compound containing one nitrogen atom. Pyrrole is present in coal-tar and bone oil and isolated in pure form from bone oil in 1857. Its structure was established in 1870. It is a colourless volatile liquid which darkens readily upon exposure to air. It turns brown over time due to accumulation of impurities such as polypyrrole and various amine oxides. Therefore, it is usually purified by distillation before use.

The 1*H*-pyrrole form and its derivatives are more stable as compared to pyrrolenines, its other tautomeric forms.

| 1*H*-Pyrrole | 2*H*-Pyrrolenine | 3*H*-Pyrrolenine |

The boiling point of pyrrole (bp 126°C) is higher than furan (bp 32°C) and thiophene (bp 84°C), due to the presence of intermolecular hydrogen bonding in pyrrole.

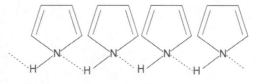

H-Bonding in pyrrole

Furan is a five membered heterocyclic aromatic compound containing one oxygen atom. The name furan originated from Latin word furfur, which means bran. Furan is a colourless, flammable, highly volatile liquid with boiling point close to room temperature. It is soluble in water and common organic solvents such as alcohol, ether and acetone Furan was first prepared by Heinrich Limpricht in 1870. Hydrogenation of furan gives **tetrahydrofuran** which is used as a solvent.

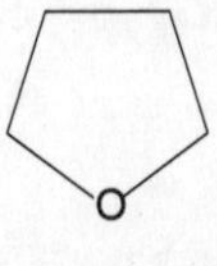

**Tetrahydrofuran
(THF)**

Thiophene is a heterocyclic five membered ring compound containing one sulfur atom. Thiophene and related compounds are found in coal tar and crude petroleum.

5.1.2 Biological importance

Pyrrolidine ring is found in a few amino acids (**proline** and **hydroxyproline**) and alkaloids (**nicotine**).

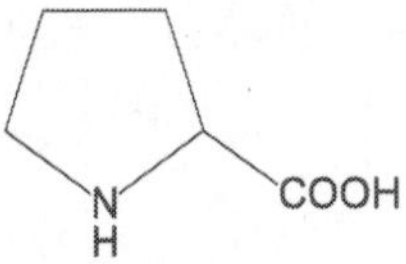

Proline 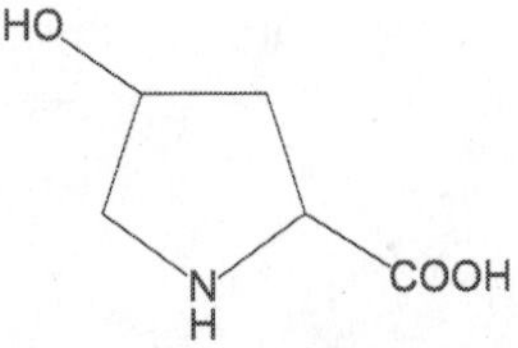   **Hydroxyproline** 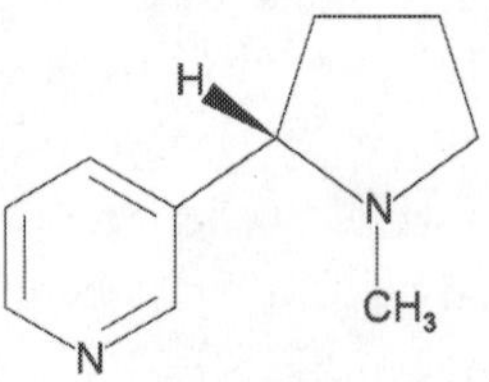   **Nicotine**

The biomolecules **chlorophyll**, **hemoglobin** and **vitamin B12** are formed from four pyrrole units joined in a larger ring system known as a porphyrin (for details see chapter 14). The bile pigments, formed during the decomposition of hemoglobin mainly in the spleen and liver, include four pyrrole rings interconnected by carbon atoms.

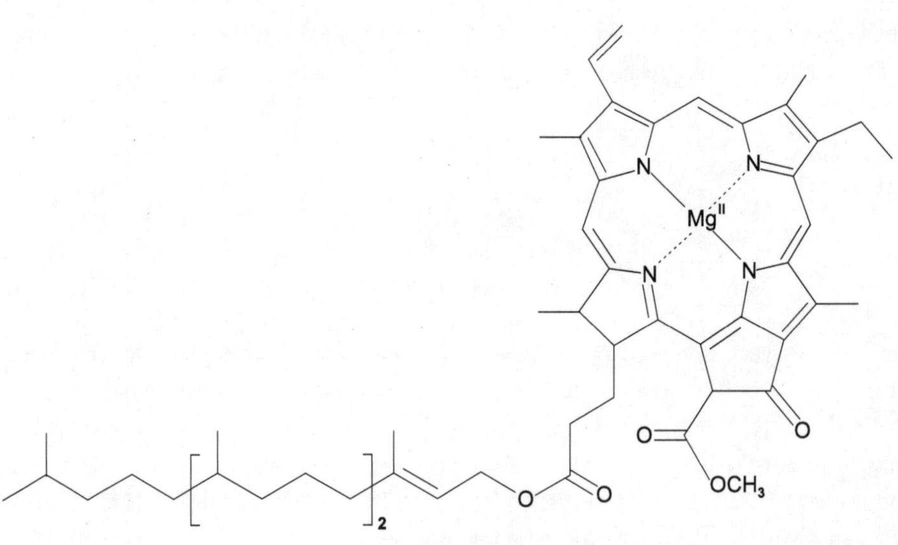

Chlorophyll a

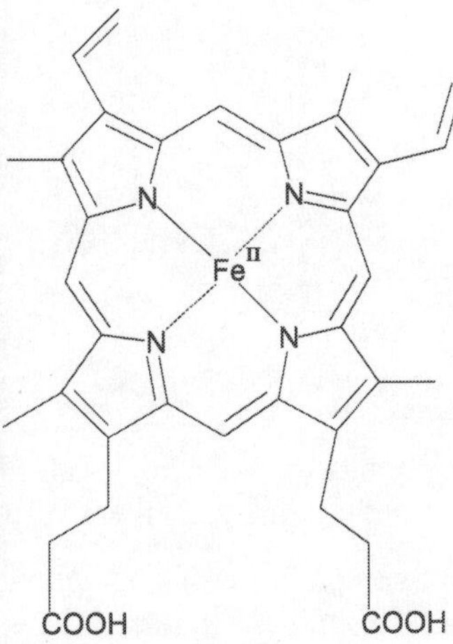

Heme group in hemoglobin

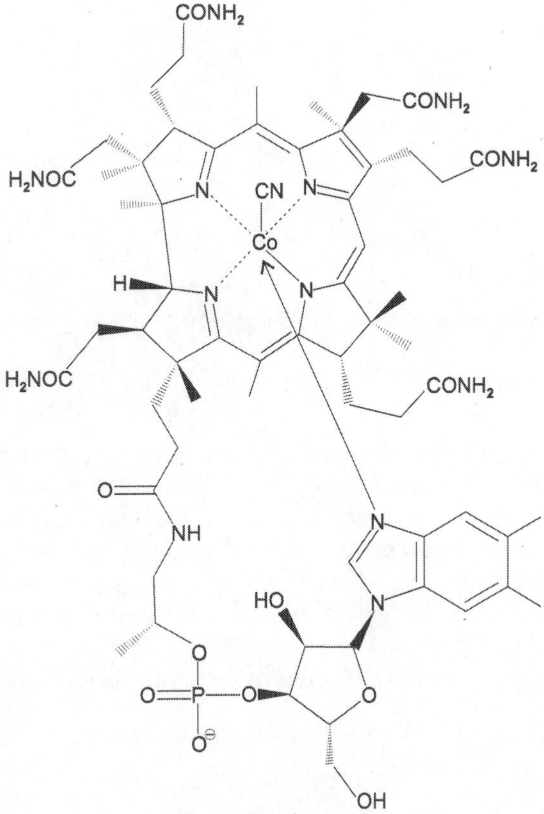

Vitamin B12

Vitamin C is a derivative of furan. The **2,5-dimethylfuran** is used as a bio-fuel. Ribose is a five carbon pentose sugar with tetrahydrofuran ring system. **Deoxyribose** and **ribose** are present in furanose form in nucleic acids (DNA and RNA).

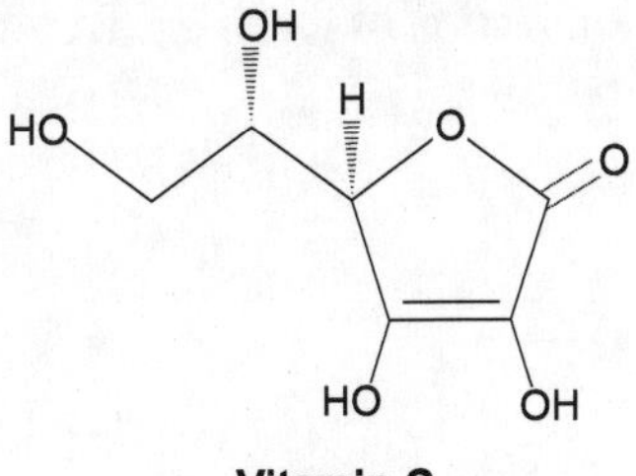

Vitamin C

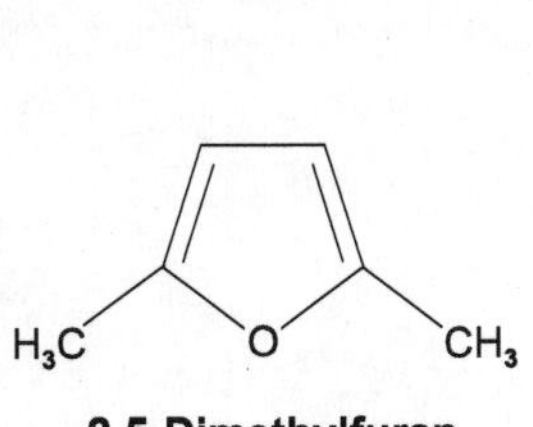

2,5-Dimethylfuran

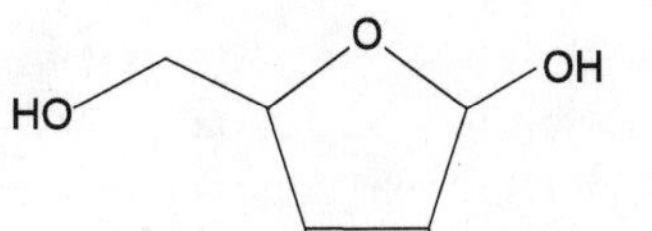

Deoxyribofuranose

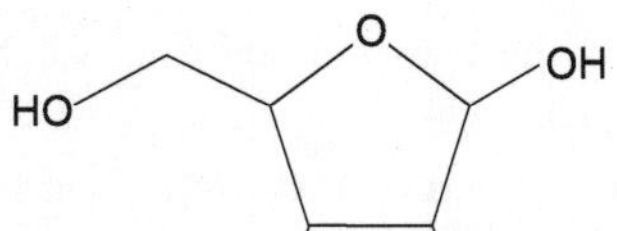

Ribofuranose

The most important biologically occurring thiophene derivative is the vitamin **biotin**.

Biotin

5.1.3 Structure and aromaticity

The most important five membered ring heterocyclic compounds with one hetero atom N, O and S along with two double bonds are pyrrole, furan and thiophene, respectively.

The hybridization of nitrogen in pyrrole is sp^2 not sp^3 as in the case of open chain amines. The three sp^2 hybrid orbitals are bonded to two carbons and one hydrogen atom. Thus in pyrrole the hydrogen atom attached to nitrogen is present in the plane of the ring. The lone pair of electrons on nitrogen is present in p-orbital. The four p-orbitals of carbon atoms, each with one electron and the p-orbital of nitrogen, with two electrons overlap and these six electrons are delocalized over the ring. This accounts for the aromatic sextet and Huckel $(4n+2)$ π-electron rule is obeyed. Thus, pyrrole being aromatic is more stable than its saturated counterpart pyrrolidine.

Lone pair in p orbital contributes
to the 6 π-e aromatic system

sp^2 hybridized

Pyrrole is a hybrid of five resonating structures. These structures do not contribute equally. In all the contributing structures carbon atoms have increased electron density. Thus carbon are said to be π-excessive. The delocalization of lone pair over the ring is clearly indicated by the resonating structures of pyrrole.

Resonance in pyrrole

In pyrrole, the C2 - C3 and C4 - C5 bonds (1.38 A°) have some single-bond character and longer than that of a true double bond (1.34 A°), whereas the C3 - C4 bond (1.42 A°) is shorter than a true single bond (1.48 A° for sp^2 - sp^2 bonds) due to some double-bond character. This confirms pyrrole to be a resonance hybrid of the five resonating structures.

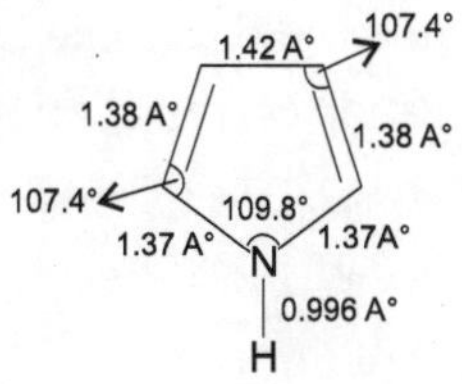

Bond parameters of pyrrole 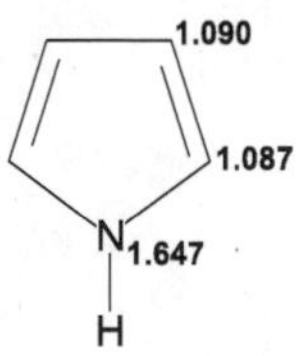   Electron density on each ring atom in pyrrole

Similarly, furan and thiophene are also aromatic and resonating structures for both the heterocyclic compounds are shown below.

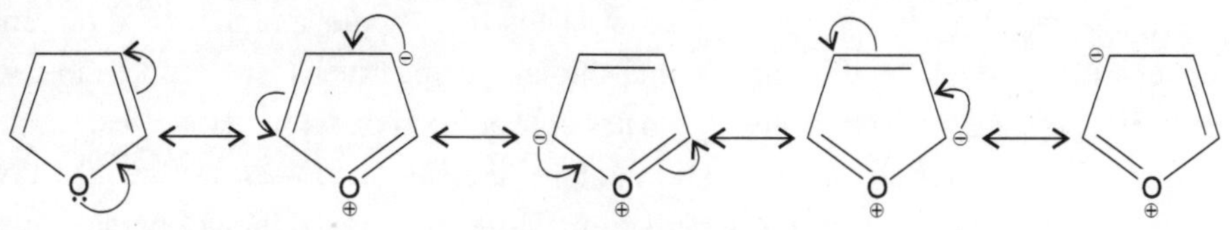

Resonance in furan

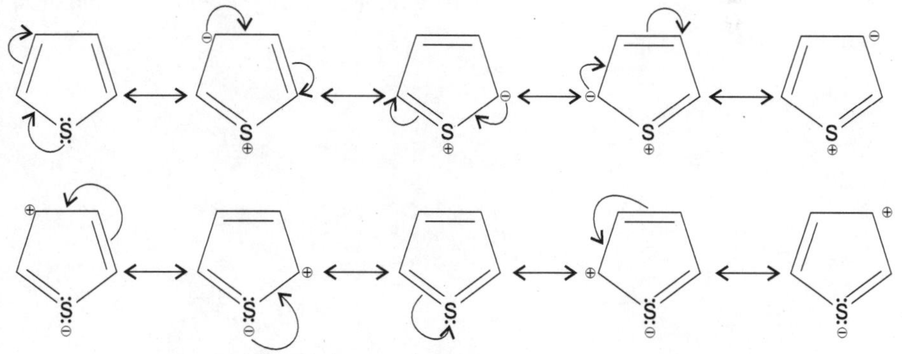

Resonance in thiophene

The resonance energy of these three heterocyclic compounds follow the order thiophene (29.1 kcal/mol or 121 kJ/mol) > pyrrole (21.6 kcal/mol or 87.7 kJ/mol) > furan (16.2 kcal/mol or 66.8 kJ/mol). The extra stabilization of these rings as indicated by resonance energy values is due to electron delocalization. Because of the lower electronegativity of sulfur (2.56) compared with nitrogen (3.04) and oxygen (3.44), the electron pair on sulfur is more effectively involved in the conjugated system. Thus delocalization in thiophene results in more resonance energy. However the resonance energies due to π-electron delocalization are less than benzene (36.0 kcal/mol). The resonance energy of thiophene (29.1 kcal/mol) is similar to that of pyridine (27.9 kcal/mol).

Therefore, order of aromaticity is **benzene > thiophene > pyrrole > furan.**

This order is consistent with the order of the electronegativity values of S, N and O, Thus electronegativity of the heteroatom is the main criteria in deciding the aromaticity of these heterocycles.

Sulfur is an element of the second period of the periodic table and thus it can expand its octet by using its d-orbitals. The highest resonance stabilization energy of thiophene is also due to the expansion of valence shells by using d-orbitals in hybridisation. Thus, it has five more resonating structures which involve d-orbital participation.

5.1.4 Dipole moment

In case of pyrrole, furan and thiophene the electrons are delocalised towards the ring. This creates a dipole from the heteroatom towards the ring. Also, an inductive effect operates towards the heteroatom. The net dipole moment is the resultant of these two effects operating in opposite direction. Thus, the dipole moment of furan and thiophene is less than tetrahydrofuran and thiolane respectively. The dipole moment of pyrrole is opposite to others due to greater contribution of mesomeric effect. In the case of pyrrolidine, tetrahydrofuran and thiolane, the saturated analogs, the direction of their dipole is from the ring to heteroatom. More over, the saturated heterocycles pyrrolidine, tetrahydrofuran and thiolane are not planar and their dipole moments do not lie in the same plane. The dipole moment values are given below and medium is given in brackets.

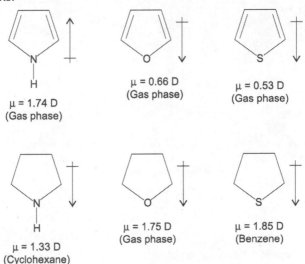

5.1.5 Spectral data

The ^{1}H NMR and ^{13}C NMR (given in brackets) chemical shift values for pyrrole, furan and thiophene are shown below:

^{1}H NMR (δ, ppm, CDCl$_3$)
^{13}C NMR (δ, ppm, CH$_2$Cl$_2$)

5.1.6 Methods of preparation of pyrrole, furan and thiophene

5.1.6.1 Methods of preparation of pyrrole

(*i*) **From furan:** Pyrrole can be synthesized by heating a mixture of furan, ammonia and steam over alumina (Al_2O_3) as catalyst.

(*ii*) **From succinic anhydride or succinimide:** Succinic anhydride on treatment with ammonia results in the formation of succinimide. Reduction of enol form of succinimide with zinc dust leads to the formation of pyrrole.

(*iii*) **Paal-Knorr synthesis of pyrroles**: The synthesis of pyrrole from 1,4-dicarbonyl compounds and ammonia or primary amine is known as Paal-Knorr synthesis. For example, hexane-2,5-dione when refluxed with ammonia in the presence of benzene leads to the formation of 2,5-dimethylpyrrole.

Hexane-2,5-dione   **2,5-Dimethylpyrrole**

Mechanism: It involves nucleophilic addition of the amine nitrogen to carbonyl carbon atom. The ring-closure followed by dehydration then yields the two double bonds and thus the aromatic π system.

(*iv*) **Hantsch pyrrole synthesis:** Hantsch pyrrole synthesis involves synthesis of pyrrole derivatives from β-ketoester, ammonia (or primary amines) and α-haloketones.

Ethylacetoacetate

Mechanism: The reaction involves the formation of ethyl β-aminocrotonate by the reaction of ammonia or amine with ethyl acetoacetate. Further reaction of α-haloketone and ethyl β-aminocrotonate gives pyrrole derivative as shown.

Ethyl β-aminocrotonate

(*v*) **Knorr pyrrole synthesis:** The reaction of α-aminoketones with carbonyl compounds containing active methylene group (electron withdrawing group at α-position) to form substituted pyrroles is known as Knorr pyrrole synthesis.

The free α-aminocarbonyl compound can undergo self condensation readily to produce dihydropyrazines. Therefore they are used in the form of their salts and the free α-aminocarbonyl can be liberated by the base present in the reaction mixture.

The original Knorr synthesis employed two equivalents of ethyl acetoacetate, one of which was converted to ethyl 2-oximinoacetoacetate by dissolving it in glacial acetic acid, and slowly adding one equivalent of saturated aqueous sodium nitrite, under external cooling. Oxime group is then reduced to the amine group by zinc dust.

Ethyl 2-oximinoacetoacetate

Mechanism: Reaction involves attack of nitrogen on more electrophilic carbonyl carbon. The reaction proceeds *via* formation of an enamine intermediate. Further cyclization leads to formation of C3-C4 bond.

(*vi*) **Piloty-Robinson pyrrole synthesis**: Piloty-Robinson pyrrole synthesis involves reaction of aliphatic carbonyl compounds with hydrazine under strongly acidic conditions to form pyrrole derivatives.

3-Pentanone **Hydrazine**

Mechanism: The reaction proceeds *via* [3+3] sigmatropic rearrangement followed by cyclisation.

$$2\ \underset{CH_3-CH_2}{\overset{CH_3-CH_2}{>}}C=O\ +\ H_2NNH_2\ \xrightarrow{\text{acid}}\ \underset{CH_3-CH_2}{\overset{CH_3-CH_2}{>}}C=N-N=C\underset{CH_2-CH_3}{\overset{CH_2-CH_3}{<}}$$

(reaction scheme showing subsequent tautomeric and cyclization intermediates)

$$\xleftarrow{H^{\oplus}}\ \underset{CH_3-CH_2}{\overset{H_3C}{>}}CH\!-\!\underset{}{C}\!-\!NH\!-\!NH\!-\!C\!-\!CH\underset{CH_2-CH_3}{\overset{CH_3}{<}}$$

$$\downarrow$$

(further intermediates)

$$\xrightarrow[-NH_3]{-H^{\oplus}}$$

$$\underset{CH_3CH_2\quad \underset{H}{N}\quad CH_2CH_3}{\overset{H_3C\qquad\qquad CH_3}{\text{pyrrole ring}}}$$

(***vii***) **From acetylene:** By passing a mixture of acetylene and ammonia over red hot tube gives pyrrole.

$$\underset{H}{\overset{H}{\underset{C}{\overset{C}{|||}}}}\ +\ NH_3\ +\ \underset{H}{\overset{H}{\underset{C}{\overset{C}{|||}}}}\ \xrightarrow{\text{Red hot tube}}\ \underset{N\ H}{\text{pyrrole}}$$

5.1.6.2 Methods of preparation of furan

(***i***) **From pentosans:** Pentosans are polysaccharides extracted from many plants, e.g corn cobs and rice husks and can be converted to xylopyranose. In the presence of acid xylopyranose losses three mole equivalents of water and gets converted to furfural. Steam distillation of furfural yields furan.

Furfural → **Furan**

(*ii*) **The Paal-Knorr synthesis:** 1,4-Dicarbonyl compound undergo cyclization and dehydration in the presence of non aqueous acidic medium usually p-toluenesulfonic acid (p-CH$_3$C$_6$H$_4$SO$_3$H) in benzene to yield furan. The other acid employed for this reaction is sulfuric acid, zinc chloride, acetic anhydride, phosphorous pentoxide or phosphoric acid.

Mechanism: The acid catalyzed furan synthesis from 1,4-diketones proceeds by protonation of one carbonyl group which is followed by ring closure and dehydration to give the resultant furan.

(*iii*) The Feist-Benary synthesis: The reaction of α-halocarbonyl compound and 1,3-dicarbonyl compound in the presence of a base to give furan, is known as Feist-Benary synthesis.

Mechanism: This method involves aldol condensation of 1,3-dicarbony compounds and 2-halocarbonyl compound which is followed by ring closure *via* intramolecular displacement of halide by enolate oxygen.

(*iv*) **From α,β-alkynic ketones:** Reaction of α,β-alkynic ketones with alcohol in the presence of light yields furans.

Mechanism: It involves the formation of 1:1 adduct between α,β-alkynic ketone and alcohol photochemically. The excited carbonyl oxygen abstracts α-hydrogen atom from alcohol which on cyclization leads to the formation of furan.

5.1.6.3 Methods of preparation of thiophene

(*i*) **From hydrocarbons:** Thiophene is synthesized by the reaction of hydrocarbons (n-butane or butadiene) and elemental sulfur at 600°C in gas phase. In case of n-butane, the sulfur first causes dehydrogenation in butane and then reacts with the unsaturated hydrocarbon *via* addition. Further dehydrogenation of the intermediate forms the aromatic system i.e., thiophene.

(*ii*) **Paal-Knorr synthesis**: The reaction of a 1,4-dicarbonyl compound with a source of sulfur such as phosphorus pentasulfide, Lawesson's reagent (LR) or bis(trimethylsilyl) sulfide gives thiophenes. This reaction is also known as Paal thiophene synthesis.

Mechanism: Thiophene synthesis is achieved *via* a mechanism very similar to the furan synthesis. The initial diketone is converted to a thioketone with a sulfurizing agent, which then undergoes the same mechanism as the furan synthesis. The phosphorous pentasulfide or Lawesson's reagent acts as sulfurizing as well as dehydrationg agent.

(*iii*) **The Hinsberg synthesis:** The condensation between a 1,2-dicarbonyl compound such as benzil and diethyl thiodiacetate in the presence of base such as sodium ethoxide in ethanol followed by acid hydrolysis gives thiophenes.

Mechanism: The mechanism of this reaction involves two consecutive aldol condensations between a 1,2-dicarbonyl compound and diethyl thiodiacetates. The immediate product is an ester-acid produced by a Stobbe type mechanism but the reactions are often worked up *via* hydrolysis to afford an isolated diacid.

(*iv*) From thioglycolates and 1,3-dicarbonyl compounds: Thioglycolates react with 1,3-dicarbonyl compounds to give thiophene-2-carboxylic acid esters.

(*v*) **From α-thiocarbonyl compounds:** 2-Ketothiols react with alkenyl phosphonium ions, to give ylides. The ylides undergo ring closure by Wittig reaction to give 2,5-dihydrothiophenes, which can be dehydrogenated to form corresponding thiophene.

2-Ketothiol · Alkenyl phosphonium ion

(*vi*) **From β-chloro-α,β-unsaturated carbonyl compounds:** β-Chloro-α,β-unsaturated carbonyl compounds on reaction with ethyl mercaptoacetate gives thiophene in the presence of triethylamine as a base.

Mechanism of this reaction is given below.

(*vii*) **From acetylene**: Thiophene can be synthesized by passing a mixture of acetylene and hydrogen sulfide through a tube containing alumina at 400°C. This method is commercially used for the synthesis of thiophene.

5.1.7 Reactions of pyrrole, furan and thiophene

Five membered ring heterocyclic compounds pyrrole, furan and thiophene undergo electrophilic substitution reactions. The delocalization pushes lone pair of electrons from heteroatom into the ring which activates the ring. This causes **faster electrophilic**

substitution than benzene. The reactions of these compounds towards electrophiles resemble that of phenol and aromatic amine. The rate limiting step of such reaction is attachment of the electrophile to the aromatic ring.

X= NH, O, S

Electrophilic substitution takes place mainly at the 2-position. It can be explained on the basis of the number of resonating structures of the intermediate formed.

The attack of electrophile at C2 (2-position) on pyrrole produces intermediate resonating structures with positive charge delocalized over three atoms (two ring carbons and one heteroatom). Thus three resonating structures are possible. But attack of electrophile at C3 produces intermediate resonance structures with two atoms sharing the charge (one carbon and one heteroatom). So only two resonating structures are formed. Thus attack of electrophile at C2 produces more number of canonical structures of the intermediate than at C3.

Other explanation is based on the stability of the intermediate formed. The intermediate resulting from C2 substitution has three resonance contributors. Two of which has a positive charge on relatively stable secondary allylic carbon. The intermediate resulting from C3 substitution has two resonance contributors. One of which has a positive charge on a secondary carbon. The stability of resonance contributor with a positive charge on a secondary allylic carbon is more than that with positive charge on a secondary carbon.

However, if both positions adjacent to the heteroatom are occupied, electrophilic substitution will take place at C3. For example, bromination of 2,5-dimethylfuran gives 3-bromo-2,5-dimethylfuran.

3-Bromo-2,5-dimethylfuran

Furan is not as reactive pyrrole in electrophilic substitution reaction. The oxygen of furan is more electronegative than the nitrogen of pyrrole so oxygen is not as effective as nitrogen in donating electron into the ring. Thiophene is less reactive than furan towards electrophilic substitution because sulfur's p electrons are in a 3p orbital which overlaps less effectively than 2p orbital of nitrogen or oxygen with 2p orbital of carbon.

Therefore, relative order of reactivity towards electrophilic aromatic substitution is : **pyrrole > furan > thiophene > benzene.**

Pyrrole   Furan   Thiophene   Benzene

Order of reactivity towards electrophiles

Pyrrole undergoes electrophilic substitution reactions about 10^5 times faster than furan. Though, its resonance energy is greater than furan, and it should react more slowly. This is because in the case of pyrrole, the σ-complex is stablilized by a carbenium-iminium mesomerism.

Of the three common five membered ring aromatic heterocyclic compounds with one heteroatom, furan is least aromatic. However furan undergoes electrophilic substitution about 10^{11} times faster than benzene under similar conditions. The furan ring has a π-electron excess.

5.1.7.1 Reactions of pyrrole

Acid-base reactions

The presence of NH group in pyrrole makes it both basic and acidic in nature. It is a very weak base ($pK_a = -3.8$ of its conjugate acid) as compared to secondary amine pyrrolidine ($pK_a = +11.3$ of its conjugate acid). This is because in pyrrole the lone pair on the nitrogen atom is delocalized over the ring to complete 6π-electron system of pyrrole. Whereas, in the case of pyrrolidine the lone pair is freely available for donation thus making it more basic.

Protonation on C2 carbon rather than on nitrogen and C3 carbon in pyrrole is more preferred because of the mesomeric stabilization of the C2 protonated pyrrole (2*H*-pyrrolium cation).

1*H*-Pyrrolium ion	2*H*-Pyrrolium ion	3*H*-Pyrrolium ion
(least stable)	(More stable)	(less stable)
	More preferred	

pK$_a$ = –3.8 (Conjugate acid)

C2 Protonation in pyrrole

The 2*H*- and 3*H*- pyrrolium cations are essentially iminium ions and as such are electrophilic. They play the key role in polymerisation and reduction of pyrroles in acid. Thus, pyrrole is unstable in strongly acidic medium and it polymerizes readily.

Neutral pyrrole is more acidic (pK$_a$ = 17.5) than pyrrolidine (pK$_a$ = 25), a saturated secondary amine because the nitrogen in pyrrole is sp^2 hybridized and thus is more electronegative than the sp^3 nitrogen of pyrrolidine. The sp^2 hybrid orbitals have greater s character than sp^3. Therefore, have greater tendency to pull the shared electrons towards themselves. This makes the release of proton from sp^2 hybrid nitrogen much easier than sp^3.

Pyrrole
pK$_a$ = 17.5

sp^3 Pyrrolidine
pK$_a$ = 25

For this reason, pyrrole is N-deprotonated by sodium, potassium or sodium hydride in inert solvents, with sodium amide in liquid ammonia and alkyllithium and alkylmagnesium halide in inert solvent to give pyrrolyl metal compounds.

The presence of active hydrogen in pyrrole can also be detected with methylmagnesium iodide by Zerewitinoff method.

N-Pyrrolyl magnesium iodide

Reaction of pyrrole with butyllithium gives N-pyrrolyl lithium.

N-Pyrrolyl lithium

The pyrrolyl anion acts as an ambident anion. This can be used to make a bond with a suitable electrophile to produce the N-substituted pyrrole, if the cation is not N-coordinated and more strongly solvated (e.g. Li^+, Na^+ and K^+) or C-substituted pyrrole, if the cation is N-coordinated (e.g. MgX^+).

N-Pyrrolyl sodium and N-pyrrolyl potassium reacts with haloalkanes, acyl halides, sulfonyl halides and chlorotrimethylsilane to yield corresponding N-substituted pyrrole.

On the other hand, 2-methylpyrrole is obtained from pyrrol-1-ylmagnesium iodide and methyliodide.

2-Methylpyrrole

Electrophilic substitution reactions

Pyrroles are resistant to nucleophilic addition and substitution, but they are attacked by electrophilic reagents very readily. Pyrrole, N-alkyl pyrrole, C-monoalkyl pyrrole and C,C′-dialkyl pyrroles are polymerised by strong acids. Hence many of the electrophilic reagents used in benzene chemistry cannot be used in case of pyrrole. If an electron withdrawing substituent such as an ester is present it prevents polymerisation and strongly acidic, nitrating and sulfonating agents can be used.

In most electrophilic substitution reactions, pyrrole is preferentially attacked at the 2-position because of the mesomeric stabilization of the intermediate cation.

Halogenation: Pyrrole on reaction with N-chlorosuccinimide (NCS) yields 2-chloropyrrole.

With SO_2Cl_2 or aq. NaOCl the product formed is 2,3,4,5-tetrachloropyrrole, from pyrrole. However, if excess of SO_2Cl_2 is used perchloro-2*H*-pyrrole is obtained.

Reaction of pyrrole with N-bromosuccinimide (NBS) and bromine forms 2-bromopyrrole and 2,3,4,5-tetrabromopyrrole, respectively.

Pyrrole with iodine in presence of potassium iodide yields tetraiodopyrrole (Iodol).

Iodol

Nitration: Nitrating mixtures suitable for benzenoid compounds cause complete decomposition of pyrrole. Pyrroles are nitrated with acetyl nitrate at low temperature, giving mainly 2-nitropyrrole. Acetyl nitrate is formed by mixing fuming nitric acid with acetic anhydride at -10°C. Nitration of pyrrole with acetyl nitrate at C2 and C3 is much faster than in benzene.

Sulfonation: Concentrated sulfuric acid causes polymerization of pyrroles. Conc. H_2SO_4 at low temperature or a mild reagent of low acidity for example, pyridine-sulfur trioxide complex at 100°C converts pyrrole into pyrrole-2-sulfonic acid.

Alkylation: Pyrroles resist direct alkylation with alkyl halides however in the presence of Lewis acid catalysts polymerization occurs.

Vilsmeier-Haack reaction: The Vilsmeier-Haack formylation leads to the formation of pyrrole-2-carbaldehyde in good yield. The actual electrophilic species is an N,N-dimethylchloromethylene iminium cation which is formed by the reaction of DMF with $POCl_3$. The final intermediate in a Vilsmeier reaction is an iminium salt which on hydrolysis produces aldehyde.

Houben-Hoesch acylation: The Houben-Hoesch acylation involves reaction of pyrroles with nitriles in the presence of hydrogen chloride to yield 2-acylpyrroles. As in case of Vilsmeier-Haack reaction, the 2-acylpyrrole is formed by the hydrolysis of iminium cation.

Reaction with diazonium salts: Pyrroles react with arene diazonium salts to give azo compounds. Pyrrole couples even faster than N,N-dimethylaniline. With 2,5-disubstituted pyrroles, coupling occurs at the 3-position.

Gattermann formylation: Reaction of pyrrole with zinc cyanide and hydrochloric acid leads to C2 formylation.

$$Zn(CN)_2 \; + \; 2\,HCl \rightleftharpoons \; ZnCl_2 \; + \; 2\,HCN$$

$$ZnCl_2 \; + \; HCl \rightleftharpoons \; \overset{\oplus}{H}Zn\overset{\ominus}{Cl_3}$$

$$HC{\equiv}N \; + \; \overset{\oplus}{H}Zn\overset{\ominus}{Cl_3} \rightleftharpoons \left[H{-}C{\equiv}\overset{\oplus}{N}{-}H\right] + \; ZnCl_2 \; + \; \overset{\ominus}{Cl}$$

$$H{-}\overset{\oplus}{C}{=}NH$$

Hydroxymethylation: Pyrroles undergo hydroxymethylation in the 2-position with carbonyl compounds in the presence of acid. The products react further to give dipyrrolylmethanes.

Michael addition: Pyrrole does not undergo Diels-Alder addition with maleic anhydride but results in an electrophilic substitution reaction. This is due to the great reactivity of pyrrole towards electrophiles. In this reaction pyrrole behaves as a 6π-electron system not as diene. This reaction can also be regarded as a Michael addition of pyrrole to maleic anhydride.

Cycloaddition reactions

Paterno-Buchi reaction involves [2+2] cycloadditions with pyrroles to form oxetanes which isomerize to give 3-(hydroxyalkyl)pyrroles under the reaction conditions.

Pyrrole undergoes cyclopropanation reaction with dichlorocarbene followed by rearrangement to give 3-chloropyridine with elimination of hydrogen chloride. The overall reaction is known as **Ciamiaci-Dennstedt rearrangement.**

Ciamician, G.L.; Dennstedt, M., Ber. 1881, 14, 1153

3-Chloropyridine

The dichlorocarbene can be generated in a weakly basic medium by heating sodium trichloroacetate and reaction proceeds *via* [2+1] cycloaddition.

$$CCl_3COONa \xrightarrow{\Delta} \overset{\ominus}{C}Cl_3\overset{\oplus}{Na} + CO_2$$
$$\overset{\ominus}{C}Cl_3\overset{\oplus}{Na} \longrightarrow \; :CCl_2 + NaCl$$

In the presence of strong base, the dichlorocarbene is electrophilically substituted on pyrrole. The main product formed is pyrrole-2-carbaldehyde. Thus similar to phenols the sodium or potassium salt of pyrrole also undergoes Reimer-Tiemann reaction to give pyrrole-2-carbaldehyde.

Oxidation

Attack of O_2 or H_2O_2 to pyrrole or N-substituted pyrrole occurs first at the 2-position and then at the 5-position, resulting finally in the formation of maleimide or N-substitued maleimide.

Ring-opening reactions

Pyrrole reacts with hydroxylamine hydrochloride and sodium carbonate in ethanol to form dioximes of 1,4-dicarbonyl compounds.

Reduction

Pyrrole can be reduced using various reducing agents. Hydrogenation of pyrroles to pyrrolidines by Raney nickel proceeds only under pressure and at high temperatures.

Hydride reducing agents or diborane do not reduce simple pyrroles. But in acidic media the protonated pyrrole can be reduced to 2,5-dihydropyrroles along with pyrrolidine as by product. Reduction with Zn/acetic acid gives 2,5-dihydropyrrole.

cis   trans

5.1.7.2 Reactions of furan

Electrophilic substitution reactions

Halogenation: The reaction of furan with chlorine and bromine is vigrous at room temperature but it does not react at all with iodine.

Chlorination of furan at -40°C produce 2-chlorofuran (major) and 2,5-dichlorofuran (minor) along with small amount of 2,3,5-trichlorofuran.

(64%)   (29%)   (7%)

Bromination with the dioxane-Br_2 complex at -5°C gives 2-bromofuran.

Under controlled conditions with bromine in dimethylformamide at room temperature furans produce 2-bromo- or 2,5-dibromo- furans.

If the bromination is conducted in an alcohol then alcoholysis of C2-bromide, produces 2,5-dialkoxy-2,5-dihydrofurans, as mixtures of *cis*- and *trans*- isomers.

(cis & trans)

Nitration: Nitrating mixtures containing concentrated acids cannot be used for carrying out nitration of furan. Nitration with fuming nitric acid in acetic anhydride (forms $AcONO_2$) at -10°C to -20°C yields 2-nitrofuran. Further nitration of 2-nitrofuran gives 2,5-dinitrofuran as the main product.

2,5-Dinitrofuran　　　　　**2-Nitrofuran**

Sulfonation: Furan and its simple alkyl-derivatives are decomposed by the usual strong acid reagents. But the pyridine-sulfur trioxide complex (pyridine-SO_3) can be used for sulfonation which converts furan into furan-2-sulfonic acid and then further into furan-2,5-disulfonic acid.

Acylation: Friedel-Crafts acylation of furans occurs with carboxylic acid anhydrides or halides normally in the presence of a Lewis acid (BF_3 or $AlCl_3$). Aluminium chloride catalysed acetylation of furan proceeds much faster at the α-position than at the β-position. 3-Alkylfurans substitute mainly at C2 but if both α-positions are occupied as in the case of 2,5-dialkylfurans acylation takes place at β-position, but generally with more difficulty.

Vilsmeier formylation: Vilsmeier formylation of furans produce furfurals.

Alkylation and alkenylation: The usual Friedel-Crafts alkylation is not possible in the furans, because the catalyst leads to polymerisation and polyalkylation. Furan can be converted to 2,5-di-tert-butylfuran in the presence of silica and sodium carbonate.

Metallation reactions

Mercuration of furan takes place very readily with replacement of hydrogen in the presence of mercury(II) chloride and sodium acetate in aqueous ethanol.

Metallation with alkyllithium proceeds selectively at α-position. The lithiation can be achieved in refluxing ether or at low temperature. More vigorous conditions can bring about 2,5-dilithiation of furan.

Addition reactions

Hydrogenation of furan with Pd/C yields tetrahydrofuran.

Furan   **THF**

Furan reacts with bromine in methanol in the presence of potassium acetate to give 2,5-dimethoxy-2,5-dihydrofuran *via* a 1,4-addition, reaction. In this reaction furan behaves as a 1,3-dienc.

Cycloaddition reactions

Furan undergoes Diels-Alder reaction with dienophiles such as maleic anhydride similar to butadiene. In the Diels-Alder reaction HOMO of furan interacts with the LUMO of maleic anhydride. The reaction is diastereoselective. The endo-adduct is formed faster than the exo-adduct in acetonitrile at 40°C (kinetically controlled). However, with long reaction time, product formed is thermodynamically controlled and the initially formed endo-compound is completely converted into the exo-compound. The exo-compound is stable by 8 kJ/mol than the endo-compound.

endo-adduct   **exo-adduct**

Furan undergoes cycloaddition with acetylenedicarboxylic ester. The adduct of furan and acetylenedicarboxylic ester isomerises in presence of acid to phenol. While in the presence of hydrogen it breaks to give alkene and substituted furan.

In the **Paterno-Buchi reaction** one of the olefnic π-bond of furan reacts with ketone photochemically.

Furan undergoes addition reaction with carbene generated from diazomethane in the presence of copper bromide and leads to cycloaddition with the insertion of carbene into C2-C3 bond.

Reaction of furan with solution of ethyl diazoacetate in the presence of ultraviolet radiation produces cyclopropane derivative which on heating at 160°C rearranges to a ring opened aldehyde.

Ring-opening reactions

Furan is stable to aqueous mineral acids. It gets decomposed by conc. H_2SO_4 or by Lewis acids ($AlCl_3$). Furan reacts slowly with HCl. Hot, dilute mineral acids leads to hydrolytic ring opening of furan.

Furan

Ring opened product

Concentrated sulfuric or perchloric acids leads to polymerization of the protonated furan. While dilute acid (perchloric acid in aqueous DMSO), leads to the protonation of furan with subsequent nucleophilic attack by water at 2-position of the protonated furan. Finally, the 2-hydroxy-2,3-dihydrofuran gives a 1,4-dicarbonyl compound (e.g. hexane-2,5-dione).

5.1.7.3 Reactions of thiophene

Thiophene mainly undergoes reaction with electrophiles. Additions and ring-opening reactions of thiophene are less important as compared to furan. Thiophene also undergoes oxidation and desulfurization reactions due to the presence of sulfur.

Nucleophilic attack on thiophene only occurs when an activating group is already present on thiophene ring.

Protonation

Many reagent leads to acid-catalysed decomposition or polymerisation of furans and pyrroles, but thiophene is stable to these.

However, with hot phosphoric acid, thiophene gets converted to a trimer. Its structure suggests that, the electrophile involved in the first C–C bonding step is the α-protonated cation.

Trimer

Electrophilic substitution reactions

Thiophene reacts more slowly than furan but faster than benzene. The S_EAr reactivity of thiophene corresponds approximately to that of anisole. Substitution is regioselective in the 2- or in the 2,5-position.

Halogenation: Halogenation of thiophene occurs very readily at low temperature (–30°C) or at room temperature. The halogenation of thiophene at 25°C is 10^8 times faster than benzene.

Thiophene is chlorinated by Cl_2, SO_2Cl_2 or N-chlorosuccinimide (NCS). 2,5-Dichlorothiophene can be obtained from thiophene by reaction with $SnCl_4$ under appropriate conditions.

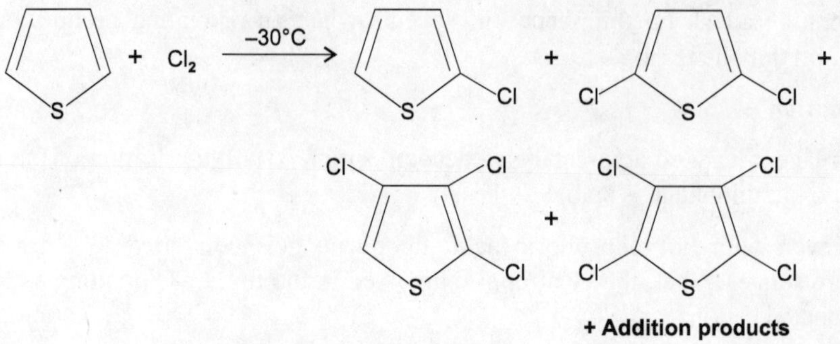

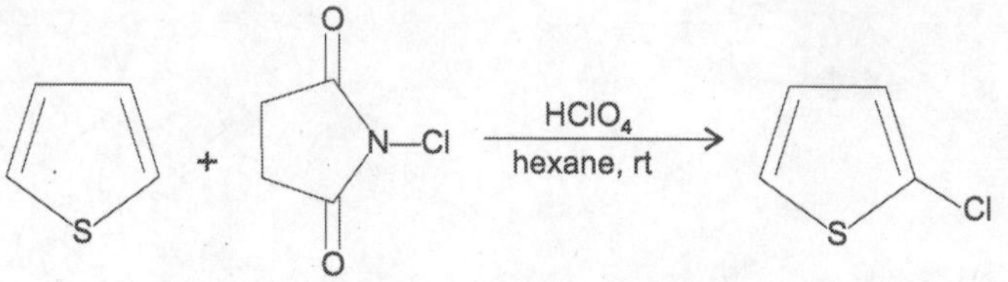

+ Addition products

NCS

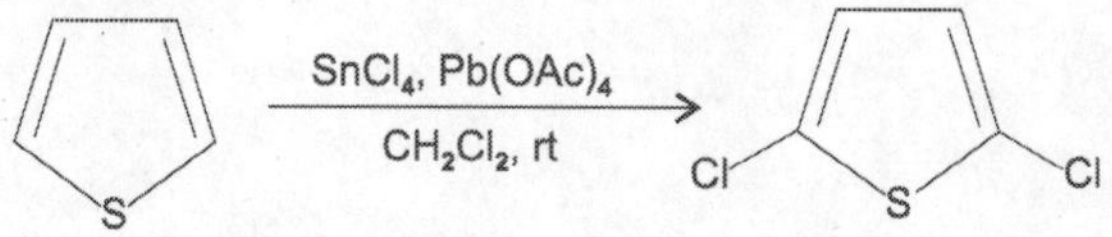

2-Bromo- and 2,5-dibromo- thiophenes can be produced under various controlled conditions. Bromination occurs at 2-position by Br_2 in acetic acid or with N-bromosuccinimide.

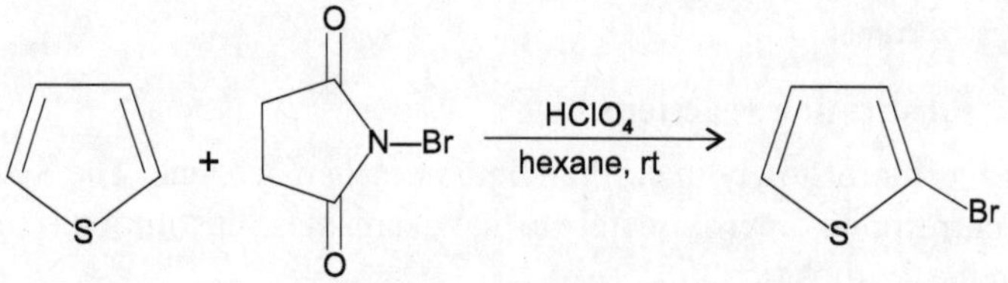

NBS

The 2,5-dibromothiophene can be obtained by treating thiophene with bromine in ether and 48% HBr.

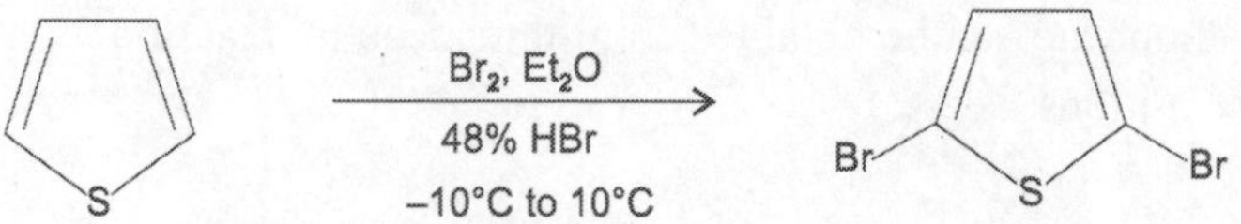

Iodination of thiophene in presence of nitric acid gives 2-iodothiophene. Iodination of thiophene in the presence of HgO gives 2-iodothiophene as major and 2,5-diiodothiophene as minor product.

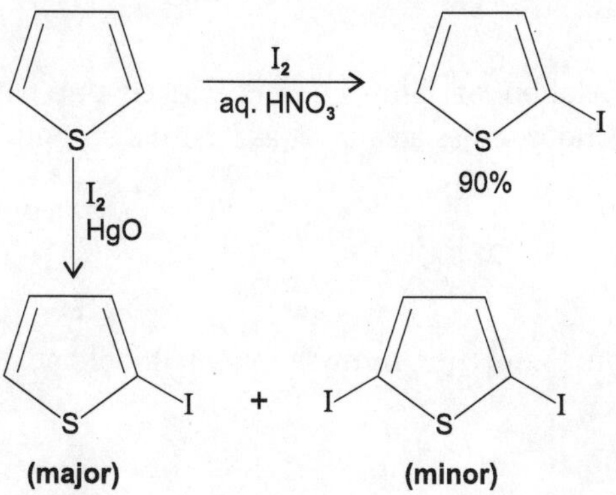

(major) + (minor)

Nitration: Nitration of thiophene with concentrated nitric acid in acetic anhydride at 10°C gives 2-nitrothiophene as major product. Nitration of 2-methylthiophene with HNO_3 gives 2-methyl-3-nitrothiophene and 2-methyl-5-nitrothiophene.

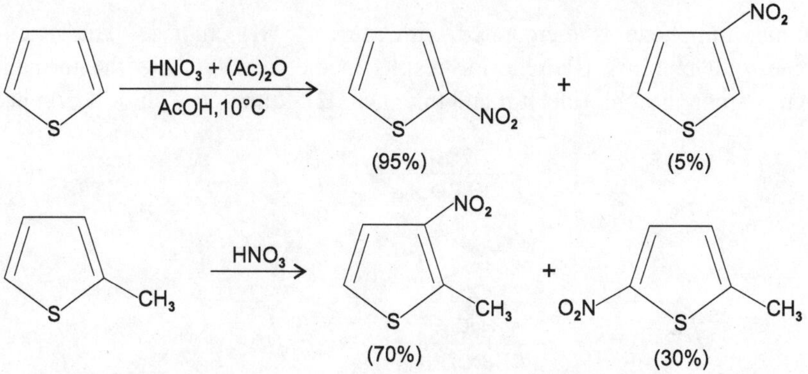

Sulfonation: Sulfonation with 96% H_2SO_4 occurs at 30°C within minutes. Benzene reacts extremely slowly under these conditions. This provides the basis for a method to remove thiophene from coal tar benzene.

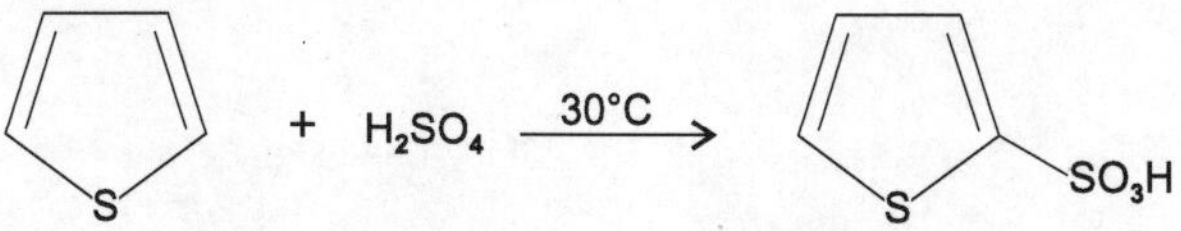

Thiophene-2-sulfonic acid

Alkylation: Alkylation of thiophene under Friedel-Crafts conditions results in polyalkylations and polymerization because of the successive protonation and electrophilic substitution. However alkylation of thiophene with alkene in the presence of phosphoric acid or boron trifluoride gives a mixture of 2-alkylated and 3-alkylated isomers.

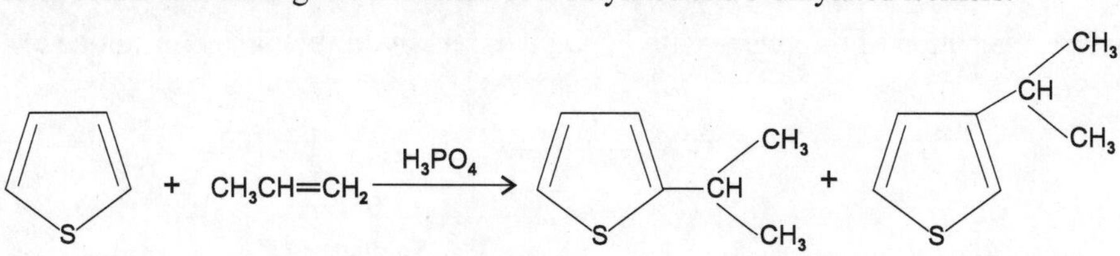

Acylation

The Friedel-Crafts acylation of thiophenes is carried out with $SnCl_4$ as catalyst. Aluminium chloride is not used as a catalyst as it forms tar with thiophene.

Vilmseier formylation: Vilmseier formylation of thiophene leads to formation of 2-formylthiophene.

Metallation

Like furan, thiophene is mercurated with mercury (II) chloride. The reaction of thiophene with mercury (II) chloride results in mercuration with the formation of thiophene-2-mercury chloride at rt and thiophene-2,5-dimercury chloride, on heating

Thiophene reacts very easily with mercury (II) acetate resulting in acetoxy mercuration of all the free nuclear positions.

Addition reactions

The hydrogenation of thiophene with Pd/C gives tetrahydrothiophene (or thiolane).

Thiolane

Diels-Alder reaction

Thiophene undergoes Diels-Alder reaction but its diene reactivity is lower than that of furan. The [4+2] cycloaddition occurs only with very reactive dienophiles (arynes and alkynes with electron withdrawing substituents) or under high pressure. The adduct of thiophene with alkyne gets converted to 1,2-disubstituted benzene after loss of sulfur.

$R = CN, COOR, Ph$

Thiophene reacts with maleic anhydride at high pressure to give exo-cycloadduct *via* [4+2] cycloaddition.

exo adduct

Thiophenes undergo [2+1] cycloaddition with carbenes across the C2-C3 bond to form cyclopropane derivative.

Ring-opening reactions

Thiophenes do not undergo polymerisation or hydrolysis by moderately concentrated Bronsted acids. Ring-opening with phenylmagnesium bromide in the presence of dichlorobis(triphenylphosphine)nickel(II) gives 1,3-diene.

Thiophene undergo reductive desulfurization with Raney nickel in ethanol to give ring opened product i.e., alkane. Raney nickel adsorbs hydrogen during its preparation, which effects the reduction. Also, hydrogen in the presence of molybdenum or tungsten as a catalyst can be used for reduction. This hydrodesulfurization is of great industrial importance for the removal of thiophenes and other sulfur compounds from petroleum.

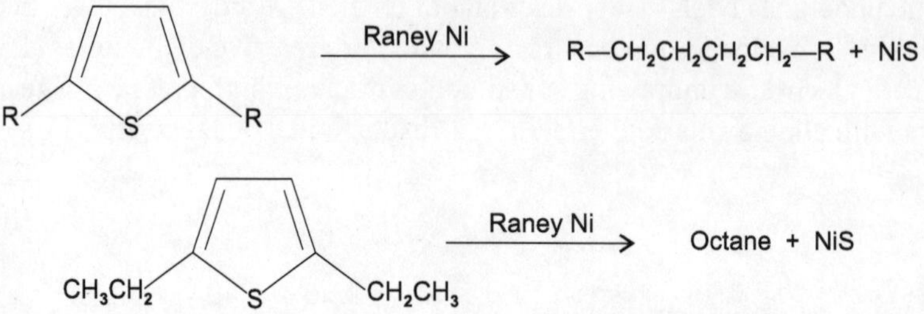

$$R-CH_2CH_2CH_2CH_2-R + NiS$$

Octane + NiS

Oxidation

The thiophene ring system is relatively stable to oxidants and side chains can be oxidized to carboxylic acid groups. Thiophene undergo oxidation with an excess of 3-chloroperoxybenzoic acid (mCPBA) to give thiophene-1-oxide, which oxidize further to thiophene-1,1-dioxide.

Both thiophene-1-oxide (Diene) and thiophene-1,1-dioxide (Dienophile) react to give Diels-Alder adduct.

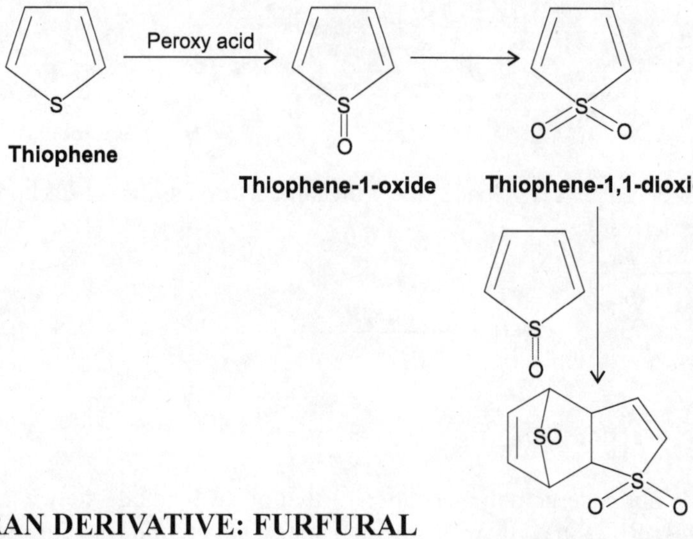

5.2 FURAN DERIVATIVE: FURFURAL

5.2.1 Introduction

Furfural or 2-furaldehyde is a heterocyclic aldehyde with odour of almonds (bp 162°C). It is a colourless oily liquid, but upon exposure to air it quickly turns yellow. Furfural is slightly soluble in water, but soluble in most polar organic solvents.

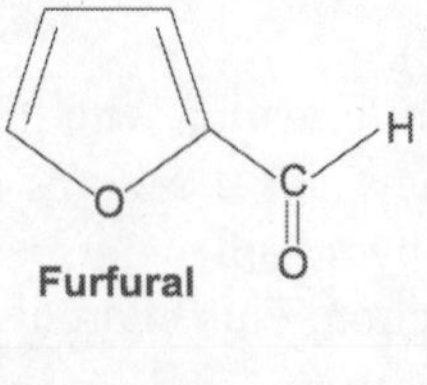

Furfural

The name furfural comes from the Latin word *furfur,* meaning bran. It is produced from the carbohydrate present in crop materials such as wheat bran, oat, corncobs and sawdust. 2-Furaldehyde is used as a solvent in the manufacture of polymers and as a starting material for syntheses. Furfural was first isolated in 1831 by Johann Wolfgang Dobereiner. In 1840 John Stenhouse determined its molecular formula, $C_5H_4O_2$ and later in 1901, Carl Harries could deduce the structure of furfural.

The ^1H and ^{13}C NMR (given in brackets) of furfural is shown below.

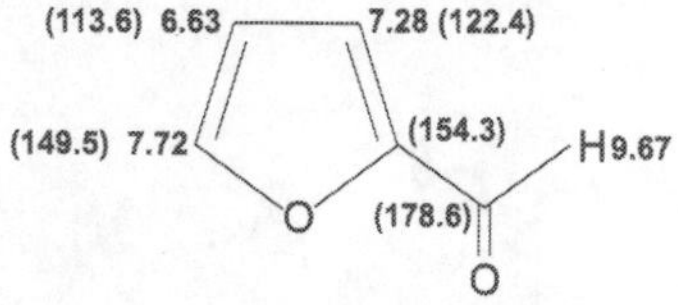

^1H NMR (δ, ppm, CDCl$_3$)
^{13}C NMR (δ, ppm, CDCl$_3$)

5.2.2 Methods of preparation of furfural

From pentosan: Industrially 2-furaldehyde (furfural) is obtained from plant residues e.g. bran which are rich in pentoses. When heated with dilute sulfuric acid the five carbon sugars undergo dehydration, losing three water molecules to yield furfural.

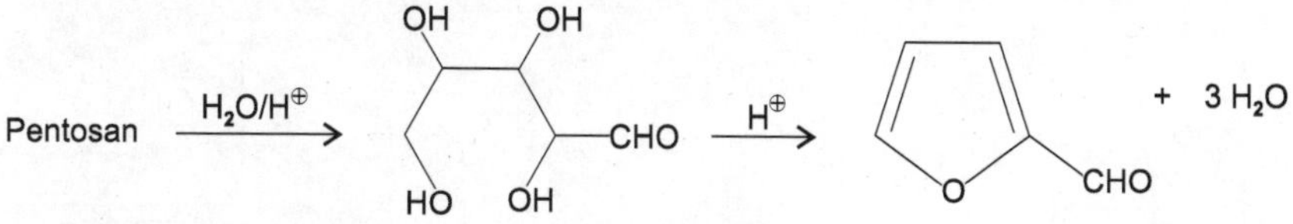

5.2.3 Reactions of furfural

Furfural is readily hydrogenated to the corresponding tetrahydrofuran derivatives. The catalytic hydrogenation of 2-furaldehyde (or furfural) yields 2-(hydroxymethyl) oxolane or tetrahydro-2-furfuryl alcohol. This compound undergoes a nucleophilic 1,2-rearrangement to yield 3,4-dihydro-2*H*-pyran (dihydropyran/DHP) in the presence of acid catalysts.

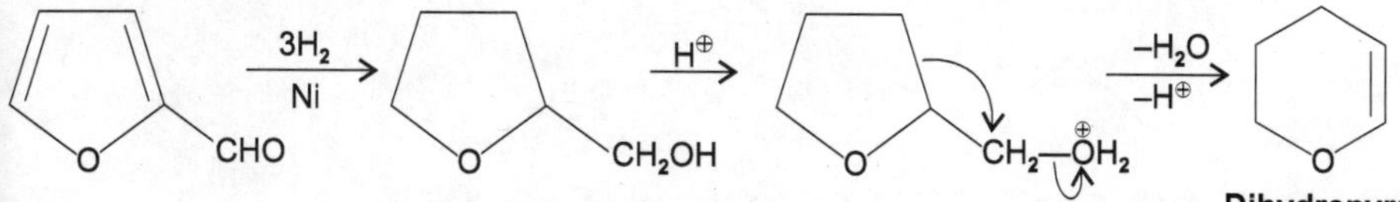

Furfural decomposes on heating above 250°C, into furan and carbon monoxide. Catalytic decarbonylation of 2-furaldehyde also produces furan.

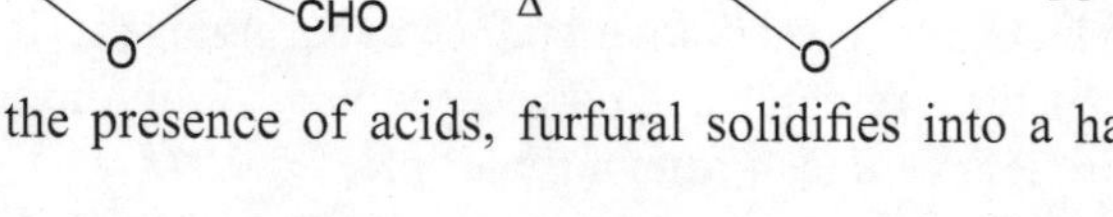

When heated in the presence of acids, furfural solidifies into a hard thermosetting resin.

Furfural undergoes reactions similar to other aldehydes and other aromatic compounds. But it is relatively less aromatic than benzene. Like benzaldehyde, it undergoes the Cannizzaro reaction, the Perkin reaction, the Knoevenagel condensation and the acyloin condensation.

Cannizzaro reaction of furfural

2-Furaldehyde undergoes ring-opening by the action of aniline and hydrochloric acid leading to the formation of coloured salt of 5-anilino-l(phenylimino)pent-2,4-dien-2-ol.

Nitration of 2-furaldehyde gives 5-nitro-2-furaldehyde (or 5-nitrofurfural).

5.3 FURAN DERIVATIVE: 2-FUROIC ACID

5.3.1 Introduction

2-Furoic acid (or furan-2-carboxylic acid) is a heterocyclic carboxylic acid, consisting of a five-membered aromatic ring. It is the first furan derivative to be reported in 1780, by Carl Wilhelm Scheele. 2-Furoic acid exists as colourless crystals (mp 132°C) and found in food products as a preservative and a flavouring agent. It is used in nylon preparation for biomedical research and optic technology. 2-Furoic acid has pK$_a$ value of 3.2 and it is a stronger acid than benzoic acid (pK$_a$ 4.2).

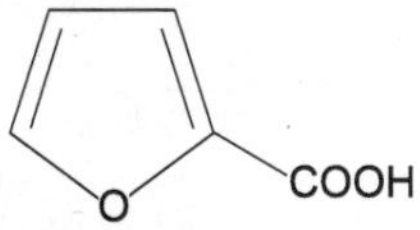

2-Furoic acid

The ^{1}H and ^{13}C NMR (given in brackets) chemical shift value of 2-furoic acid are shown below.

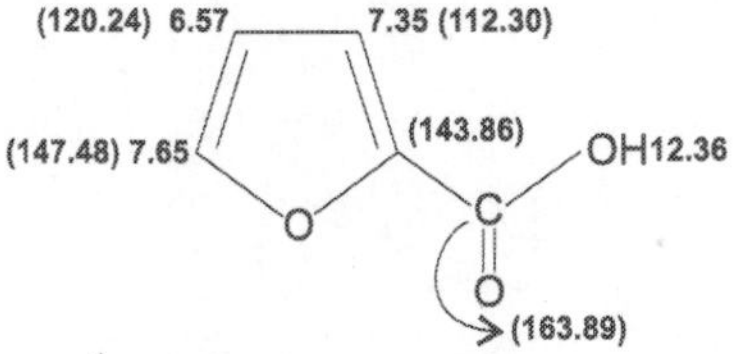

^{1}H NMR (δ, ppm, CDCl$_3$)
^{13}C NMR (δ, ppm, CDCl$_3$)

5.3.2 Methods of preparation 2-furoic acid

(*i*) **From D-galactaric acid:** Furoic acid is obtained by dry distillation of D-galactaric acid (mucic acid).

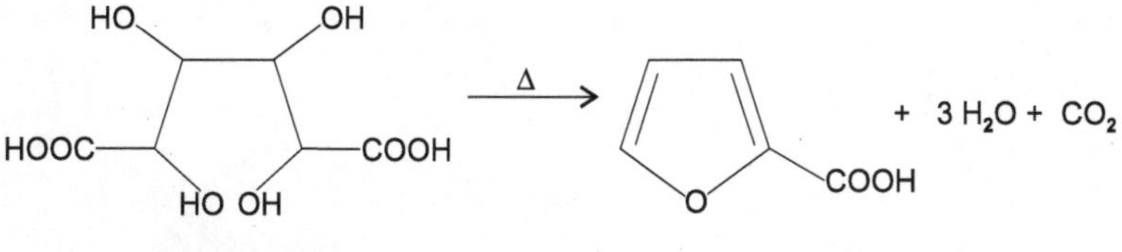

D-Galactaric acid
(Mucic acid)

(*ii*) **Oxidation of 2-furfuryl alcohol and 2-furanaldehyde:** 2-Furoic acid can be synthesized by the oxidation of 2-furfuryl alcohol and 2-furanaldehyde using a microbial biocatalytic preparation with *Nocardia corallina*.

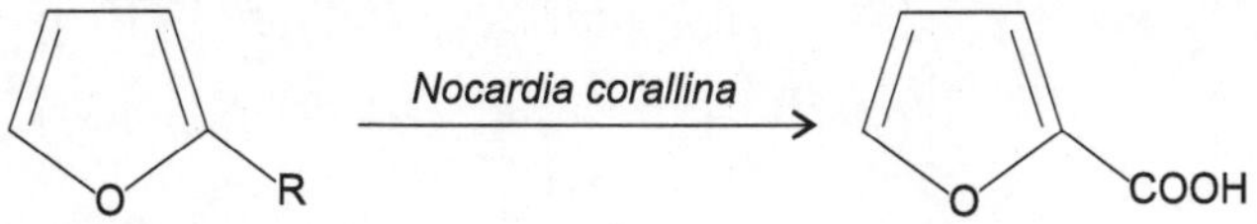

R = CH$_2$OH
R = CHO

5.3.3 Reactions of 2-furoic acid

2-Furoic acid reacts with lithium diisopropylamide (LDA) at low temperature to form lithium 5-lithio-2-carboxylate. It reacts with n-butyllithium to produce lithium 3-lithio-2-carboxylate.

2-Furoic acid is reduced by metal/ammonia to give 2,5-dihydro derivative.

Decarboxylation of 2-furoic acid with copper powder in quinoline produces furan.

Furan

EXERCISE

Q.1   Compare the basicity of pyrrole and pyrrolidine.

[Hint. Pyrrolidine is more basic than pyrrole] (Refer sec. 5.1.7.1)

Q.2   Write a method of preparation of thiophene from 1,4-diketone.

[Hint. Paal-Knorr synthesis]

Q.3   Compare the aromaticity of pyrrole, furan and thiophene.

[Hint. order of aromaticity thiophene > pyrrole > furan] (Refer sec. 5.1.3)

Q.4   Discuss electrophilic substitution reactions of pyrrole.

[Hint. attack of electrophile at C2 produces more number of canonical structures of the intermediate than at C3.] (Refer sec. 5.1.7)

Q.5   Give an industrial method for the synthesis of furan.

[Hint. From pentosans]

Q.6   Arrange thiophene, pyrrole, furan and benzene in order of their reactivity towards electrophile.

Ans.   benzene < thiophene < furan < pyrrole (Refer sec. 5.1.7)

Q.7   Suggest a method for the synthesis of iodol.

[Hint. Pyrrole + I_2/KI]

Q.8   Explain aromatic character of furan on the basis of resonance.

Q.9   Write the mechanism of Hantsch synthesis for pyrrole derivatives

Q.10   How will you carry out the following transformation:

(a) Pyrrole to 3-chloropyridine

(b) Furan to 2-furoic acid

Q.11   What happens when

(a) Pyrrole is treated with dichlorocarbene under strongly basic and weakly basic conditions.

(b) Thiophene is treated with a mixture of nitric acid and acetic anhydride

(c) Furan is reduced by hydrogen in presence of nickel.

(d) Furfural is treated with acetic anhydride in presence of sodium acetate.

Ans.   (a) strongly basic condition gives 1*H*-pyrrole-2-carbaldehyde and weakly basic conditions gives 3-chloropyridine.

(b) 2-Nitrothiophene

(c) Tetrahydrofuran

(d) Furfural undergoes Perkin reaction to yield furan-2-acrylic acid

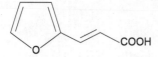

Q.12 Write the mechanism of reaction of pyrrole with chloroform in the presence of alkali.

Q.13 Explain why

(a) Electrophilic substitution in furan occurs preferentially at the 2-position.

(b) Thiophene is more aromatic than furan.

(c) Electrophilic substitution in pyrrole takes place at 2-position, whereas in pyridine at 3-position.

(d) Furan and pyrrole polymerizes in the presence of strong acids but pyridine remains unaffected.

(e) Pyrrole is weaker acid than diethyl amine.

(f) Furan is least aromatic of the five membered heterocycles.

(g) Pyrrole is more reactive than furan towards electrophiles.

Q.14 Complete the following reactions:

(a) Pyrrole + $CHCl_3$ + KOH $\longrightarrow$

(b) Thiophene + H_2 + Raney Ni $\longrightarrow$

(d) Pyrrole + aq. K_2CO_3 $\longrightarrow$

(e) Furan + $(CH_3CO)_2O$ + $BF_3.Et_2O$ $\longrightarrow$

Ans. (a) 1*H*-pyrrole-2-carbaldehyde

(c) n-Butane and NiS

(d) 3-Chloropyridine

(e) 2-Acetylfuran

Q.15 Mention the conditions for the following conversions:

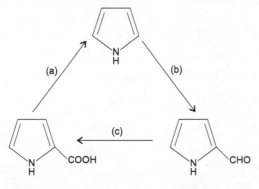

Ans. (a) heat, 200°C

(b) DMF/POCl$_3$ and aq. Na$_2$CO$_3$

(c) aq. Alkaline KMnO$_4$

Q.16 Write the product formed in the following Paal-Knorr synthesis:-

(a)

$$\xrightarrow[\text{H}^{\oplus} \text{catalyst}]{\text{PhCH}_2\text{NH}_2}$$

(b)

$$\xrightarrow{\text{conc. HCl}}$$

Ans. (a)   (b)

Q.17 Which is more basic out of pyrrolidine and phospholane?

Ans. Pyrrolidine is more basic than phospholane as the lone pair of electrons present on phosphorous of phospholane have less repulsion than in the case of pyrrolidine. In pyrrolidine due to greater electronegativity of nitrogen the lone pair electrons are more towards nitrogen. Because pyrrolidine lone pairs experience more repulsion so protonation leads to greater stability in it, making it comparatively more basic.

+ H$^{\oplus}$ ⟶

more basic   less acidic

+ H$^{\oplus}$ ⟶

less basic   more acidic

Q.18 Mention the product formed in the following reactions

(a) Pyrrole + NH_2OH/reflux $\longrightarrow$

(b) Thiophene + H_2/Raney Ni $\longrightarrow$

Ans.

(a) 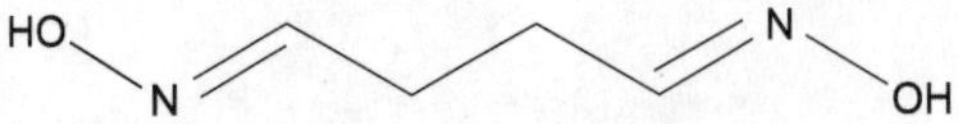 OH and NH_3

(b) n-Butane and NiS

Q.19 What products will be obtained when pyrrole, furan and thiophene are treated with

(a) Maleic anhydride

(b) Ethyl diazoacetate ($N_2CHCOOEt$) in presence of light

Ans. (a) Furan undergo Diels-Alder reaction, thiophene undergo Diels-Alder reaction at 100°C and high pressure, while pyrrole undergo electrophilic substitution reaction

(b)

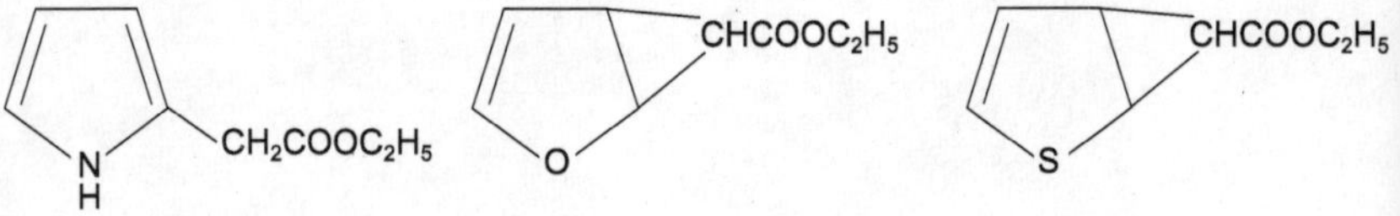

Q.20 Give the products of reaction of furan with the following reagents:

(a) CH_3COCl in presence of $SnCl_4$ as a catalyst

(b) n-Butyl lithium

(c) $C_6H_5N_2{}^+Cl^-$ in presence of sodium hydroxide

Ans (a) 2-Acetylfuran

(b) 2-Furyl lithium

(c) 2-Phenylfuran (furan undergo phenylation rather than diazotization)

Q.21 Complete the following reaction:

(a)

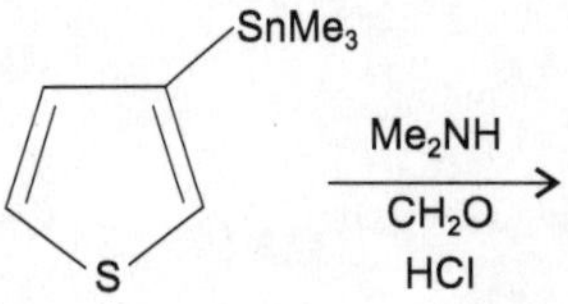

$\xrightarrow[\substack{CH_2O \\ HCl}]{Me_2NH}$

(b)

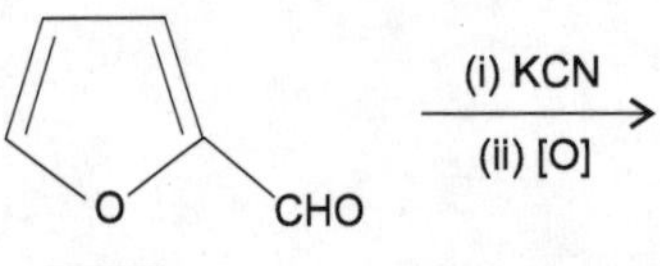

$\xrightarrow[\text{(ii) [O]}]{\text{(i) KCN}}$

(c)

$\xrightarrow[NaOEt]{CH_3COOEt}$

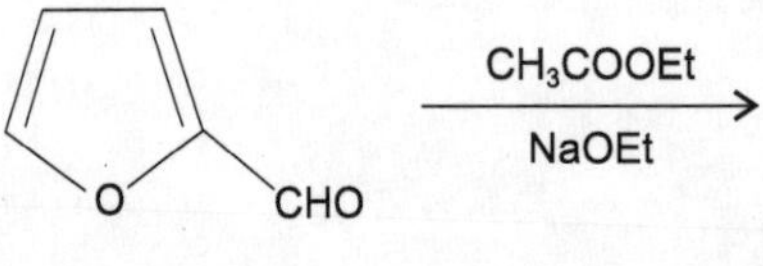

Ans. (a)

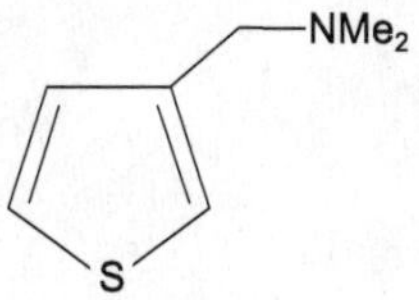

(Mannich reaction)

(b)

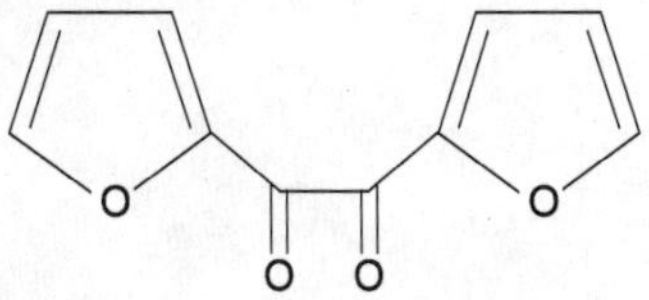

(Benzoin condensation)

(c)

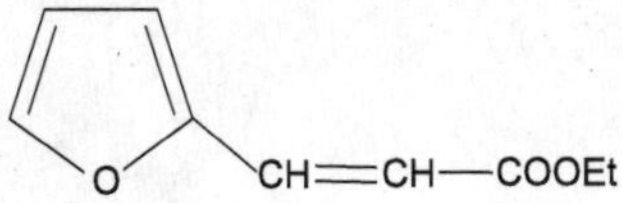

(Perkin condensation)

Q.22 Complete the following reactions:

(a)

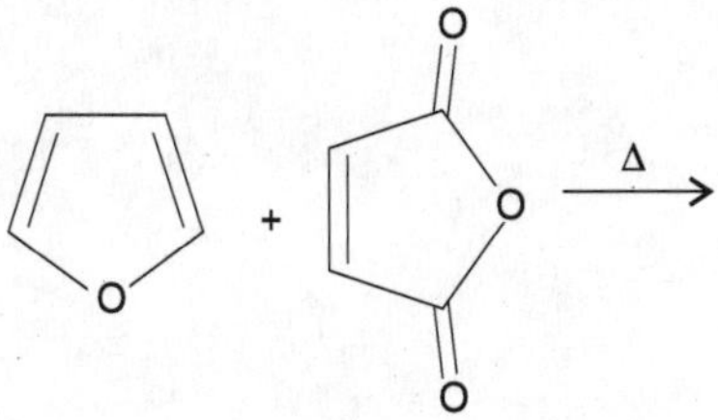

(b)

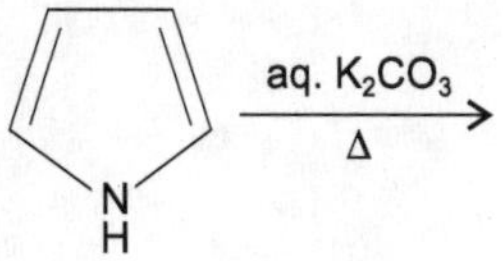

$\xrightarrow[\Delta]{\text{aq. K}_2\text{CO}_3}$

(c)

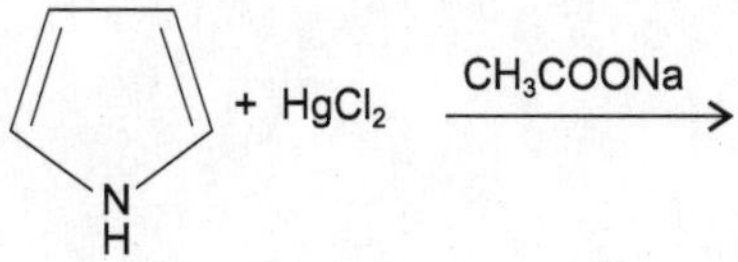

+ HgCl$_2$ $\xrightarrow{\text{CH}_3\text{COONa}}$

(d)

$\xrightarrow{\text{CH}_3\text{COONO}_2, \text{Ac}_2\text{O}}$

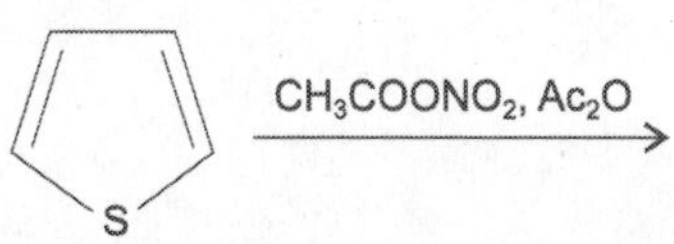

Q.23 Write the product formed in the following reactions:

(a)

$\xrightarrow[\text{Catalyst}]{\text{H}_2\text{S, H}^{\oplus}}$

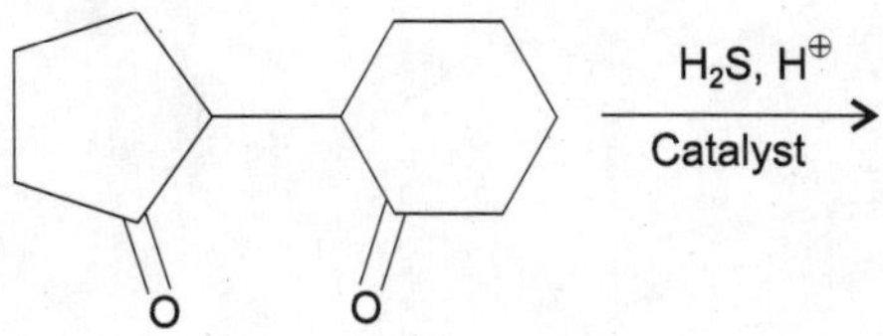

(b)

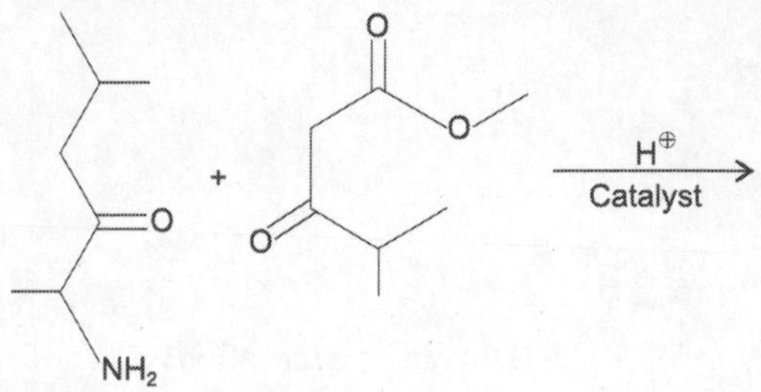

Ans. (a)

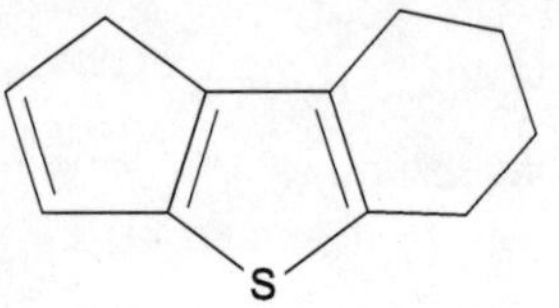

(b)

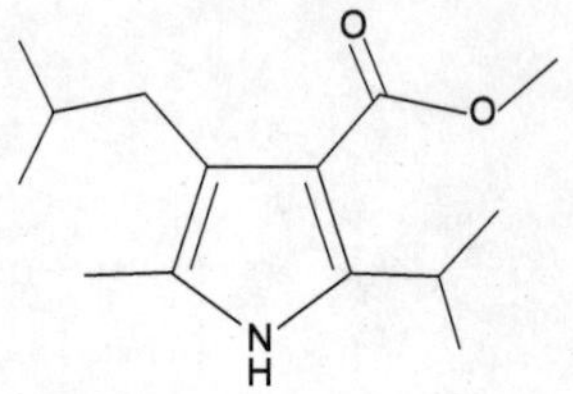

Q.24 Write the product formed A, B and C in the following reactions:

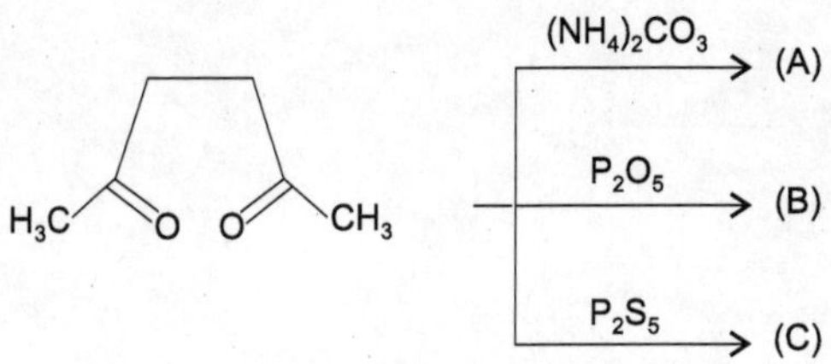

$(NH_4)_2CO_3 \longrightarrow$ (A)

$P_2O_5 \longrightarrow$ (B)

$P_2S_5 \longrightarrow$ (C)

Q.25 The following conversion involves:-

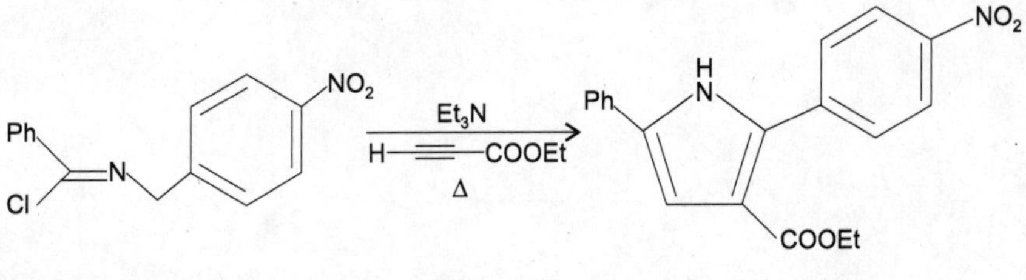

Ans. A 1,3-dipolar species as reactive intermediate, and a cycloaddition

6

Bicyclic Ring Systems Derived from Pyrrole, Furan and Thiophene

Benzo[b]pyrrole (Indole), Benzo[b]furan, Benzo[b]thiophene, Benzo[c]pyrrole (Isoindole), Benzo[c]furan and Benzo[c]thiophene

The benzene ring and the five membered heterocyclic pyrrole, furan or thiophene can fuse together to produce two possible aromatic structures differing in their position of fusion. These are namely **benzo[b]pyrrole, benzo[b]furan, benzo[b]thiophene, benzo[c]pyrrole, benzo[c]furan** and **benzo[c]thiophene.**

X		
NH	Benzo[b]pyrrole (Indole)	Benzo[c]pyrrole (Isoindole)
O	Benzo[b]furan	Benzo[c]furan
S	Benzo[b]thiophene	Benzo[c]thiophene

6.1 BENZO[b]PYRROLE (INDOLE)

6.1.1 Introduction

Indole is the trivial name for benzo[b]pyrrole. Indole is a planar aromatic heterocyclic molecule in which benzene ring is fused at the 2 and 3 position of pyrrole ring. It is solid at room temperature (mp 52-54°C, bp 253-254°C). Solubility of indole in water is low but it is freely soluble in many organic solvents. Indole itself has naphthalene like odour.

The name indole is a combination of the words indigo and oleum, since indole was first isolated by treatment of the indigo blue dye with oleum. In 1866, Adolf von Baeyer converted indigo to isatin and then to oxindole, which was reduced to indole using zinc dust. The work on indole was continued by Baeyer and Emmerling who proposed the structure of indole.

Indigo — **Isatin** — **Oxindole** — **Indole**

6.1.2 Biological importance

Indole has significance due to its presence in many natural products, pharmaceuticals, dyes etc. **Skatole** (3-methylindole), a degradation product of **tryptophan** having an indole unit contributes to odour of feces. Derivatives of **tryptamine** are biologically active compounds including neurotransmitters and psychotropic drugs. Indolic compounds of great importance include amino acid **tryptophan**, neurotransmitter **serotonin** (or 5-hydroxytryptamine) and **melatonin**, plant hormone **auxin** (indolyl-3-acetic acid), anti-inflammatory drug **indomethacin**, beta-blocker **pindolol**, naturally occurring hallucinogen **dimethyltryptamine** (N,N-DMT), naturally occurring dyes **tyrian purple** (royal purple, imperial purple or imperial dye) and **indigo**.

Skatole   **Tryptamine**   **Tryptophan**

Serotonin   **Melatonin**

Auxin   **Indomethacin**

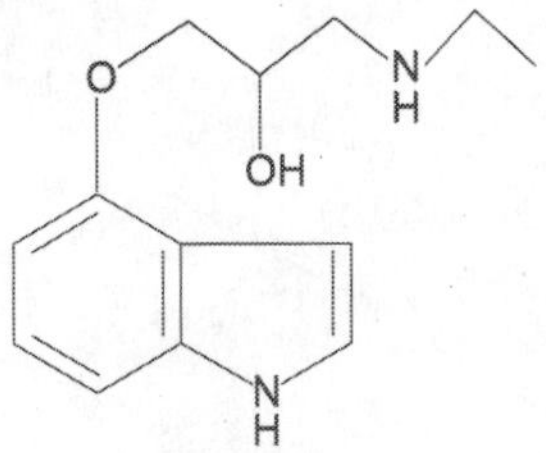

Pindolol

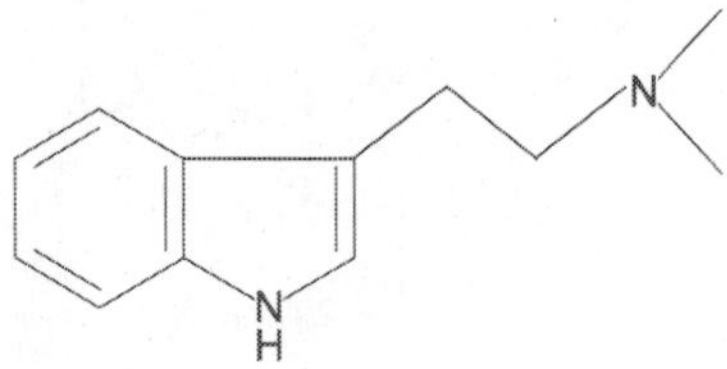

Dimethyltryptamine

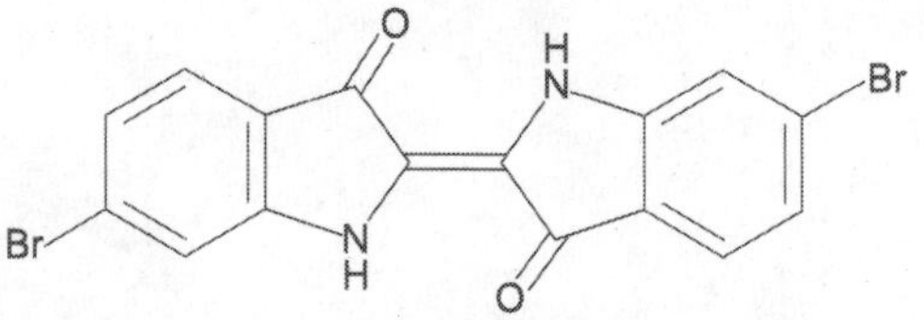

Tyrian purple

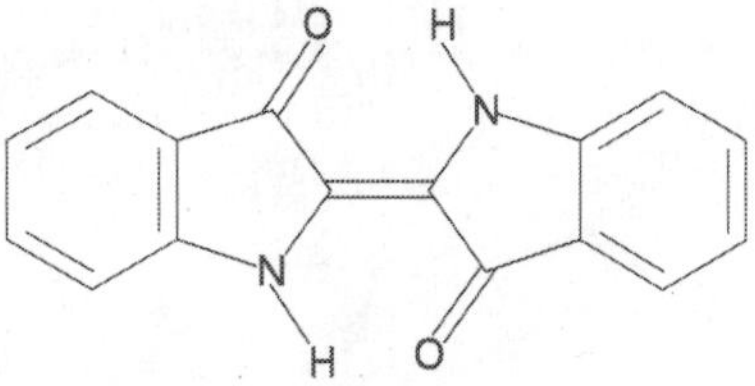

Indigo

The medicine such as **sumatriptan, rizatriptan** and **eletriptan** used for migraine are indole derivatives. **Lysergic acid diethylamide (LSD)** is a potent psychoactive compound which is prepared from lysergic acid, an alkaloid natural product of the ergot fungus. The mushroom hallucinogens **psilocin** and **psilocybin,** the drugs **yohimbine** and **reserpine**, and the poison **strychnine** all contain indole nucleus.

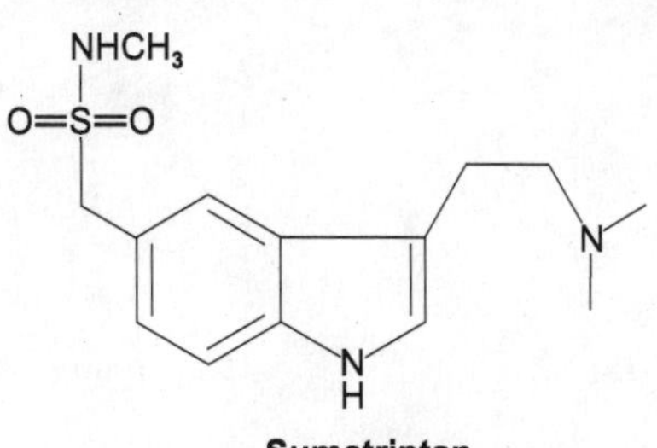

Sumatriptan

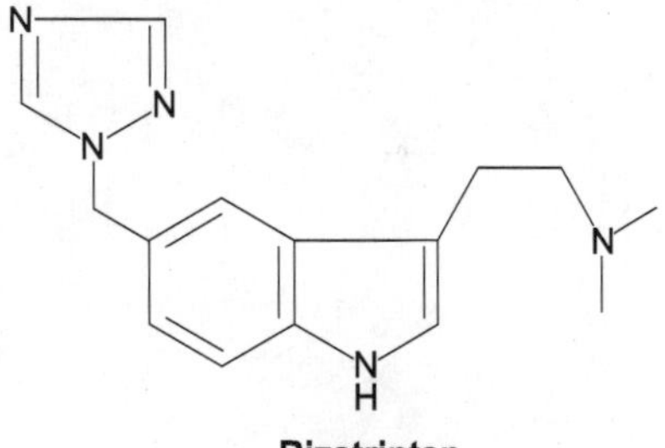

Rizatriptan

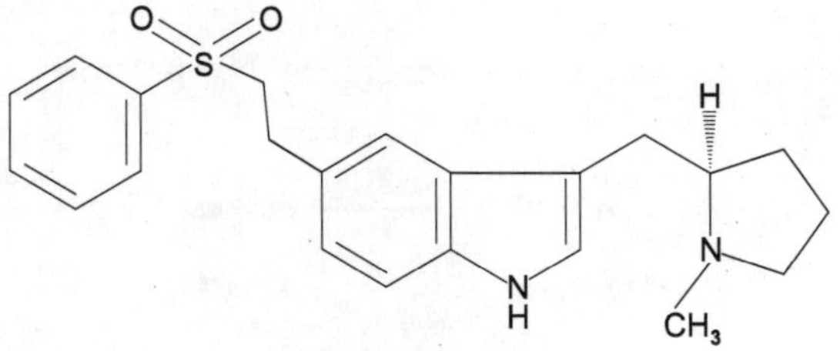

Eletriptan

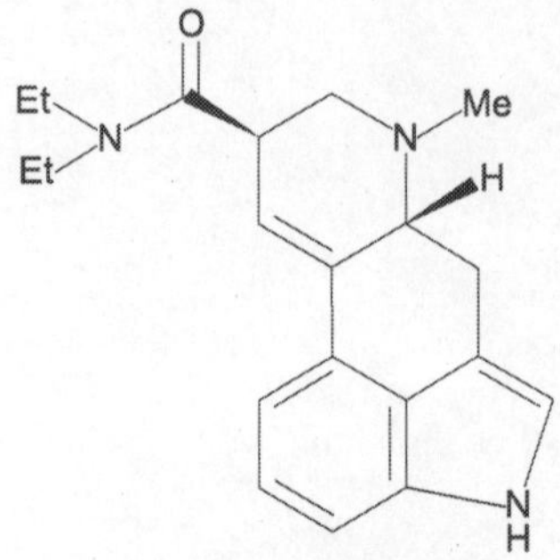

Lysergic acid diethylamide (LSD)

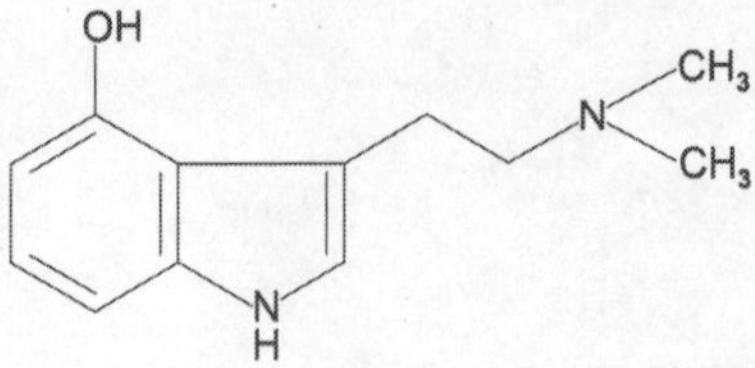

Psilocin

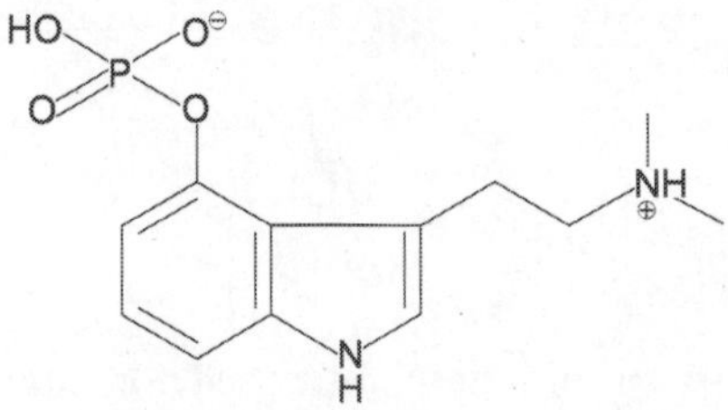

Psilocybin

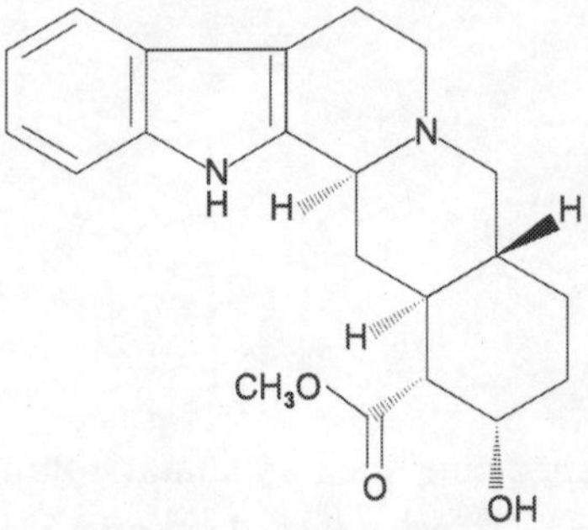

Yohimbine

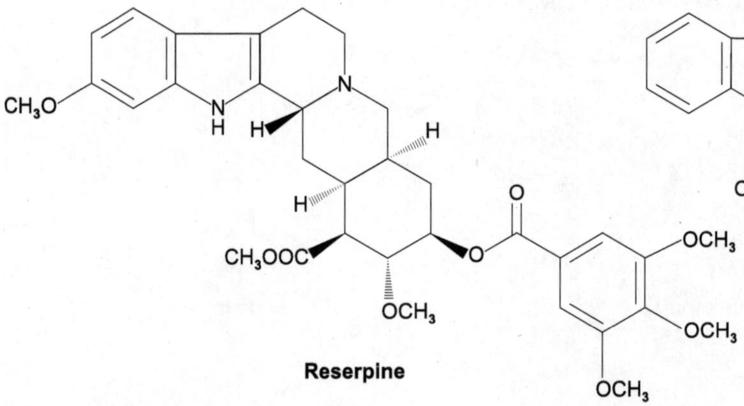

Reserpine

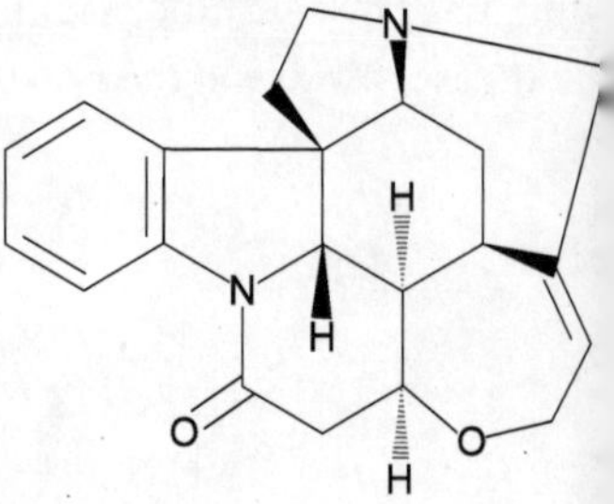

Strychnine

6.1.3 Spectral data

The ^{1}H and ^{13}C NMR (given in brackets) of indole is shown below:

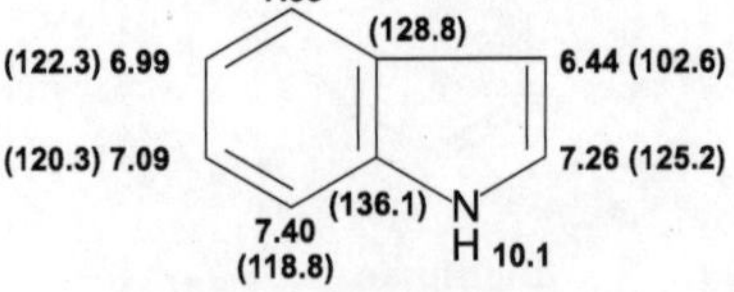

^{1}H NMR (δ, ppm, acetone-d$_6$)

^{13}C NMR (δ, ppm, CDCl$_3$)

6.1.4 Structure

Indole is a planar aromatic heterocyclic molecule. The resonating structures of indole have negative charge on carbon atoms and nitrogen atom carries positive charge.

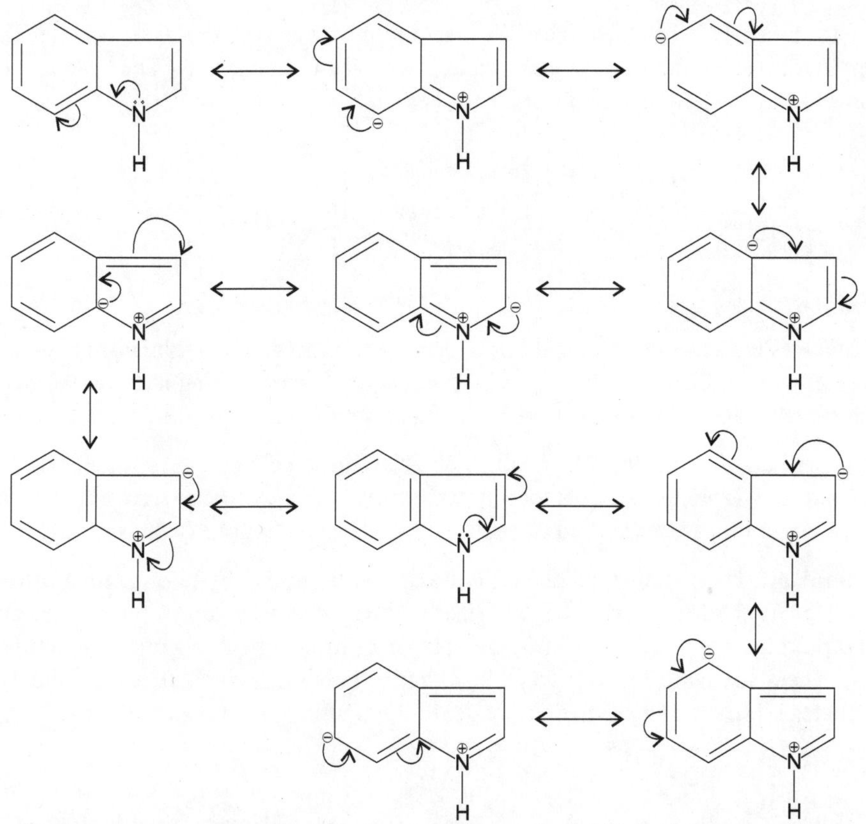

The dipole moment value of indole in excited state is 5.6 D. Indole exists entirely in the 1*H*- form, its 3*H*- tautomeric form (**indolenine**) is also known.

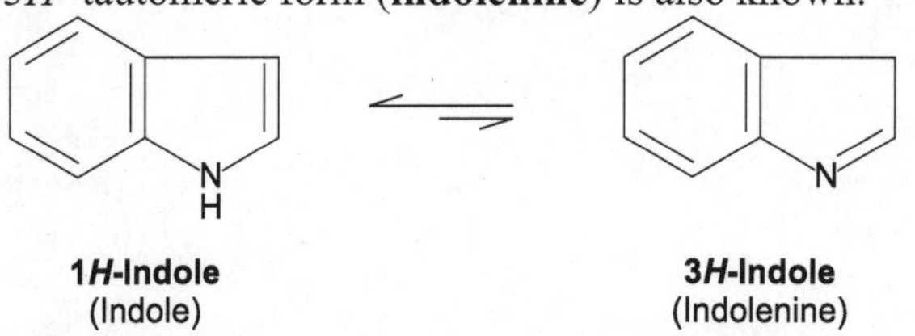

1*H*-Indole
(Indole)

3*H*-Indole
(Indolenine)

The 3*H*-indole (**indolenine**) can be generated in ethereal solution from 2,3-dihydro-N-indolylacetophenone on irradiation. However it is stable only at -100°C and tautomerizes to 1*H*- form at -50°C.

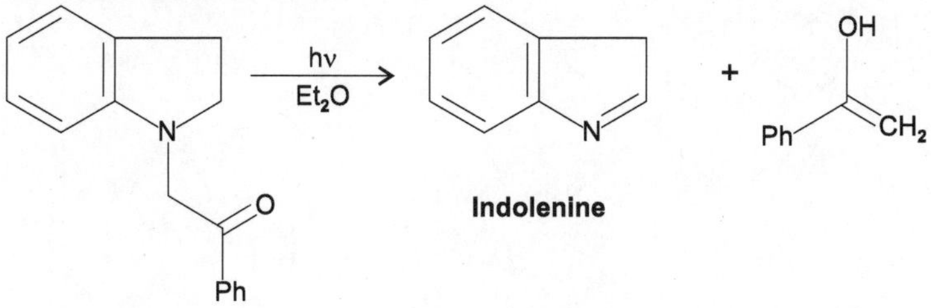

Indolenine

6.1.5 Methods of preparation of indole

Indole is obtained by fractional distillation of coal-tar at 220-260°C. Various other methods for the preparation of indole are as follows.

(*i*) **The Fischer synthesis:** The Fischer indole synthesis produces indole from phenylhydrazine and carbonyl compound under acidic conditions. The reaction was discovered in 1883 by Hermann Emil Fischer.

Phenyl hydrazine   **Phenyl hydrazone**   **Indole**

It involves the acid or Lewis acid catalysed rearrangement of an arylhydrazone with the loss of ammonia. Electron releasing substituents on the benzene ring increase the rate of Fischer cyclization, whereas electron withdrawing substituents slow the process.

Bronsted acids such as HCl, H_2SO_4, polyphosphoric acid and para-toluenesulfonic acid and Lewis acids such as boron trifluoride, zinc chloride, iron chloride and aluminium chloride can be used as catalyst in the Fischer indole synthesis.

Mechanism: The reaction of phenylhydrazine with an aldehyde or ketone initially forms a phenylhydrazone which isomerizes to the respective enamine (ene-hydrazine). After protonation, a cyclic [3,3]-sigmatropic rearrangement occurs producing an imine. The resulting imine forms a cyclic amino compound, which under acid catalysis eliminates NH_3, resulting in the energetically favorable aromatic indole.

enamine

[3,3]-Sigmatropic rearrangement

(*ii*) **Madelung synthesis:** The Madelung synthesis produces substituted or unsubstituted indoles from N-phenyl amides using strong base.

Mechanism: The reaction involves the intramolecular cyclization of N-phenylamides in presence of strong base, followed by dehydration.

Indole

(*iii*) **Grandberg synthesis of tryptamines**: Grandberg synthesis is modified Fischer indole synthesis that involves the reaction of 4-halobutanals or their acetals with phenylhydrazine in ethanol to give tryptamines.

Similarly, 2-methyltryptamine can be synthesised by the reaction of phenyl hydrazine with 5-chloropentan-2-one.

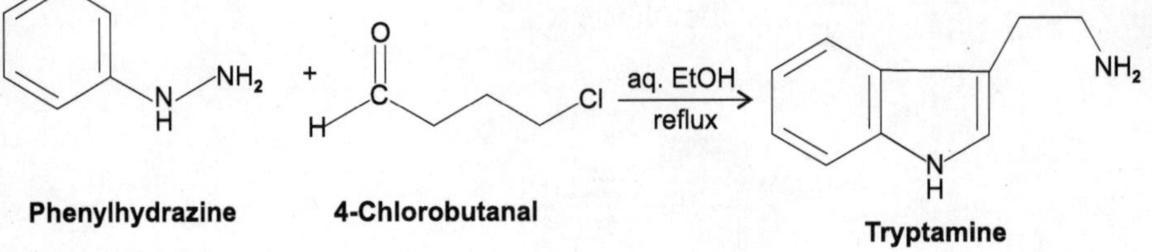

Phenylhydrazine   **4-Chlorobutanal**   **Tryptamine**

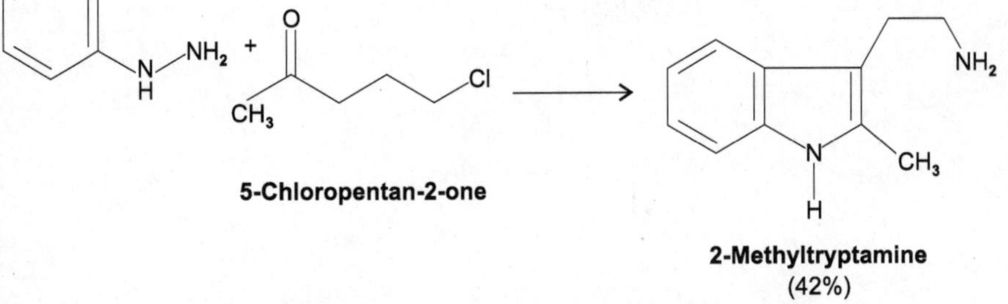

5-Chloropentan-2-one

2-Methyltryptamine
(42%)

Mechanism: The mechanism of this reaction involves the formation of the hydrazone from phenylhydrazine and the 5-chloropentan-2-one, which after cyclization and isomerization from iminium to enamine form followed by the rearrangment afford the tricyclic intermediate. Loss of a proton provides the 2-methyltryptamine with 42% yield.

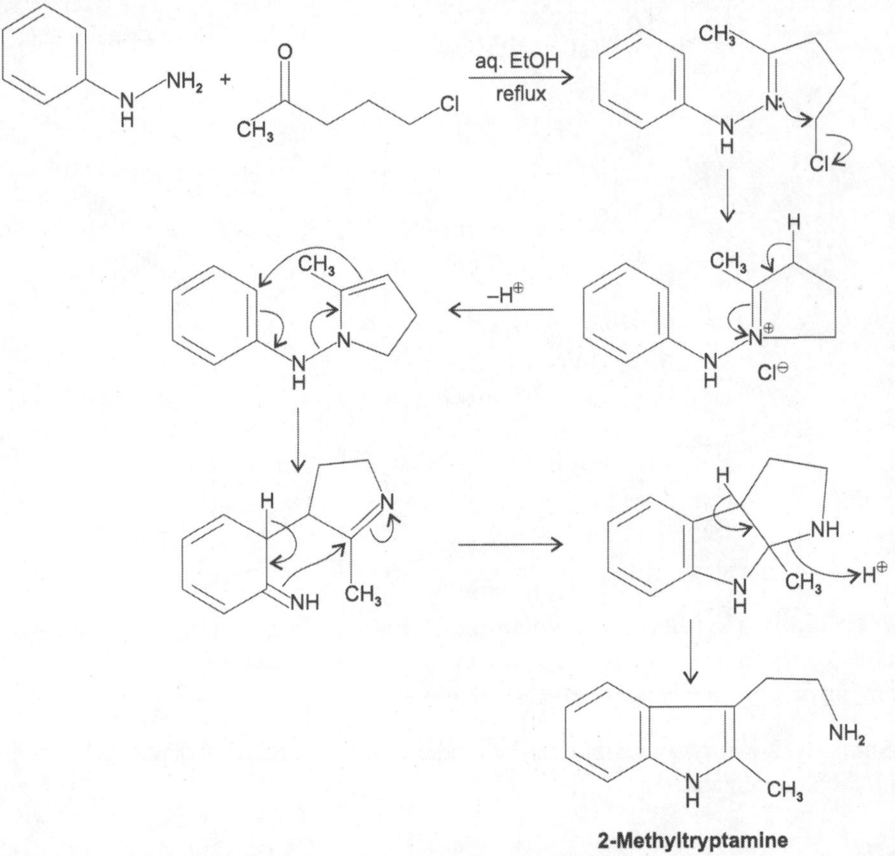

2-Methyltryptamine

(*iv*) **The Reissert synthesis:** The Reissert synthesis of indole or substituted indoles involves the reaction of ortho-nitrotoluene and diethyl oxalate.

o-Nitrotoluene   Diethyloxalate   o-Nitrophenyl pyruvate

Indole   Indole-2-carboxylic acid

Mechanism: The first step of the Reissert indole synthesis is the condensation of ortho-nitrotoluene with a diethyl oxalate to give ethyl ortho-nitrophenyl pyruvate. The reductive cyclization with zinc in acetic acid gives ethyl ester of indole-2-carboxylic acid which on hydrolysis followed by decarboxylation with heat gives indole. It has been found that the hydrogenation of 2-nitrophenylpyruvic ester with 5% Pd/C in EtOH is the best reduction condition for this reaction. Cyclization of ortho-(2-oxoalkyl)-anilines by simple intramolecular condensation with loss of water occurs spontaneously. Potassium ethoxide has been shown to give better results than sodium ethoxide.

(*v*) **Batcho-Leimgruber indole synthesis:** The Batcho-Leimgruber indole synthesis produces indoles from ortho-nitrotoluenes and N,N-dimethylformamide dimethyl acetal and pyrrolidine. The first step is the formation of an enamine. Indole is then formed in a second step by reductive cyclization. The reductive cyclization can be carried out by using Raney Ni/hydrazine, Pd/C/H$_2$, stannous chloride, sodium dithionite or iron/acetic acid.

(*vi*) **Bischler indole synthesis:** In this reaction indoles are synthesized by reaction of aniline with α-haloketones in the presence of acid.

Mechanism: The α-halocarbonyl compound reacts with aniline to form α-arylamino carbonyl compound in the presence of an acid, which undergoes electrophilic intramolecular cyclization followed by aromatization to give indole derivative.

(*vii*) **Aniline and ethylene glycol:** Indole and substituted derivatives can be synthesized *via* vapor-phase reaction of aniline with ethylene glycol in the presence of catalyst such as Cu/SiO$_2$. Among all kinds of catalysts, the silver-based catalyst prepared by the co-precipitation method is regarded as a good catalyst because the yield of indole can be up to 78%.

| Aniline | Ethylene glycol |

6.1.6 Reactions of indole

Basicity and protonation

Unlike most amines, nitrogen of indole is not basic (pK$_a$ of conjugate acid = –3.6). The bonding situation is completely analogous to that in pyrrole thus indoles like pyrroles (pK$_a$ of conjugate acid = –3.8), are very weak bases.

Very strong acids such as hydrochloric acid are required to protonate indole giving 1*H*-indolium and 2*H*-indolium cations. In presence of dilute acids, the 3*H*-indolium cation is formed. It is the thermodynamically stablest cation, and retains benzene aromaticity (in contrast to the 2*H*-indolium cation).

1*H*-Indolium cation   **2*H*-Indolium cation**

3*H*-Indolium cation

2-Methylindole (pK$_a$ of conjugate acid –0.3) is a stronger base than indole (pK$_a$ of conjugate acid –3.6) due to stabilisation of the cation by electron release from the methyl group while 3-methylindole (pK$_a$ of conjugate acid –4.6) is a weaker base than indole.

| pK$_a$: | – 0.3 | – 3.6 | – 4.6 |

Reactions of 3-protonated indoles

$3H$-Indolium cations are electrophilic species in contrast to neutral indoles. The $3H$-indolium cation add bisulfite at pH = 4 to give crystalline product sodium salt of indoline-2-sulfonic acid. The salt can be converted back to indole on adding water. It can be N-acetylated and the resulting acetamide on halogenation and nitration at C5 followed by hydrolysis with loss of bisulfite affords the 5-halogenated and 5-nitro indole, respectively.

3H-Indolium cation   **Indole**

Electrophilic substitution reactions

In contrast to pyrrole electrophilic substitution occurs preferably at the 3-position in indole. Because attack of the electrophile on the 3-position leads to the formation of a low energy intermediate structure of the σ-complex. A simple explanation is that reaction at the 3-position involves an intermediate in which the aromaticity of the benzene ring is not disturbed. The positive charge at 2-position is stabilized by the donation of electrons from nitrogen.

While attack on the 2-position, results in a high energy orthoquinonoid enamine structure. As the reaction in the 2-position disturbs the aromaticity of the benzene ring, therefore 2-position is less preferred for electrophilic substitution.

More favourable
(Six membered ring is aromatic)

Since the pyrrole ring is the most reactive portion of indole, electrophilic substitution of the benzene ring can take place only after N1, C2, and C3 are substituted. If the 3-position is substituted, attack of electrophile occurs first on the 2- and subsequently on the 6-position. In most electrophilic substitution reactions, indole reacts more slowly than pyrrole but faster than benzofuran. Indole is less reactive than pyrrole because of the fact that the share of negative charge by each carbon of indole is less than pyrrole due to greater delocalization as evident from the resonating structures of indole.

Pyrrole　　>　　**Indole**　　>　　**Benzo[b]furan**

Order of reactivity towards electrophiles

Vilsmeier reaction: The Vilsmeier reaction of indole with DMF and POCl$_3$ gives indole-3-carboxaldehyde.

DMF　　　　　　　　　　　　**Indole-3-carboxaladehyde**

Mechanism: The mechanism for Vilsmeir formylation of indole is shown below.

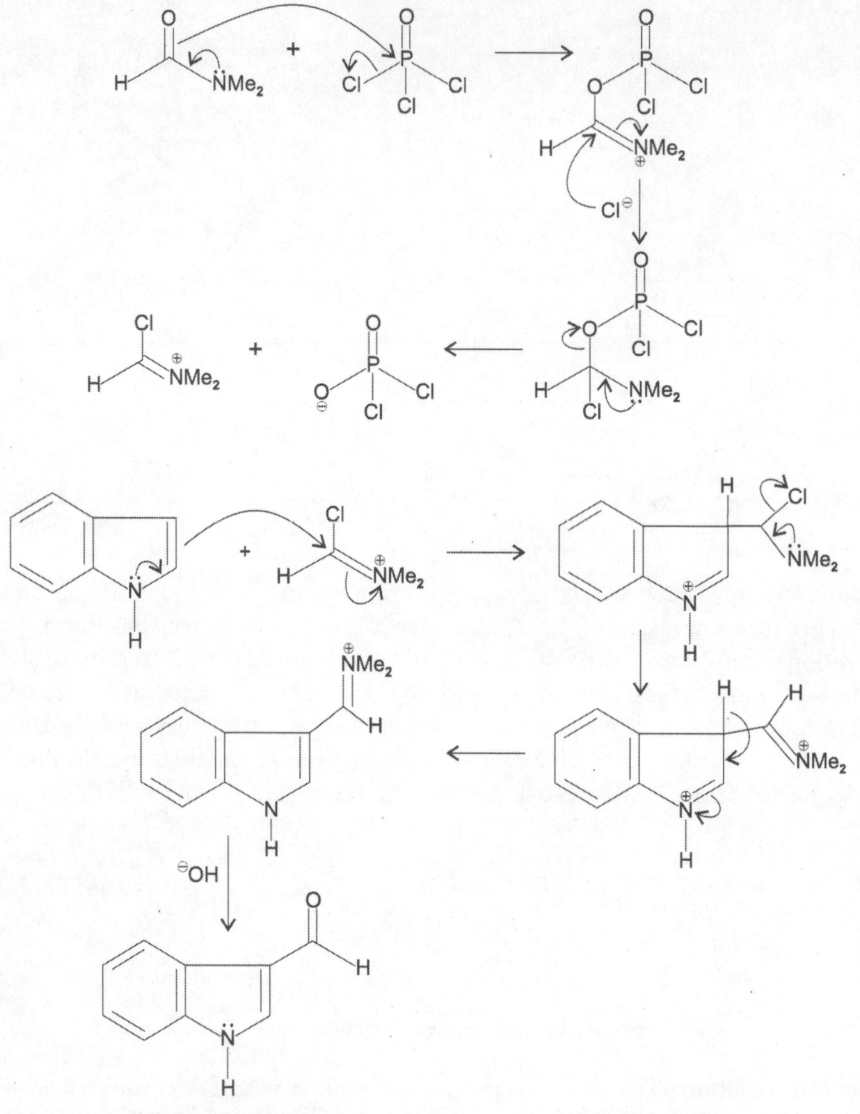

Indole-3-carboxaldehyde

Mannich reaction: Gramine, a useful synthetic intermediate, is produced *via* Mannich reaction of indole with dimethylamine and formaldehyde. It is the precursor to indole acetic acid and synthetic tryptophan.

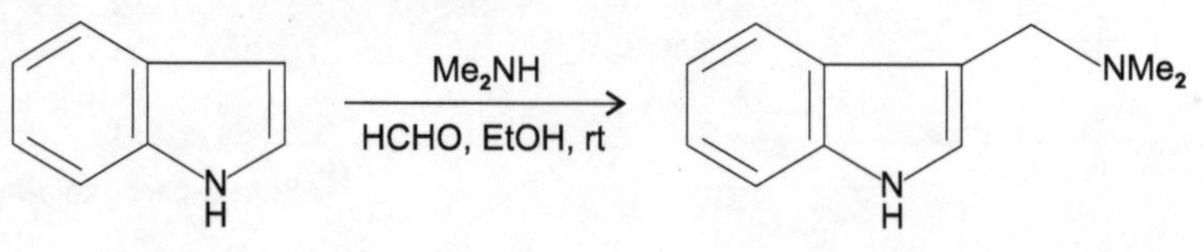

Gramine

Mechanism: It involves nucleophilic attack of dimethylamine on formaldehyde resulting in the formation of imine intermediate, which reacts with indole to give 3-substituted product.

Gramine

Nitration: The usual nitrating mixture leads to acid-catalysed polymerisation. Action of HNO_3 on indole causes oxidation of the pyrrole ring followed by polymerization.

Indole can be nitrated using the non acidic nitrating agents such as benzoyl nitrate (PhCOCl and $AgNO_3$) or acetyl nitrate (conc. HNO_3 and acetic anhydride) at low temperature.

Indoles substituted in the 2-position react with HNO_3 in acetic acid to give 3,6-dinitroindole. 2-Methylindole gives a 3-nitro derivative with benzoyl nitrate. It can also be nitrated successfully with concentrated nitric/sulfuric acids to give 5-nitro derivative. Under these conditions attack on the heterocyclic ring does not take place due to the complete protonation of 2-methylindole.

If an acetyl group is present at C3, nitration with nitronium tetrafluoroborate and tin(IV) chloride takes place at either C5 or C6 depending on the temperature of the reaction.

Sulfonation: Sulfonation of indole, at C3, occurs using the pyridine-sulfur trioxide complex in hot pyridine.

Indole-3-sulfonic acid

Gramine is sulfonated in oleum to give the corresponding 5- and 6- sulfonic acids.

Halogenation: With sodium hypochlorite, N-chlorination of indole occurs first which rearranges to form 3-chloroindole. Chlorination of indole may also be carried out with sulfuryl chloride which gives 3-chloroindole. However, if excess sulfuryl chloride is used 2,3-dichloroindole may be obtained. 3-Halo- and 2-halo- indoles are unstable and used soon after preparation.

Reaction of indole with bromine in DMF or iodine in DMF gives very high yields of corresponding 3-halogenated indoles.

Acylation: Indole reacts with acetic anhydride above 140°C, to yield 1,3-diacetylindole. Initially attack takes place at 3-position of give 3-acetylindole, which is converted into 1,3-diacetyl indole. If acetylation is carried out in the presence of sodium acetate, N-acetylindole is formed exclusively *via* the indolyl anion. With trifluoroacetic anhydride, indole gets acylated at C3 in dimethylformamide and at nitrogen in dichloromethane.

Alkylation: Indoles do not react with alkyl chlorides and bromides at room temperature. Indole reacts with iodomethane in dimethylformamide (or MeOH) at about 80°C to form **skatole**. On increasing the temperature to 110°C further methylation occurs to form 1,2,3,3-tetramethyl-3H-indolium iodide.

Cycloaddition reactions

Indole do not undergo Diels-Alder reaction.

Reactions with bases

As in pyrroles (pK$_a$ = 17.5), the N-hydrogen in indoles is much more acidic (pK$_a$ = 16.2) than that of an aromatic amine.

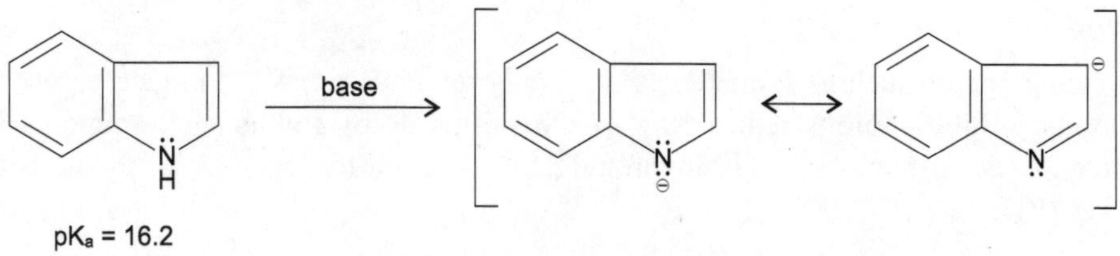

pK$_a$ = 16.2

Very strong base such as alkyl Grignard reagent, sodium hydride and n-butyllithium converts an N-unsubstituted indole into the corresponding indolyl anion.

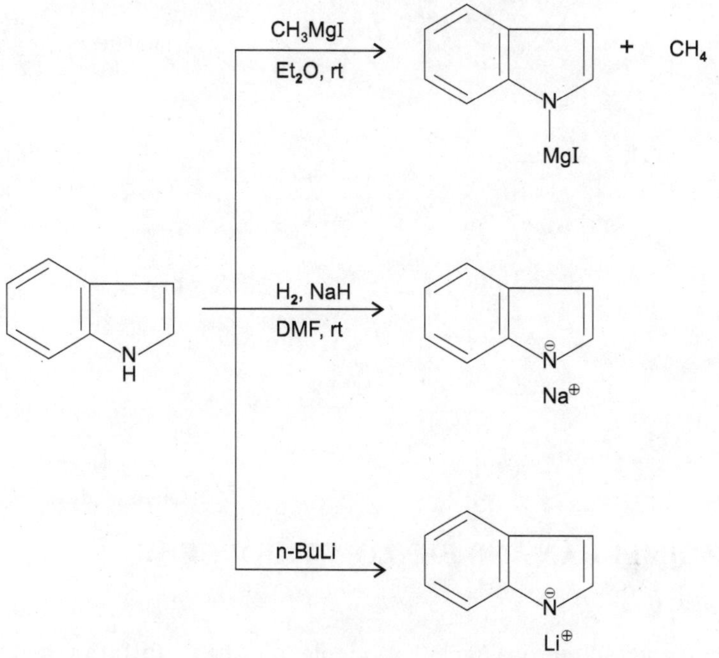

Oxidation

Due to the electron-rich nature of indole, it is easily oxidized. Simple oxidants such as N-bromosuccinimide will selectively oxidize indole to oxindole.

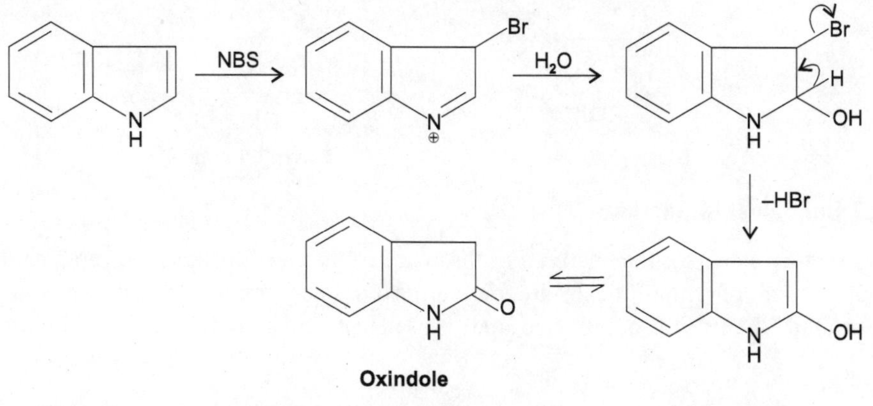

Oxindole

Reduction

Reducing agents such as Raney Ni, Zn/phosphoric acid or tin/HCl reduce indole to indoline. While indole is reduced to 4,7-dihydroindole by sodium/ liq. ammonia in methanol (Birch reduction). Octahydroindole is obtained by reduction of indole with nickel/HCl.

6.2 BENZO[b]FURAN AND BENZO[b]THIOPHENE

6.2.1 Introduction

The oxygen and sulfur analogues of indole are **benzo[b]furan** and **benzo[b] thiophene.** Benzo[b]furan and benzo[b]thiophene are also known as **benzofuran** and **benzothiophene**, respectively. Benzofuran is a water insoluble oily colourless liquid (bp 173-175°C, mp < -18°C). Benzothiophene occurs naturally as a constituent of petroleum related deposits such as lignite tar. It is colourless solid (mp 32°C and bp 221°C), with a smell similar to that of naphthalene.

Benzo[b]furan   **Benzo[b]thiophene**

6.2.2 Biological importance

Psoralen is a benzofuran derivative that occurs in several plants. Several natural products and pharmaceuticals are derived from benzofuran e.g., **amiodarone**, a substituted 3-benzoyl-2-butylbenzofuran, is used in the treatment of cardiac arrhythmia.

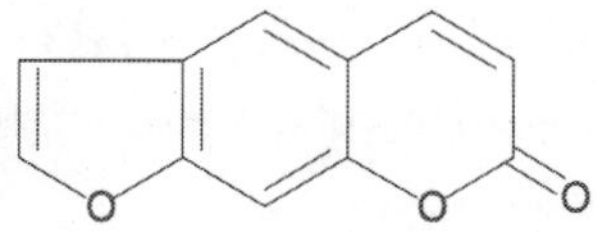

Psoralen or
7H-Furo[3,2-g]chromen-7-one

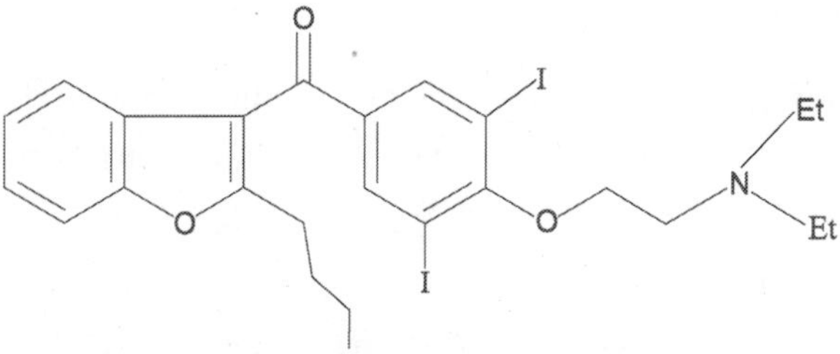

Amiodarane

Benzothiophene occurs in roasted coffee beans. It is found within the chemical structures of pharmaceutical drugs such as **raloxifene, zileuton, sertaconazole** and **benzothiophenylcyclohexylpiperidine (BTCP)**. It is also used in the manufacturing of dyes such as **thioindigo**. **Benzothiophene-3-acetic acid** a benzothiophene derivative promotes plant growth.

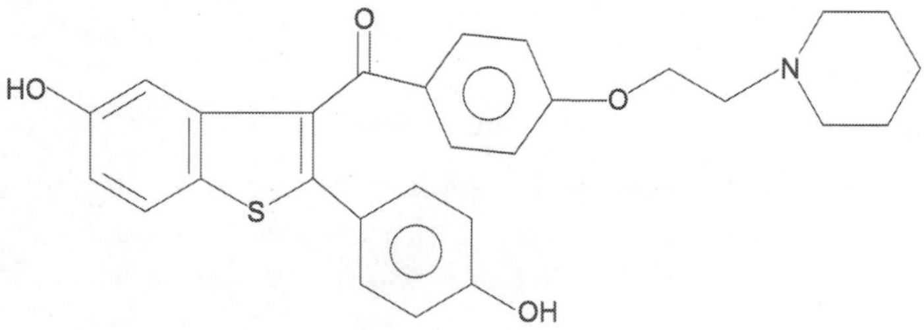

Raloxifene

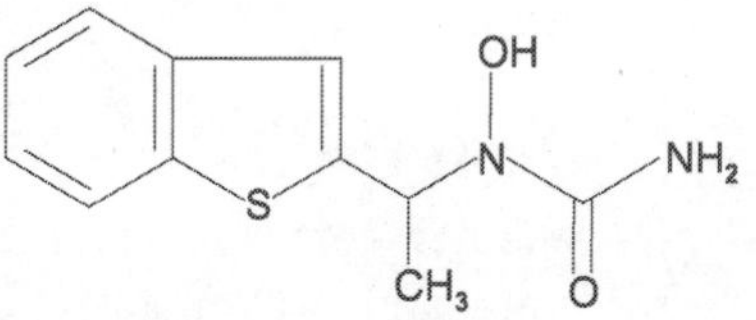

Zileuton

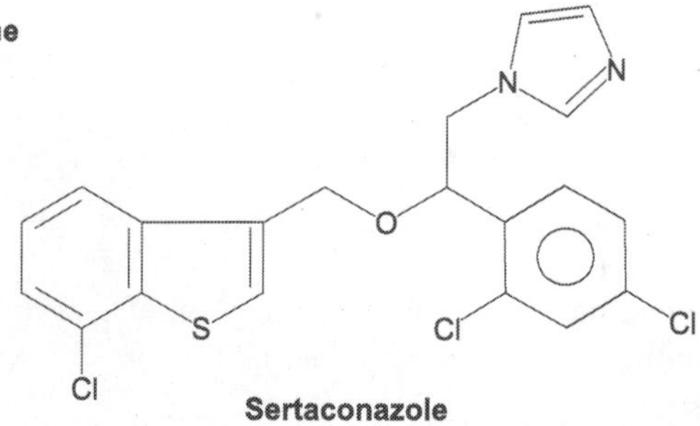

Sertaconazole

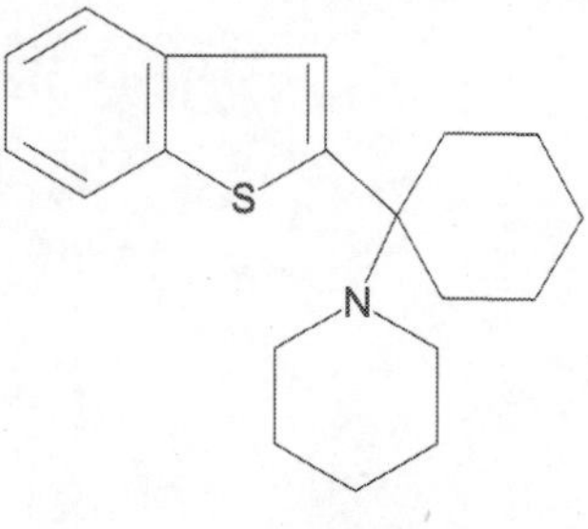

Benocylidine
BTCP

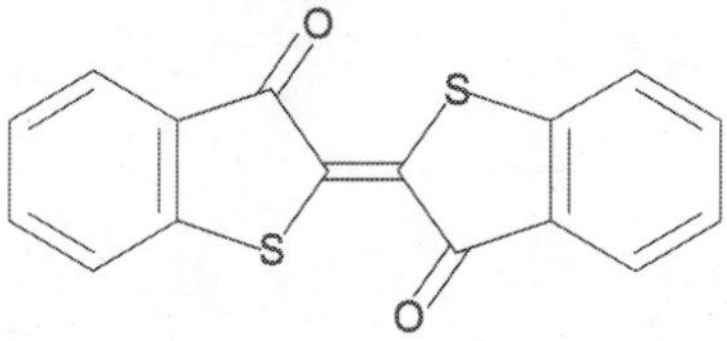

Thioindigo

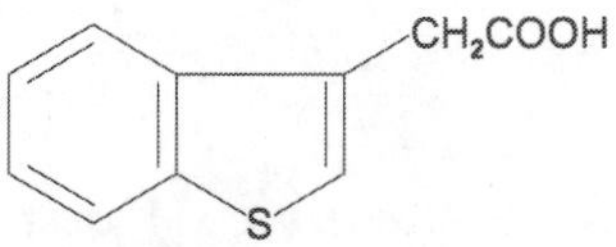

Benzothiophene-3-acetic acid

6.2.3 Structure

Several resonating structures are possible for benzo[b]furan and benzo[b]thiophene as in case of indole. The canonical structures in which aromaticity of benzene ring is disturbed contribute less towards resonance energy.

6.2.4 Spectral data

The ^1H and ^{13}C NMR (given in brackets) chemical shift values for benzo[b]furan and benzo[b]thiophene are shown below.

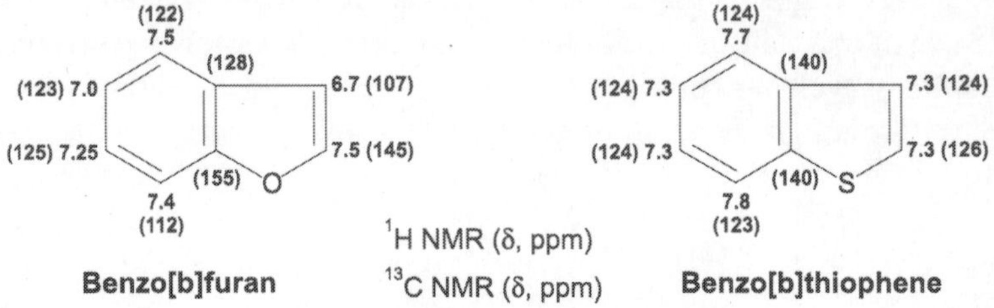

^1H NMR (δ, ppm)
^{13}C NMR (δ, ppm)

Benzo[b]furan   **Benzo[b]thiophene**

6.2.5 Methods of preparation of benzofuran and benzothiophene

(*i*) **From 2-aryloxy- or 2-arylthiocarbonyls:** Intramolecular cyclization of 2-aryloxy- or 2-arylthiocarbonyls with loss of water produce benzofurans and benzothiophenes, respectively. In this reaction H_2SO_4, PPA, $ZnCl_2$ or $POCl_3$ can be used as dehydrating agents.

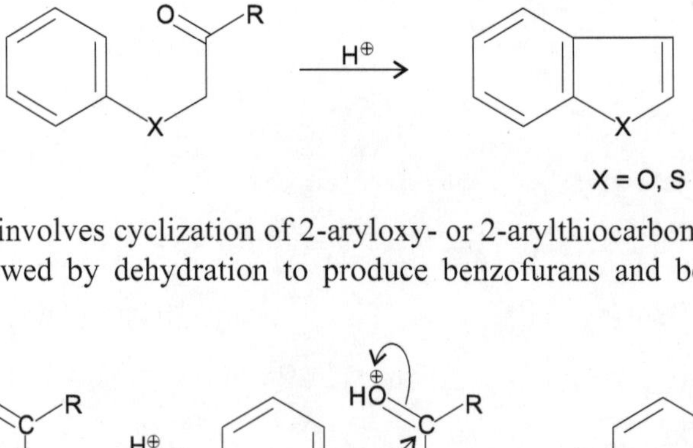

X = O, S

Mechanism: It involves cyclization of 2-aryloxy- or 2-arylthiocarbonyls under acidic conditions followed by dehydration to produce benzofurans and benzothiophenes, respectively.

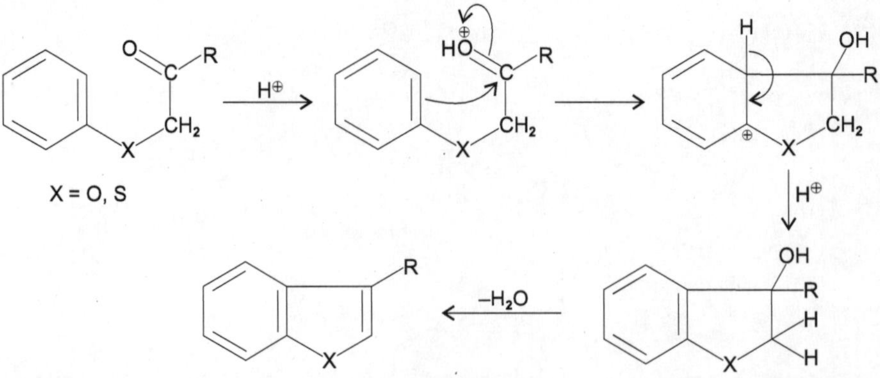

(*ii*) **From aryl 2-chloroprop-2-enyl ethers (or sulfides):** The aryl 2-chloroprop-2-enyl ethers (or sulfides) undergo rearrangement to provide and 2-methylbenzofurans (or 2-methylbenzothiophenes).

Mechanism: It involves **Claisen** rearrangement (or **thio-Claisen** rearrangement) of the starting molecule to give the product.

(*iii*) **From ortho-formylaryloxyacetic acids:** The intramolecular condensation of ortho-formylaryloxyacetic acid yields benzofuran.

o-Formyl aryloxyacetic acid

The synthesis can be performed in one pot in which ortho-hydroxyaryl aldehydes or ketones, are O-alkylated with α-halo ketones, then intramolecular aldol condensation *in situ* produces 2-acyl or 2-aroyl benzofurans.

6.2.6 Reactions of benzofuran and benzothiophene

Electrophilic substitution reactions

Unlike indole where electrophilic substitution takes place at 3-position, electrophilic substitution in benzofuran occurs preferentially at 2-position. This is because the intermediate formed due to attack of electrophile at 2-position carries positive charge at carbon, while intermediate for attack at 3-position has positive charge on oxygen in one of the resonating structure. As, positive charge on carbon is more stable than on the electronegative oxygen, therefore, electrophilic substitution is favourable at 2-position in benzofuran.

But, benzothiophene undergoes electrophilic substitution at 3-position similar to indole.

The electrophilic substitution in benzofuran and benzothiophene is much less regioselective than that of indole. Because of this reason electrophilic substitution reactions of these systems produce mixture of products where substitution occurs at both heterocyclic as well as benzene ring.

Nitration: Nitration of benzofuran with nitric acid in acetic acid gives 2-nitrobenzofuran. However, with N_2O_4 in benzene, mixture of 2-nitro and 3-nitro benzofuran is obtained.

Nitration of benzothiophene in presence of nitric acid gives a mixture in which, although the main product is the 3-nitro derivative, lesser quantities of 2-nitro-, 4-nitro-, 6-nitro- and 7-nitro benzothiophenes are also produced. However, ceric ammonium nitrate in acetic anhydride at room temperature produces a high yield of 3-nitrobenzothiophene.

(major)

Halogenation: Treatment of benzofuran with halogens results in 2,3-addition products. This on reaction with sodium ethoxide undergoes hydrogen halide elimination to yield 3-monohalo benzofuran. But the 2,3-addition product on heating in presence of acetic acid gives 2-halo product.

(*cis & trans*)

Controlled reaction of benzothiophene with chlorine (or bromine) produces 3-chloro (or 3-bromo) benzothiophene in moderate yield. Reaction of benzothiophene with chlorine in the presence of one mole of iodine yields 2,2,3,3,4,5,6,7-octachloro-2,3-dihydrobenzothiophene. It reacts with iodine in the presence of mercuric oxide to give 3-iodobenzothiophene.

Formylation: Benzofuran displays a lesser selectivity for 3-substitution, therefore, formylation of benzofuran gives only the 2-formylderivative.

Friedel-Crafts acetylation: Benzofuran undergoes acetylation with acetic anhydride in presence of $SnCl_4$ or H_3PO_4 to yield 2-acetyl derivative.

Friedel-Crafts acetylation of benzothiophene gives a mixture of 3- and 2- acetyl derivatives in a ratio of 7:3.

7:3

Oxidation

Benzofuran on oxidation with acidified potassium permangnate yields 2-hydroxybenzoic acid.

Benzothiophen when oxidised with mCPBA affords benzothiophene-1,1-dioxide (sulfone).

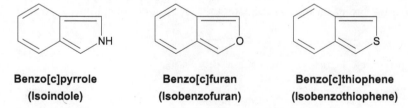

Reduction

Both benzofuran and benzothiophene when treated with sodium in ethanol gets converted to their 2,3-dihydroderivatives. However, treatment of benzofuran and benzothiophene with sodium in liquid ammonia results into the formation of 2-ethylphenol and 2-ethylbenzenethiol, respectively.

$$X = O, S$$

6.3 BENZO[c]PYRROLE (ISOINDOLE), BENZO[c]FURAN AND BENZO[c]THIOPHENE

6.3.1 Introduction

The benzofused heterocyclic compounds **benzo[c]pyrrole (isoindole)**, **benzo[c]furan** and **benzo[c]thiophene** have benzene ring fused at 3 and 4 position of pyrrole, furan and thiophene, respectively. Benzo[c]furan and benzo[c]thiophene are also known as **isobenzofuran** and **isobenzothiophene**, respectively.

| **Benzo[c]pyrrole** | **Benzo[c]furan** | **Benzo[c]thiophene** |
| (Isoindole) | (Isobenzofuran) | (Isobenzothiophene) |

6.3.2 Biological importance

The isoindole structure can be found in several natural products (e.g. alkaloids) and pharmaceutical compounds. Isoindoles are building blocks for phthalocyanines which are relevant as dyes. Some potent isoindole ring containing compounds are **thalidomide** (teratogenic), **(S)-pazinaclone** (sedative) and **lenalidomide** (anticancer, multiple myleloma).

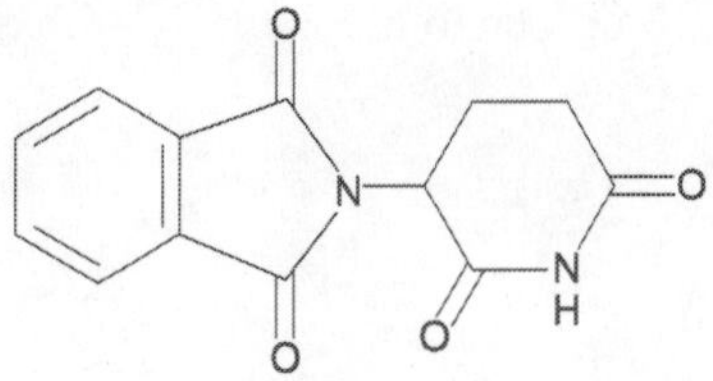

Thalidomide

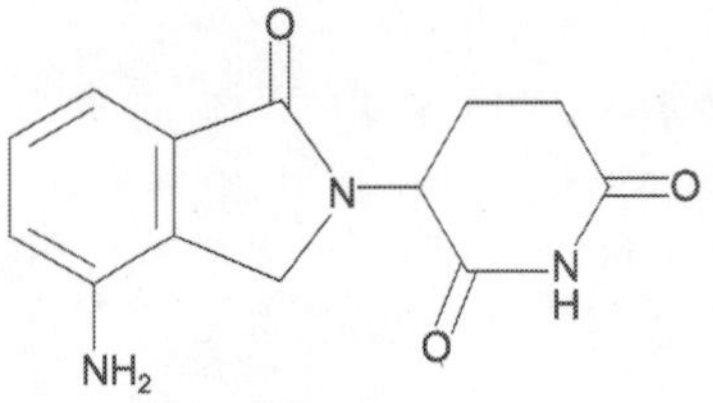

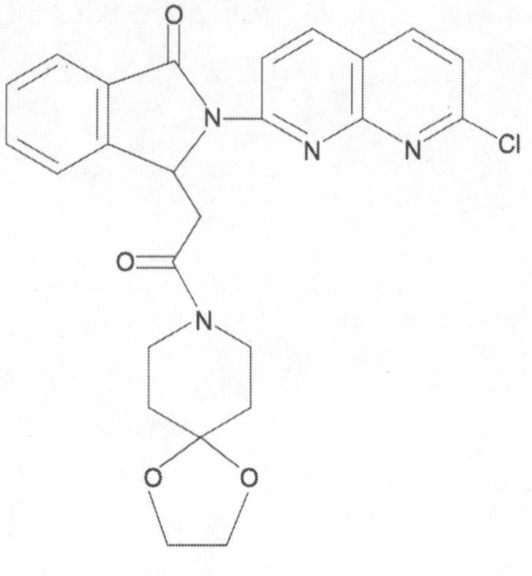

(S)-Pazinaclone

Lenalidomide

6.3.3 Stuctrure

Isoindole, isobenzofuran and isobenzothiophene are much less stable than their isomers, indole, benzothiophene and benzofuran. This is because, in these compounds the six membered ring is not a complete benzenoid unit and have lower aromaticity.

Isoindole (2*H*-isoindole) is an isomer of indole which exists in the ortho-quinoid form. The reduced form is known as **isoindoline** (2,3-dihydro-1*H*-isoindole). The benzo[c]pyrrole (2*H*-isoindole) shows tautomerism with its imine form known as **isoindolenine** (**1*H*-isoindole**). In solution, the 2*H*-isoindole tautomer is the predominant form and therefore the compound resembles a pyrrole more than a simple imine. The degree to which the 2*H* predominates depends on the solvent, and for substituted isoindole-isoindolinine systems it depends upon the nature of substitutents. The electron releasing substituent stabilize 2*H*-isoindole tautomers, whereas electron withdrawing group stabilizes isoindolenine (1*H*-isoindole) tautomeric form.

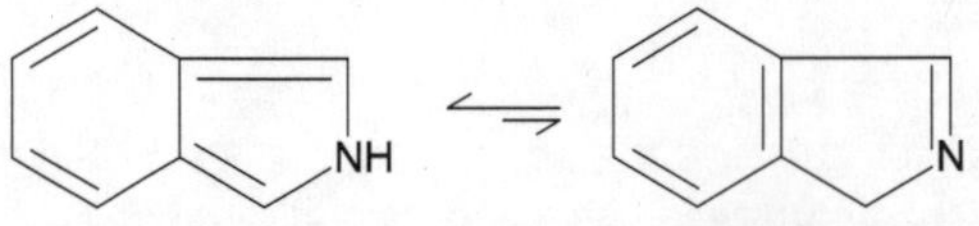

2*H*-Isoindole   **1*H*-Isoindole**   **Isoindoline**
                          (Isoindolenine)

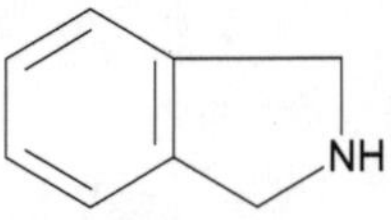

Tautomerism in isoindole

6.3.4 Spectral data

The ^{1}H NMR of isoindole is given below:

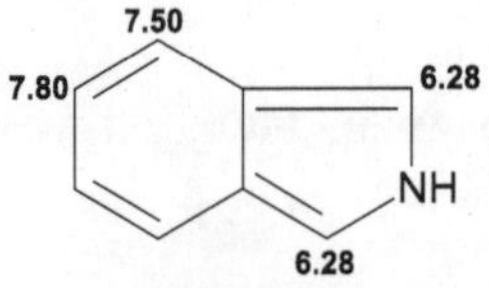

Isoindole

^{1}H NMR (δ, ppm, $(CD_3)_2CO$)

6.3.5 Methods of preparation of benzo[c]pyrrole, benzo[c]furan and benzo[c]thiophene

(*i*) **From ortho-diacylbenzene:** Isoindole can be synthesized by the condensation of primary amines with ortho-diacylbenzene. Isomerization followed by sequential dehydration–hydration process gives isoindole derivative.

(*ii*) **From α-azido carbonyl compounds:** Isoindole derivatives can be synthesized from α-azido carbonyl compounds bearing a 2-alkenylaryl moiety at the α-position. The azides undergo 1,3-dipolar cycloaddition onto alkenes to give substituted isoindoles.
Hui, B.W.Q.; Chiba, S., Org. Lett., ***2009****, 11, 729-732.*

(*iii*) **From N-substituted isoindolines:** Cyclic hydroxylamine acetate (N-substituted isoindolines) on pyrolysis eliminate acetic acid to yield isoindoles which is trapped at -196°C.

(trap at −196°C)

(*iv*) **Synthesis of 1,3-diphenylisoindoles:** The 1,3-diphenylisoindoles can be prepared from 1,2-diaroyl benzenes such as 1,2-phenylenebis(phenylmethanone) by reaction with methyl amine and a reducing agent sodium borohydride.

(*v*) **Synthesis of benzo[c]furan derivative:** Purification of 2-[chloro(phenyl) methyl]-N,N-diethylbenzamide by flash chromatography leads to the formation of 1-(diethylamino)-3-phenylbenzo[c]furan hydrochloride.

(*vi*) **Synthesis of benzo[c]furan:** Isobenzofuran can be isolated by trapping on a cold finger, (laboratory equipment used to generate cold surface) by flash vacuum thermolysis of 1,4-epoxy-1,2,3,4-tetrahydronaphthalene.

(100%)

(*vii*) **Synthesis of benzo[c]thiophene:** Dihydrobenzo[c]thiophene S-oxides undergo elimination in presence of alumina to yield benzo[c]thiophene.

6.3.6 Reactions of benzo[c]pyrrole, benzo[c]furan and benzo[c]thiophene

Four molecules of 1,3-diiminoisoindoline eliminates ammonia in presence of tetralin on heating to produce phthalocyanine. The phthalocyanine macrocyclic system is the basis for many blue dyestuffs.

Phthalocyanine

Isoindole gives Diels-Alder reaction, whereas indole does not.

1,3-Diphenyl derivative of benzo[c]furan undergoes cycloaddition with ortho-thiobenzoquinonemethide.

1,3-Dithienylbenzo[c]furan gets converted to benzo[c]selenophene in 67% yield on interaction with **Woollins reagent** (0.25 eq.) at room temperature.

EXERCISE

Q.1 What happens when indole is treated with acetyl chloride in presence of $SnCl_4$?

Q.2 Write mechanism of the following reactions:

(a) Reissert indole synthesis

(b) Fischer indole synthesis

(c) Madelung synthesis of 2-substituted indole derivatives

Q.3 How will you synthesize gramine from indole.

Ans. Mannich reaction of indole with Me_2NH and HCHO give gramine.

Q.4 Give the synthesis of 2-methyl indole by Fischer-indole method. Give the mechanism of the reaction.

Q.5 Explain why:-

(a) Electrophilic substitution is preferred at the 3-position in indole.

(b) Indole is less reactive than pyrrole in electrophilic substitution reactions.

(c) In benzofuran electrophilic substitution takes place at 2-position whereas in indole and benzothiophene at 3-position. (Refer sec. 6.2.6)

(d) Isoindole is less stable than indole (Refer sec. 6.3.3)

Q.6 Explain the aromaticity and basicity of indole on the basis of its resonating structures.

[Hint. Indole is an aromatic heterocyclic compound. It has 10 π-electrons and all the atoms are sp^2 hybridized. Both the rings of indole are coplanar.]

However, indole is not basic as the lone pair on nitrogen are involved in delocalization in the aromatic system and not free to react.

Q.7 Complete the following reaction:

(a) Indole + $CHCl_3$ + NaOH $\longrightarrow$

(b) Indole + CH_3COCl + $SnCl_4$ $\longrightarrow$

(c) Benzo[b]furan + $(CH_3CO)_2$ + $SnCl_4$ $\longrightarrow$

(d) Benzo[b]furan + DMF + $POCl_3$ $\longrightarrow$

(e) Benzo[b]furan + n-BuLi + ether at -10°C $\longrightarrow$

(f) Benzo[b]furan + $KMnO_4$ + H_2SO_4 $\longrightarrow$

(g) Benzo[b]thiophene + I_2 + HgO $\longrightarrow$

(h) Benzo[b]thiophene + HCHO + HCl $\longrightarrow$

(i) Benzo[b]thiophene + CH_3COCl + $SnCl_4$ or $AlCl_3$ $\longrightarrow$

Ans. (a) Indole-3-carbaldehyde; (b) 3-Acetylindole; (c) 2-Acetylbenzofuran; (d) Benzo[b]furan-2-carbaldehyde; (e) 2-Lithiobenzo[b]furan; (f) Salicylic

acid; (g) 3-Iodobenzo[b]thiophene; (h) 3-Chloromethylbenzo[b]thiophene;
(i) 3-Acetylbenzo[b]thiophene

Q.8 What product is obtained in the following reaction:

(a)

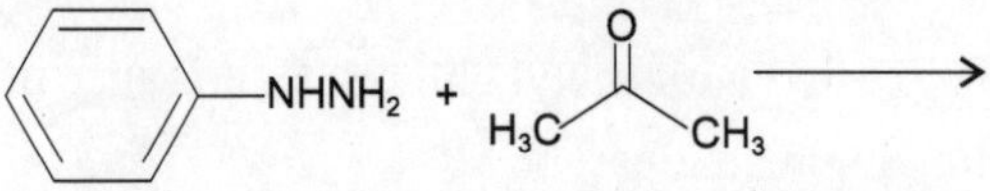

(b)

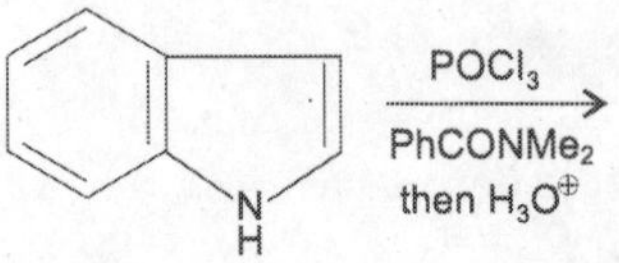

(c)

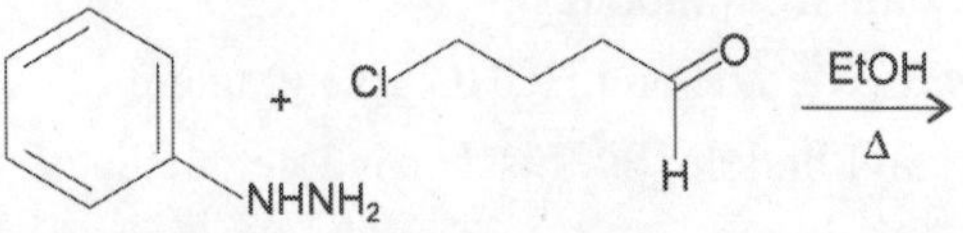

(d)

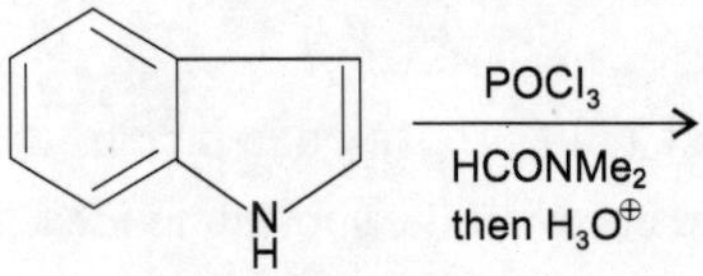

Ans. (a) 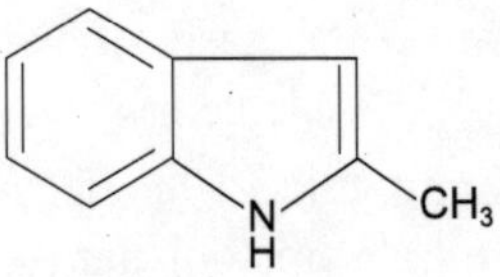   (b)

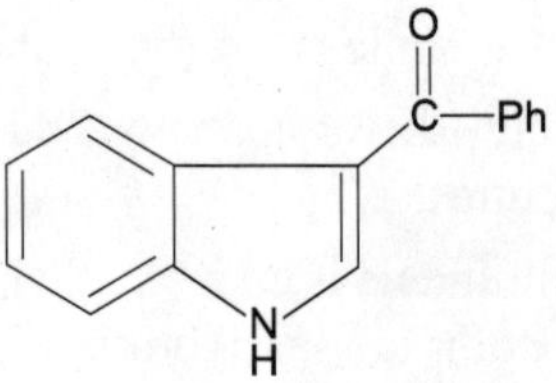

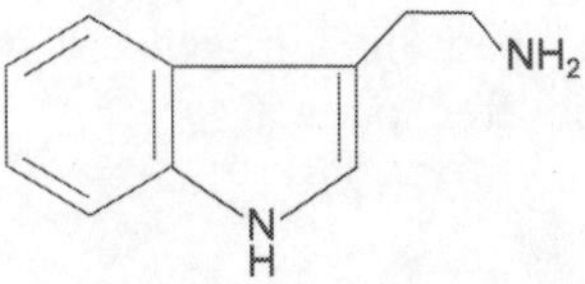

(c) 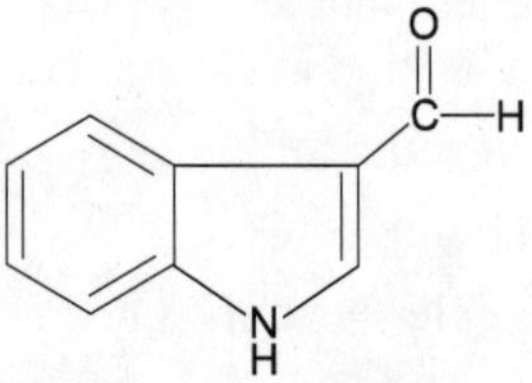   (d)

Q.9 The major product formed in the following reaction is

(a)

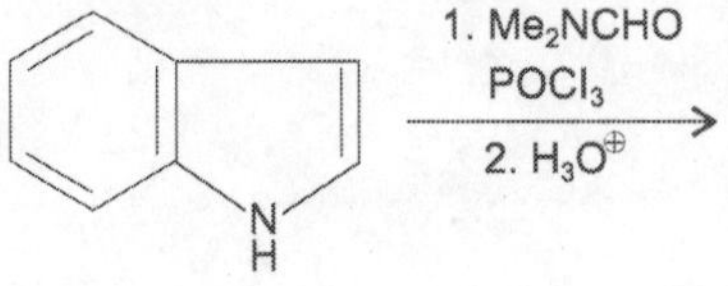

(b) Indole + Sn/HCl $\longrightarrow$

(c) 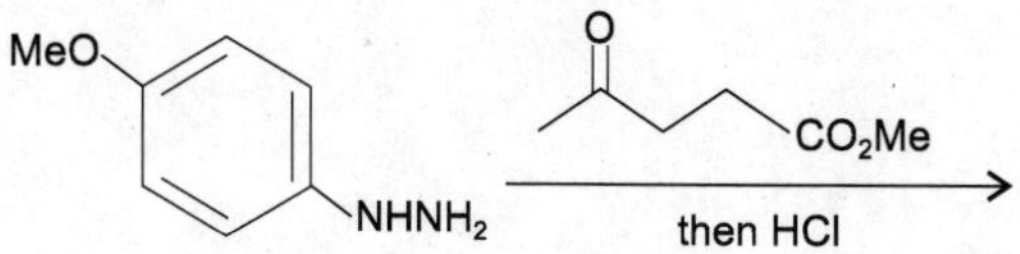
MeO
NHNH2
O
CO2Me
then HCl

Ans. (a) 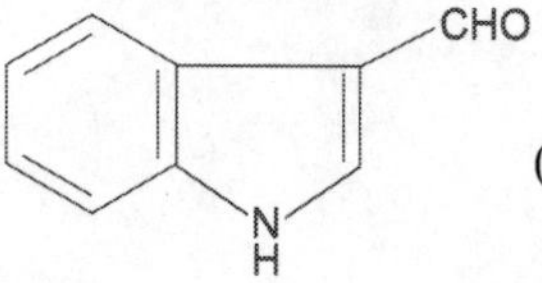
CHO
N
H

(b) 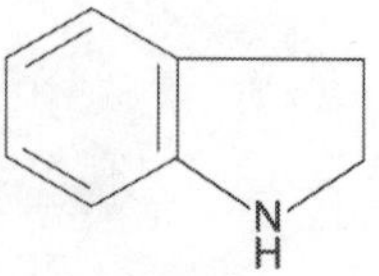
N
H

(c) 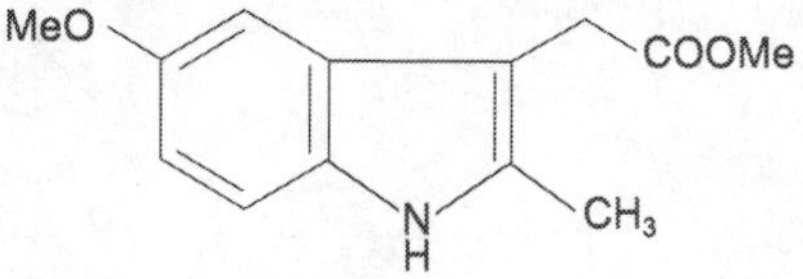
MeO
COOMe
N
H
CH3

Five Membered Heterocyclic Compounds with Two Heteroatoms

1,2-Azoles: Pyrazole, Isoxazole and Isothiazole

1,3-Azoles: Imidazole, Oxazole and Thiazole

7.1 1,2-AZOLES: PYRAZOLE, ISOXAZOLE AND ISOTHIAZOLE

7.1.1 Introduction

The three main five membered heterocycles with two heteroatoms at adjacent positions, one of which is nitrogen, are pyrazole, isoxazole and isothiazole. These belong to the azole class of compounds.

| Pyrazole | Isoxazole | Isothiazole |

Pyrazole is a heterocyclic compound with a five membered ring of three carbon and two adjacent nitrogen atoms. Isoxazole is an azole with an oxygen atom next to the nitrogen. Isothiazole, or 1,2-thiazole contains a five membered aromatic ring that consists of three carbon atoms, one nitrogen atom and one sulfur atom.

The corresponding dihydro heterocycles of pyrazole, isoxazole and isothiazole are named as pyrazoline, isoxazoline and isothiazoline, respectively.

1-Pyrazoline
4,5-Dihydro-3*H*-pyrazole

2-Pyrazoline
4,5-Dihydro-1*H*-pyrazole

4-Pyrazoline
2,3-Dihydro-1*H*-pyrazole

2-Isoxazoline
4,5-Dihydroisoxazole

3-Isoxazoline
2,5-Dihydroisoxazole

4-Isoxazoline
2,3-Dihydroisoxazole

2-Isothiazoline
4,5-Dihydroisothiazole

3-Isothiazoline
2,5-Dihydroisothiazole

4-Isothiazoline
2,3-Dihydroisothiazole

Pyrazoline has only one endocyclic double bond and is basic in nature. 1-Pyrazoline, 2-pyrazoline and 3-pyrazoline are the three partially reduced forms of the pyrazole structure having double bonds at different positions. These three forms exist in equilibrium. 2-Pyrazoline exhibits the monoimino character and hence more stable than the rest even though all the three types have been synthesized. 2-Pyrazolines (4,5-dihydro-1*H*-pyrazolines), can be considered as a cyclic hydrazine moiety, seem to be the most frequently studied pyrazoline type compounds. 2-Pyrazoline is insoluble in water but soluble in propylene glycol because of its lipophilic character.

The tetrahydro-1,2-azoles are called pyrazolidine, isoxazolidine and isothiazolidine.

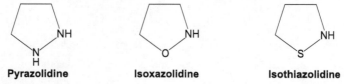

Pyrazolidine   **Isoxazolidine**   **Isothiazolidine**

7.1.2 Importance of 1,2-azoles

In 1959, the first natural pyrazole, 3-(**1-pyrazolyl**)-**alanine**, was isolated from seeds of watermelons. Since then many synthetic pyrazoles of importance as drugs and dyes are known. Among them is **phenylbutazone (butazolidin)**, an anti-inflammatory drug used for the treatment of arthritis. **Antipyrine** is a well-known pyrazole based drug used as analgesic, antipyretic and anti-inflammatory agent. **Celebrex** is another pyrazole based drug which is a sulfonamide non-steroidal anti-inflammatory drug (NSAID). **Tartrazine** most commonly used as a yellow dye for food is also pyrazole derivative.

Isoxazolyl group is found in many semisynthetic β-lactamase-resistant antibiotics, such as **cloxacillin, dicloxacillin, flucloxacillin** and **oxacillin**. Isoxazoles also form the basis for a number of drugs such as **valdecoxib (Bextra)**, a non-steroidal anti-inflammatory drug (NSAID). Isoxazole ring is found in some natural products, such as **ibotenic acid**, powerful neurotoxin that is used as a brain-lesioning agent. Some other drugs containing isoxazole nucleus are **drazoxolon** (antibacterial), **cycloserine** (antibacterial), **zonisamide** (anticonvulsant), **leflunomide** (rheumatoid arthritis).

The zwitterionic compound 5-(aminomethyl)isothiazol-3-ol (**thiomuscimol**) containing isothiazole ring is active in the central nervous system as an agonist of receptors for gama-aminobutyrate (GABA).

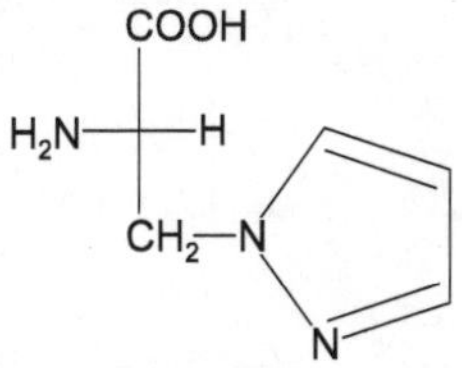

3-(1-Pyrazolyl)-alanine

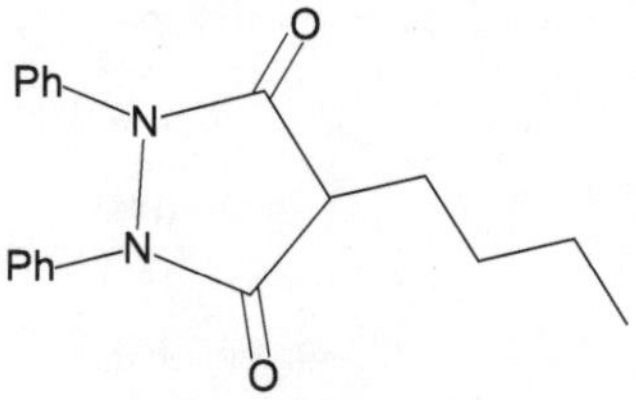

Phenylbutazone

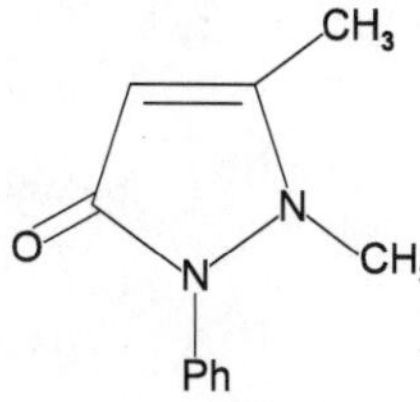

Antipyrine

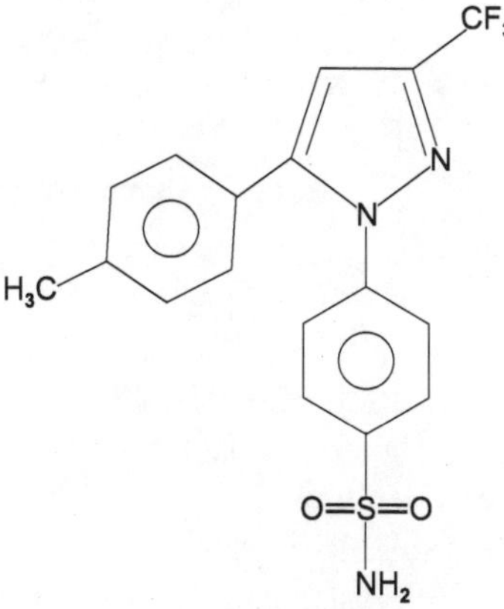

Celebrex

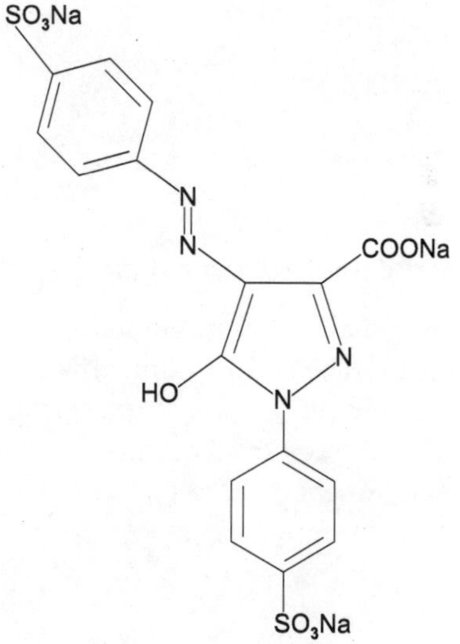

Tartrazine

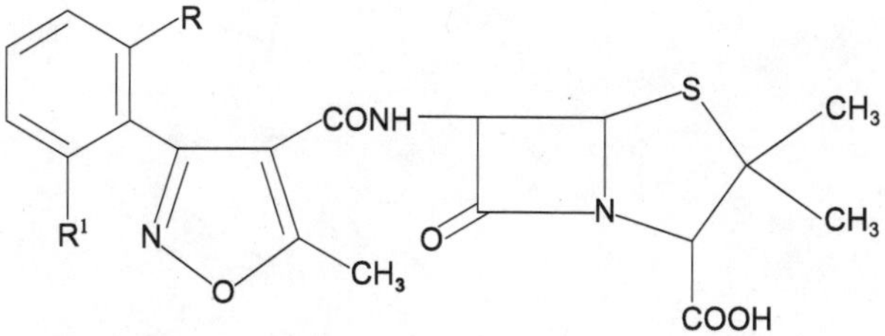

R	R¹	
Cl	H	**Cloxacillin**
Cl	Cl	**Dicloxacillin**
Cl	F	**Flucloxacillin**
H	H	**Oxacillin**

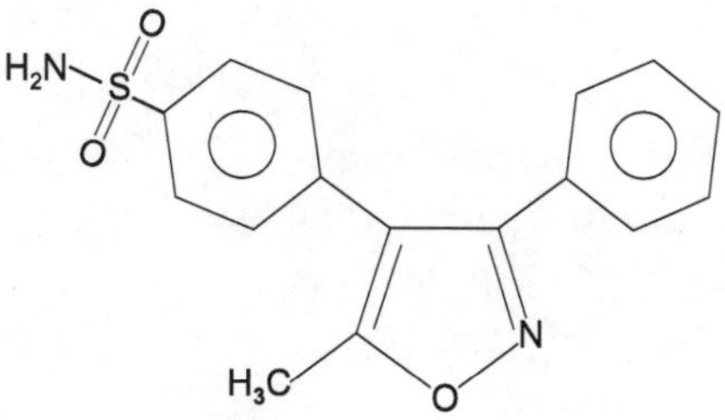

Valdecoxib

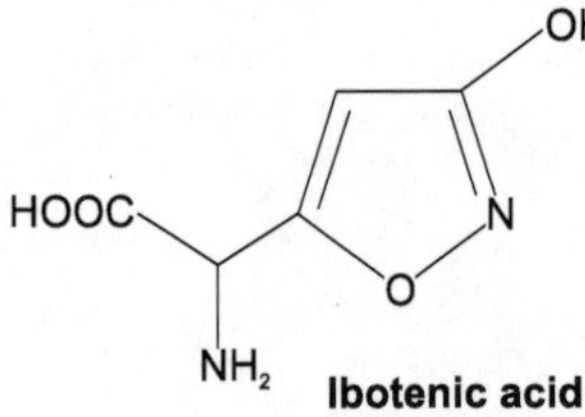

Ibotenic acid 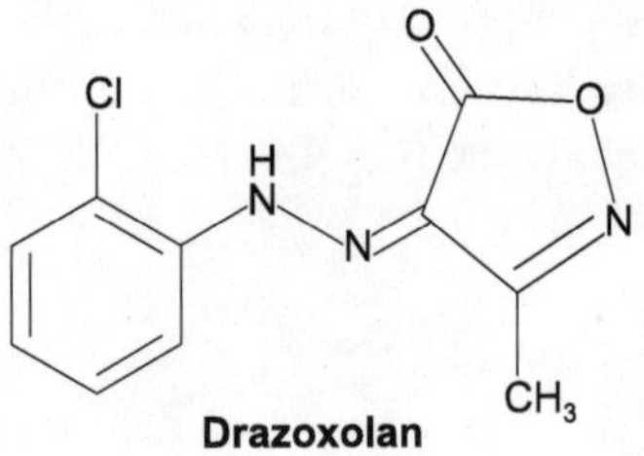   **Drazoxolan** 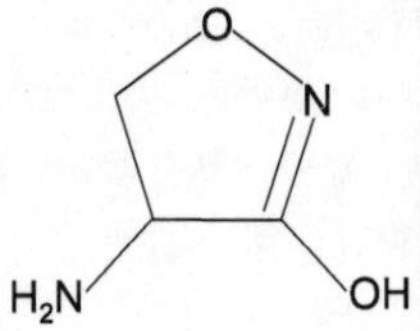   **Cycloserine**

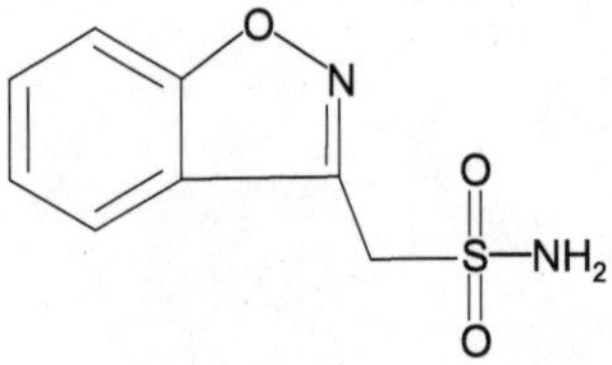

Zonisamide 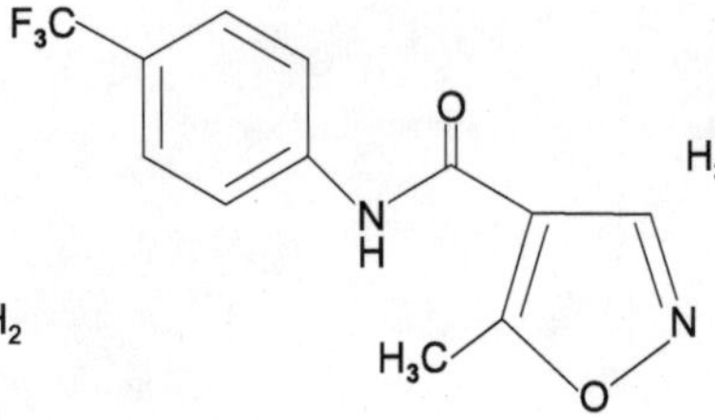   **Leflunamide** 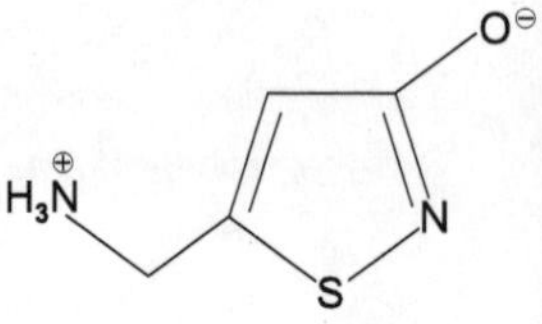   **Thiomuscimol**

7.1.3 Structure and aromaticity

1,2-Azoles have a pyridine-like odour. Pyrazole is solid (mp 67-70°C) at room temperature. The boiling point (187°C) of pyrazole is higher than isothiazole (114°C) and isoxazole (95°C) due to intermolecular hydrogen bonding. This association is in the form of dimers, trimers and oligomers. Pyrazole is soluble in water. Isothiazole has a solubility of about 3.5% in water and is miscible with most organic solvents.

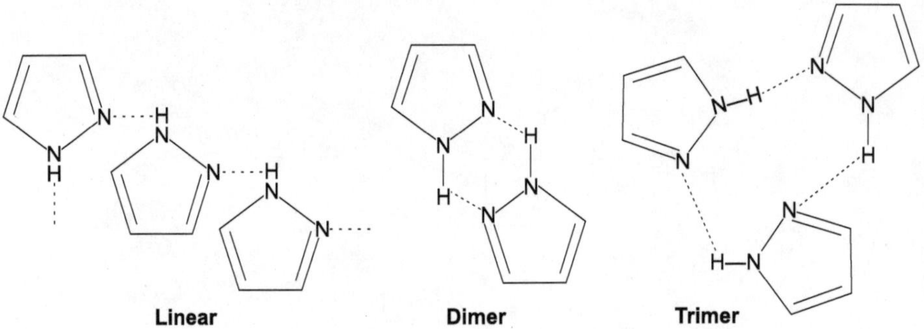

Linear   **Dimer**   **Trimer**

Hydrogen bonding in pyrazole

The boiling point of isoxazole (95°C) is higher than oxazole (69°C) and furan (bp 32°C) due to intermolecular association by the hydrogen bond between nitrogen atom and hydrogen at position-3.

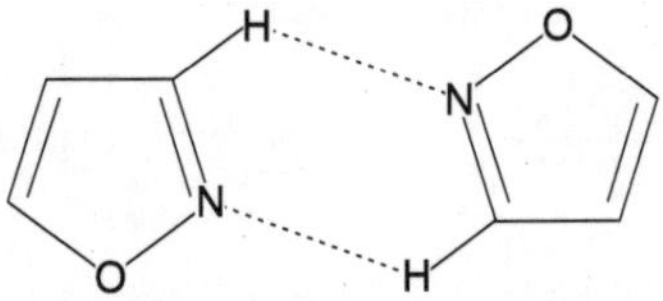

Hydrogen bonding in oxazole

Pyrazole undergoes rapid tautomerism which involves switching of hydrogen atom from one nitrogen to the other. Thus substituted pyrazoles such as 3-methylpyrazole is a mixture of two inseparable isomers, 3-methyl and 5-methyl pyrazoles hence named as 3(5)-methylpyrazole.

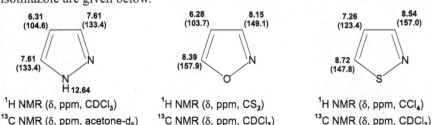

3-Methylpyrazole　　　　**5-Methylpyrazole**

3(5)-Methylpyrazole

Tautomerism in pyrazoles

1,2-Azoles are planar having (4n+2) π-electrons, thus aromatic in nature. The order of aromaticity of the 1,2-azoles is **isothiazole > pyrazole > isoxazole.**

The resonance energy of pyrazole is 29.3 kcal/mol or 123 kJ/mol. It is more aromatic than imidazole (resonance energy 14 kcal/mol or 59 kJ/mol). The aromaticity of isothiazole is similar to that of thiazole. The resonating structures of isoxazole and isothiazole have negative charge on carbon and nitrogen atoms while positive charge resides on oxygen and sulphur respectively.

X = NH, O, S

Resonance in 1,2-azoles

The dipole moment of pyrazole (μ = 2.64 D in dioxane) was found by Huckel et al. The dipole moment of isoxazole and isothiazole is 3.07 and 2.4 D, respectively.

μ = 2.64 D　　　　μ = 3.07　　　　μ = 2.4 D

7.1.4 Spectral data

The ¹H NMR and ¹³C NMR (given in brackets) data for pyrazole, isoxazole and isothiazole are given below.

6.31 (104.6)		7.61 (133.4)
7.61 (133.4)	N	
	N H 12.64	

¹H NMR (δ, ppm, CDCl₃)
¹³C NMR (δ, ppm, acetone-d₆)

6.28 (103.7)　8.15 (149.1)
8.39 (157.9)　O, N

¹H NMR (δ, ppm, CS₂)
¹³C NMR (δ, ppm, CDCl₃)

7.26 (123.4)　8.54 (157.0)
8.72 (147.8)　S, N

¹H NMR (δ, ppm, CCl₄)
¹³C NMR (δ, ppm, CDCl₃)

7.1.5 Methods of preparation of 1,2-azoles

(*i*) **From 1,3-dicarbonyl compounds:** Pyrazoles can be synthesized by the reaction of 1,3-dicarbonyl compounds with hydrazine.

Mechanism: This method is based on the double nucleophilic character of hydrazine, allowing them to react in turn with each carbonyl group of a 1,3-dicarbonyl compounds.

The 3-methyl-1-phenyl-5-pyrazolone can be prepared similarly by the condensation of phenylhydrazine with ethyl acetoacetate.

Similarly, isoxazoles can be synthesized from 1,3-dicarbonyl compound which may be masked as enol, ether, acetal, imine or enamine by the reaction with hydroxylamine.

Synthesis of sildenafil (Viagra): Sildenafil is a drug used for relaxing and dilating the blood vessels in the lungs. The synthesis of sildenafil involves the reaction of 1,3-dicarbonyl compound with hydrazine to form pyrazole. The pyrazole formed is then functionalised to get sildenafil.

Sildenafil

(*ii*) **By 1,3-cycloaddition:** Pyrazoles are synthesized by the cycloaddition of nitrile imines to alkynes.

Nitrile imines

3,5-Substituted isoxazoles are produced by the dipolar cycloaddition of nitrile oxides to alkynes. Nitrile oxides can be prepared by the elimination of chlorooximes (PhC(Cl)=NOH) or the dehydration of nitroalkanes, (PhCH$_2$NO$_2$).

Nitrile oxides   **(7:3)**

(*iii*) **From oximes:** Reaction of ketoximes containing an α-hydrogen with two mole equivalents of n-butyllithium leads to O- and C- lithiation (*syn* to the oxygen). Further reaction of this dianion with dimethylformamide as electrophile allows C-formylation and ring closure *in situ* to form corresponding isoxazole derivative.

Oxime

(*iv*) **From alkenes:** A mixture of alkene, sulfur dioxide and ammonia on passing over alumina as catalyst at high temperature yields isothiazole.

$$CH_3-CH=CH_2 \ + \ SO_2 \ + \ NH_3 \ \xrightarrow[200°C]{Al_2O_3} \quad + \ H_2O \ + \ H_2S$$

(*v*) **From γ-iminothiols:** Isothiazoles are prepared by the reaction of γ-iminothiols with halogens.

γ-Iminothiol

7.1.6 Reactions of 1,2-azoles

Basicity

The direct linking of two heteroatoms in 1,2-azoles decreases their basicity. The pK_a values of the conjugate acids of 1,2-azoles are: pyrazole (2.5), isoxazole (-3.0) and isothiazole (-0.5). The higher basicity of pyrazole as compared to other 1,2-azoles is due to the symmetry of the cation. The pyrazolium cation has two equivalent contributing resonance structures. Isoxazole is less basic than isothiazole due to greater electronegativity of N than S. The order of basicity is **pyrazole > isothiazole > isoxazole.**

	Pyrazolium cation	Isoxazolium cation	Isothiazolium cation
pK$_a$ of conjugate acid:-	2.5	–3.0	–0.5

Reaction with electrophiles

Azoles are less reactive than monohetero aromatic five membered heterocycles pyrrole, thiophene and furan due to the additional nitrogen atom present in the ring. The additional heteroatom (N-atom) in azoles has electron withdrawing effect. Moreover, under acidic conditions the nitrogen in 1,2-azoles get protonated to form corresponding azolium cation which is less susceptible towards electrophilic attack. Thus, 1,2-azoles undergo electrophillic substitution reaction with less ease than pyrrole, furan and thiophene, but more than benzene.

The order of reactivity of the 1,2-azoles towards electophiles is **pyrazole > isothiazole > isoxazole**.

1,2-Azoles (pyrazole, isoxazole and isothiazole) undergo electrophilic attack preferentially at 4-position. The nitrogen present at 2-position deactivates position-3 and -5 towards the attack of electrophile. However, position-4 is activated due to electron release from heteroatom at position-1.

Also, when an electrophile attacks position-3 or -5 the intermediate formed has positively charged sextet nitrogen (azomethine) which is highly unfavourable. Whereas no such unfavourable resonating structure is formed when electrophile attacks the position-4 of these 1,2-azoles.

X = NH, O, S

(Unfavourable)

(attack at C–5)   **(Unfavourable)**

Nitration: Pyrazole, isoxazole and isothiazole undergo nitration at C4 under appropriate conditions. For example, nitration of pyrazole with concentrated nitric acid and sulfuric acid gives 4-nitropyrazole.

Similarly, isoxazole and isothiazole also give 4-nitro product with concentrated nitric acid and sulfuric acid.

However, with acetyl nitrate, 1-nitropyrazole is formed, which can be rearranged to 4-nitropyrazole in the presence of acid at low temperature.

Sulfonation: Both pyrazole and isothiazole can be sulfonated at C4. But sulfonation of isoxazole is difficult. However it can be sulfonated under drastic conditions with oleum.

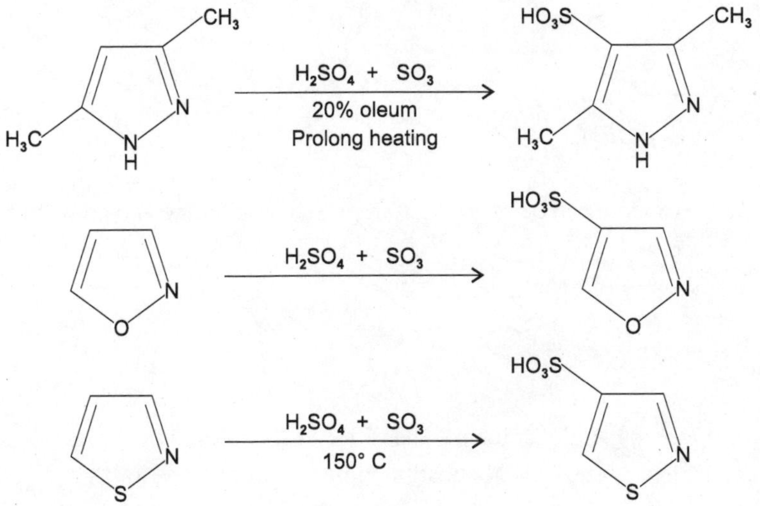

$H_2SO_4 + SO_3$
20% oleum
Prolong heating

$H_2SO_4 + SO_3$

$H_2SO_4 + SO_3$
150° C

Halogenation: Halogenation of pyrazole gives 4-monohalopyrazoles under controlled conditions. Chlorination of pyrazole can be carried out by using sulphuryl chloride in chloroform or chlorine in carbon tetrachloride (or acetic acid). Pyrazole on treatment with bromine in chloroform yields 4-bromopyrazole.

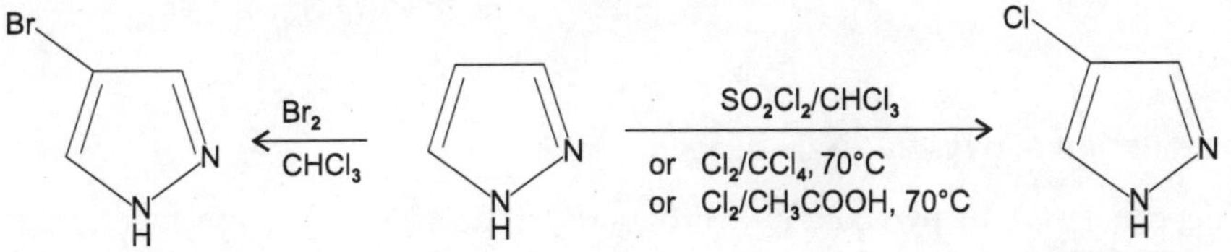

Br_2
$CHCl_3$

$SO_2Cl_2/CHCl_3$
or Cl_2/CCl_4, 70°C
or Cl_2/CH_3COOH, 70°C

4-Bromopyrazole

4-Chloropyrazole

4-Bromination of pyrazoles and isoxazoles can also be achieved using N-bromosuccinimide and microwave irradiation.

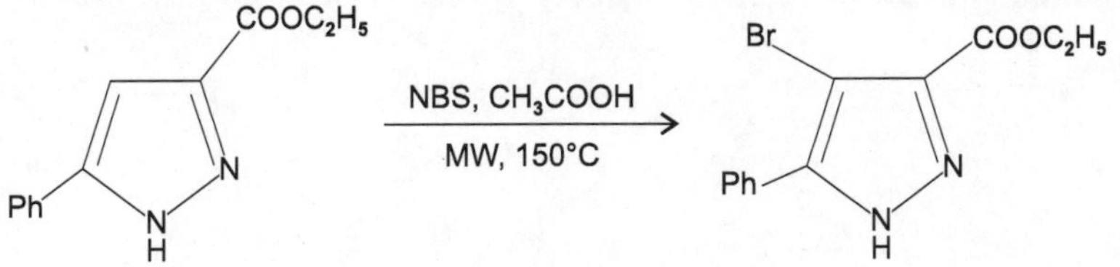

NBS, CH_3COOH
MW, 150°C

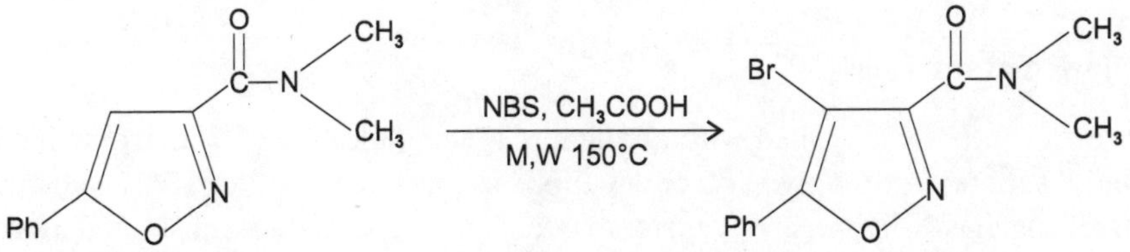

NBS, CH_3COOH
M,W 150°C

Isoxazoles undergo halogenation more readily than pyridine, but less than pyrrole and furan. Chlorinating agents used for isoxazole, are sodium hypochlorite, sulfuryl chloride, aq. $CHCl_3$ in 20% HCl, N-chlorosuccinamide (NCS). Isoxazole gives 4-chloro or 4-bromo- derivative with chlorine or bromine, respectively.

Iodination of isoxazole can be carried out in presence of iodine and conc. nitric acid.

Acylation: Only pyrazole amongst the three 1,2-azoles, undergoes acylation at 4-position and, even here, only N-substituted pyrazoles react well.

Deprotonation of pyrazole N-hydrogen

The deprotonation of pyrazole N-hydrogen by strong bases is more facile than for pyrrole, imidazole or indole. The pK_a for loss of the N-hydrogen of pyrazole is 14.2, thus it is an appreciably stronger acid than imidazole (NH pK_a = 14.5), indole (NH pK_a = 16.2) and pyrrole (NH pK_a = 17.5). This is because the anion formed after deprotonation of pyrazole has two equal contributing resonance structures.

pK_a = 14.2

Alkylation at nitrogen

Pyrazoles can be N-alkylated with alkylating agents (alkyl halides, diazomethane or dimethyl sulfate) in strongly basic or under phase-transfer conditions. 3(5)-Substituted pyrazole that has N-hydrogen gives rise to two isomeric N-alkyl-3-substituted pyrazole and N-alkyl-5-substituted pyrazole.

N-Methyl-3-methylpyrazole   **N-Methyl-5-methylpyrazole**
(65%)   (35%)

Pyrazole under Mannich conditions produces an N-protected pyrazole, presumably *via* attack at the imine nitrogen, followed by loss of proton from the other nitrogen. Similarly, a N-tetrahydropyranyl pyrazole is prepared.

Isoxazole undergoes N-alkylation with alkyl sulfates or alkyl iodides.

If a hydroxyl group is present at position-3 or -5 in isothazole it tautomerises to NH and it is alkylated by diazomethane, dimethyl sulfate or alkyl halide.

Acylation at nitrogen

An acyl or phenylsulfonyl group can be introduced into a pyrazole nitrogen in the presence of a weak base, such as pyridine. Since acylation, is reversible, the more stable product is obtained.

Reactions with oxidizing agents

The 1,2-azole ring systems are relatively resistant to oxidizing agents. But the alkyl or acyl substituents present in pyrazole ring may be oxidized to carboxylic acid.

The unsaturated side chains and oxygenated functional groups in isoxazole are oxidized to carboxylic acid group.

Ozone cleaves the substituted isoxazole ring at 4,5-double bond to give acyl derivatives of α-diketone monoximes.

**Acyl derivatives of
α-diketone monoxime**

An isothiazole unsubstituted at position-3 gets oxidized to 1,2-thiazol-3(2*H*)-one-1,1-dioxides with hydrogen peroxide in acetic acid.

However, a trisubstituted isothiazole gets oxidized to its 1-oxide first and then further to 1,1-dioxide with peracids.

Reactions with reducing agents

Pyrazoles are relatively stable to catalytic and chemical reduction when there is no substituent on nitrogen. Catalytic reduction can be achieved in acidic medium.

Pyrazole when heated with sodium/alcohol or palladium/hydrogen yields 2-pyrazoline.

2-Pyrazoline

Catalytic hydrogenolysis of the N-O bond in isoxazoles takes place readily with H_2/Ni and $LiAlH_4$, irrespective of the substituents

Direct ring C-metallation

C-5-Lithiation of N-substituted pyrazoles takes place with n-BuLi.

However, C-lithiation of isoxazoles with hydrogen at C-3 leads to ring opening, with the oxygen as anionic leaving group.

Substituted isothiazole on reaction with n-butyl lithium gives 5-lithioisothiazole.

When reacted with n-butyllithium and then with iodine, 4-methylisothiazole yields 5-iodo-4-methylisothiazole as well as a small amount of the acyclic sulfide 3-(butylsulfanyl)-2-methylprop-2-enenitrile.

In case of pyrazoles and isoxazoles mercuration takes place at position-4 with mercury(II)actetate.

Ring contraction

5-Alkoxyisoxazoles undergo ring contraction with iron(II) chloride, producing azirine esters.

7.2 1,3-AZOLES: IMIDAZOLE, OXAZOLE AND THIAZOLE

7.2.1 Introduction

The 1,3-azoles, namely imidazole, oxazole and thiazole contain two heteroatoms in a five membered ring at 1,3-positions. All are very stable compounds that do not auto-oxidize. Imidazole with two nitrogen atoms is a colourless solid that dissolves in water to give mildly basic solution. Oxazole is an 1,3-azole with an oxygen and a nitrogen atom. Thiazole is a pale yellow heterocyclic compound that contains sulfur and nitrogen. Oxazole and thiazole are water miscible liquids with pyridine like odour.

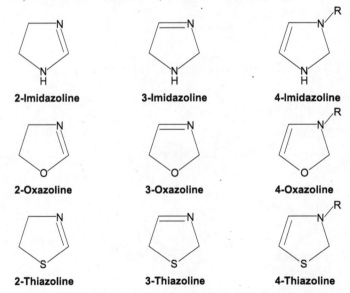

Imidazole   **Oxazole**   **Thiazole**

The dihydro-1,3-azoles are imidazoline, oxazoline and thiazoline.

2-Imidazoline   **3-Imidazoline**   **4-Imidazoline**

2-Oxazoline   **3-Oxazoline**   **4-Oxazoline**

2-Thiazoline   **3-Thiazoline**   **4-Thiazoline**

Tetrahydro-1,3-azoles are imidazolidine, oxazolidine and thiazolidine.

Imidazolidine   **Oxazolidine**   **Thiazolidine**

7.2.2 Importance of 1,3-azoles

Imidazole ring system is present in important biological building blocks, such as **histidine** and **histamine**. Imidazoles are present in many antifungal, antiprotozoal and antihypertensive medicines. Many antifungal drugs such as **ketoconazole, miconazole, clotrimazole** and the sedative **midazolam** contain an imidazole ring. Imidazole has been used extensively as a corrosion inhibitor on certain transition metals, such as copper.

Histidine   **Histamine**

Ketoconazole

Miconazole   **Clotrimazole**   **Midazolam**

Ditazole containing oxazole ring is an inhibitor of platelet aggregation. **Darglitazone** has a variety of insulin-sensitizing effects, such as improving glycemic and lipidemic control and is used in the treatment of metabolic disorders such as type II diabetes. **Tilmacoxib** a COX-2 inhibitor and NSAID also contains oxazole moiety.

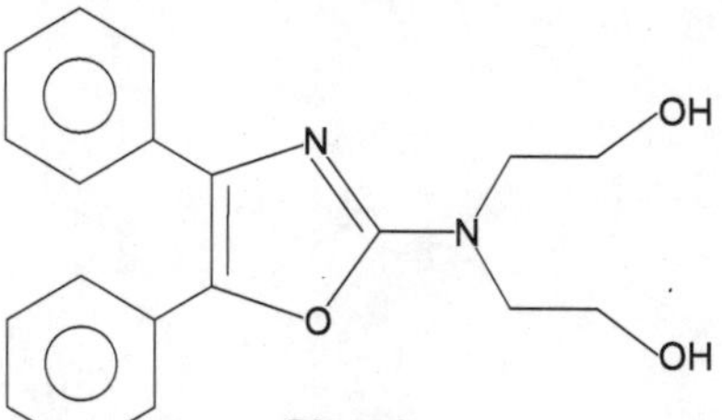

Ditazole

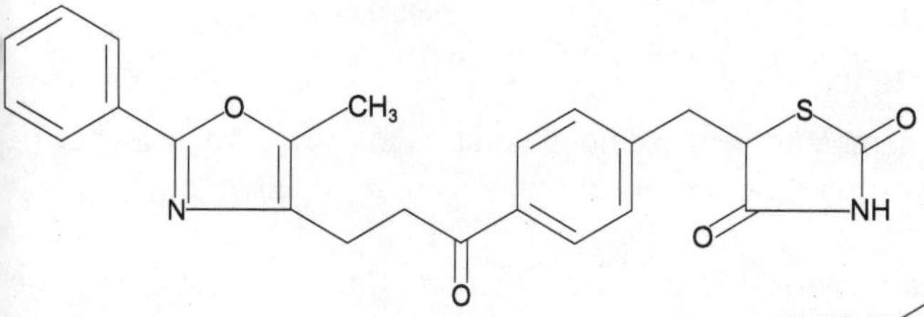

Darglitazone

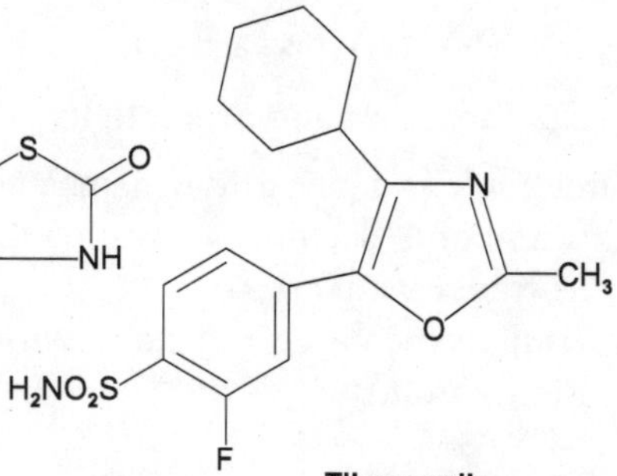

Tilmacoxib

The thiazole ring is present in **thiamine** (vit. B1), **epohthilones** and **luciferin** (firefly chemical). Commercial significant thiazoles include mainly dyes and fungicide. **Thifluzamide** and **thiabendazole** are marketed for control of various agricultural pests. Another widely used thiazole derivative is the non-steroidal anti-inflammatory drug (NSAID) **meloxicam**.

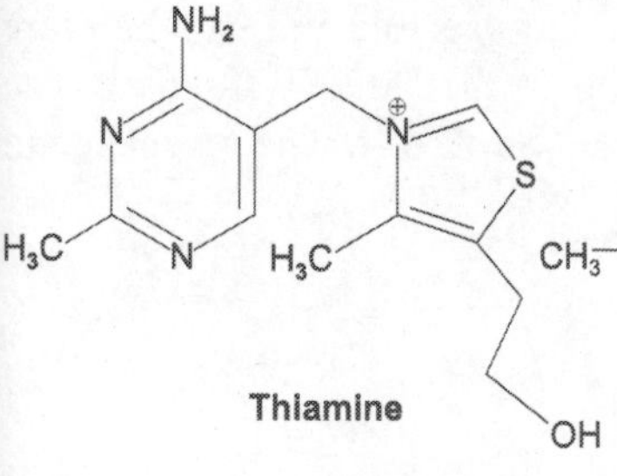

Thiamine

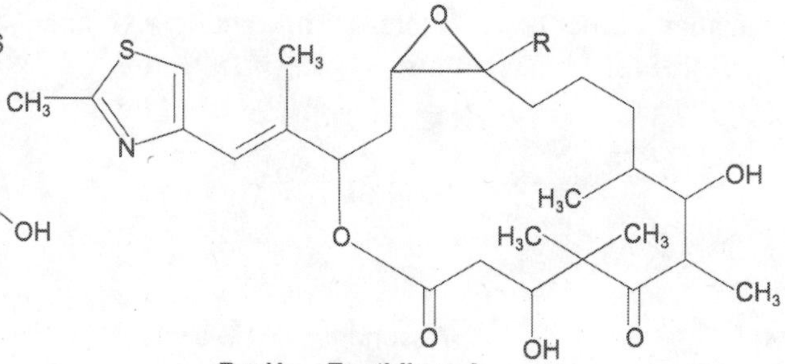

R = H,   Epothilone A
R = CH₃, Epothilone B

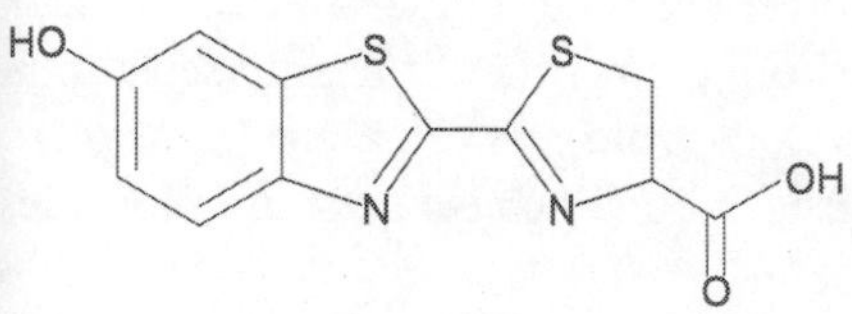

Luciferin

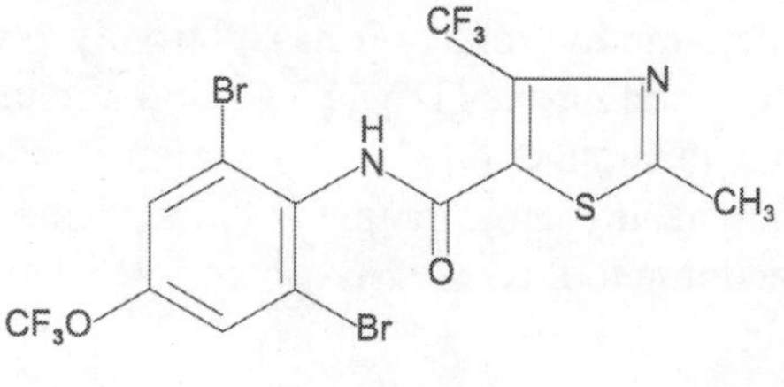

Thifluzamide

Thiabendazole   **Meloxicam**

7.2.3 Structure and aromaticity

Imidazole is a planar five membered ring aromatic compound due to the presence of a sextet of π-electrons, consisting of a pair of electrons from nitrogen atom (NH) and one from each p orbital of the remaining four atoms of the ring. The lone pair on the pyridine type nitrogen is not involved in delocalization and thus does not contribute towards aromaticity.

Lone pair involved in aromatic sextet

X = NH, O, S

Oxazole and thiazole rings are also planar and aromatic. Thiazole is characterized by larger π-electron delocalization than the corresponding oxazole and has therefore greater aromaticity. Oxazole is least aromatic amongst the three 1,3-azoles. The aromaticity order of 1,3-azoles is: **imidazole > thiazole > oxazole.** The resonance in 1,3-azole is shown as follows:

Resonance in 1,3-azoles (X = NH, O, S)

Imidazole boils at high temperature (256-257°C) due to intermolecular H-bonding, while 1-methylimidazole has relatively low boiling point (198-199°C). Oxazole (69-70°C) and thiazole (116-117°C) boil at much lower temperature than imidazole. This is due to stronger dipolar association resulting from the permanent charge separation in imidazole (dipole moment of imidazole is 5.6 D; oxazole is 1.4 D; thiazole is 1.6 D). In addition to this extensive intermolecular hydrogen bonding also exist in imidazole.

μ = 5.6 D   μ = 1.4 D   μ = 1.6 D

Imidazole, is a good donor and a good acceptor of hydrogen bonds, the imine nitrogen is donar and the N-hydrogen, is an acceptor.

H-bonding in imidazole

Imidazoles with a ring N-hydrogen exists in two equivalent tautomeric forms, because the proton can be located on either of the two nitrogen atoms. It is evident in unsymmetrically substituted compounds such as the methylimidazole. 4-Methylimidazole is in rapid tautomeric equilibrium with 5-methylimidazole. All such tautomeric pairs are inseparable and named as 4(5)-methylimidazole.

4-Methylimidazole   **5-Methylimidazole**

4(5)-Methylimidazole

Tautomerism in imidazoles

7.2.4 Spectral data

The chemical shift of the ring protons of 1,3-azoles in ^1H NMR spectroscopy is between δ 7.25 and 8.95 ppm due to a strong diamagnetic ring current. The ^1H and ^{13}C NMR (given in brackets) values of imidazole, oxazole and thiazole are typical for aromatic hydrogens and aromatic carbons. The downfield shift for H-2 in all the 1,3-azoles is due to the deshielding caused by the combined electron-withdrawing effects of two heteroatoms.

^1H NMR (δ, ppm, CDCl$_3$)
^{13}C NMR (δ, ppm, CDCl$_3$)

7.2.5 Methods of preparation of 1,3-azoles

7.2.5.1 Methods of preparation of imidazole

(*i*) **From α-halocarbonyl compounds:** The reaction of α-halocarbonyl compounds with amidines gives imidazoles.

X = Halogen   **Amidine**

Mechanism:

Reaction of α-halocarbonyl compound with guanidine results in the formation of 2-aminoimidazoles.

α-**Halocarbonyl**   **Guanidine**
compound

(*ii*) Imidazole itself can be prepared efficiently from bromoacetaldehyde ethylene acetal, formamide and ammonia.

(*iii*) **Debus-Radziszewski synthesis of imidazole**: In 1822 Radziszewski synthesised imidazoles by the condensation of α-diketones with aldehyde and ammonia. He synthesised lophine from benzil, ammonia and benzaldehyde.

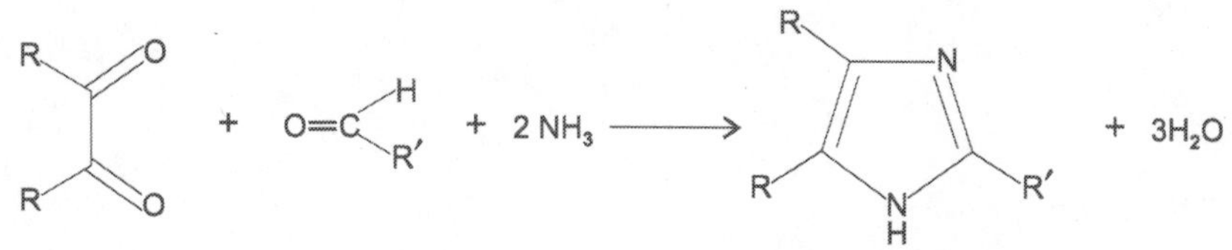

R = Ph, Benzil   **R = R′ = Ph, Lophine**

Mechanism: It involves condensation of 1,2-dicarbonyl compound with two moles of ammonia to form diimine, which reacts with aldehyde to give imidazole.

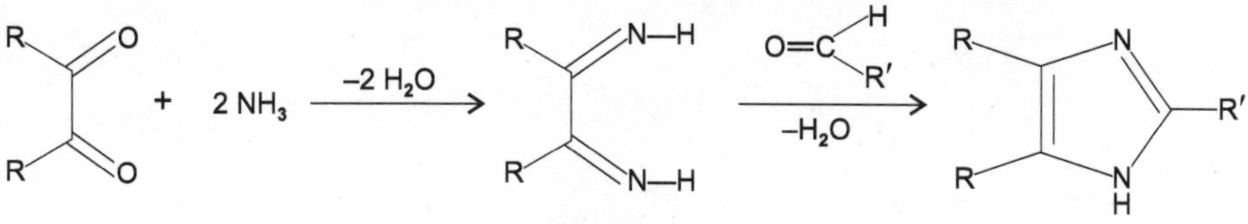

Diimine

Later Debus synthesised imidazole from glyoxal, formaldehyde and ammonia.

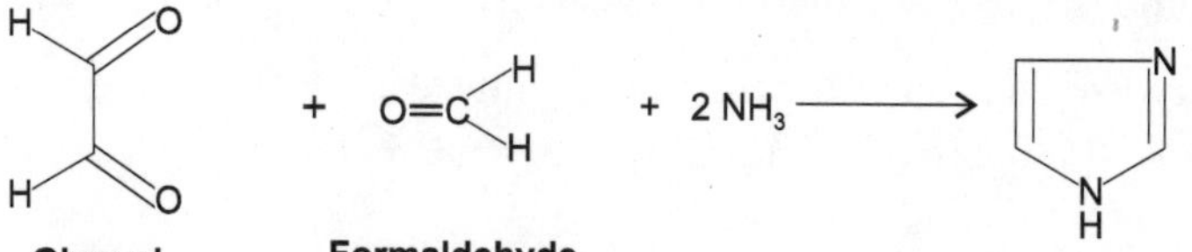

Glyoxal   Formaldehyde

1-Methylimidazole is synthesized on industrial scale by this method from glyoxal formaldehyde and a mixture of ammonia and methylamine

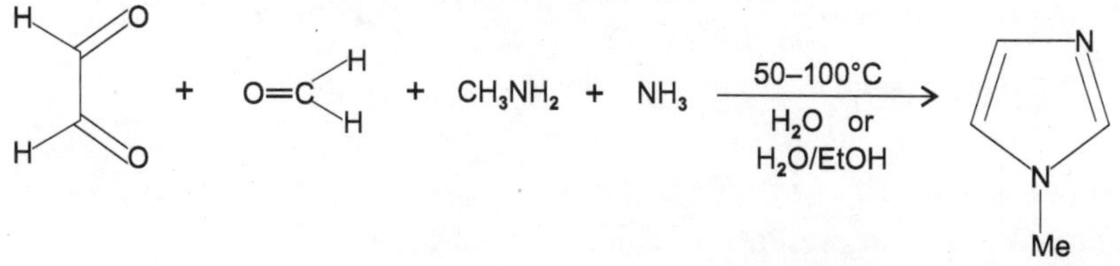

The α-diketones when treated with formamide also results in the formation of imidazoles.

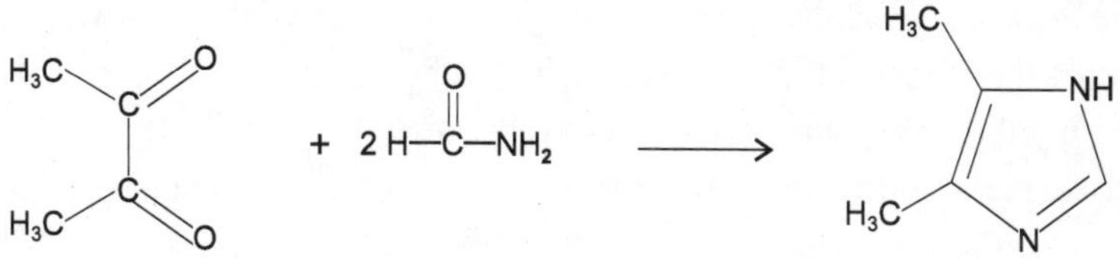

The stepwise formation of imidazoles from α-diketones is shown below.

7.2.5.2 Methods of preparation of oxazole

(*i*) **From 2-hydroxyketones:** Oxazoles unsubstituted at 2-position are readily prepared by the reaction of 2-hydroxyketones with formamide in the presence of sulphuric acid.

2-Hydroxyketone   **Formamide**

(*ii*) **Robinson-Gabriel synthesis of oxazoles:** It involves cyclization of α-acylaminoketones in presence of acid followed by dehydration. Cyclodehydration reagents which allow this transformation include concentrated sulfuric acid, phosphorus pentachloride, phosphorus pentoxide, phosphoryl chloride, thionyl chloride, phosphoric acid/acetic anhydride, polyphosphoric acid and anhydrous hydrogen fluoride. Polyphosphoric acid has been found to be the most suitable. This reaction has been applied for the synthesis of 2,5-disubstituted (especially for the synthesis of 2,5-diaryloxazoles) and 2,4,5-trisubstituted oxazole derivatives. This method can be also be used for the synthesis of thiazole or imidazole.

α-**Acylaminoketone** R^1, R^2, R^3 = **Alkyl, aryl, heteroaryl**

Mechanism:

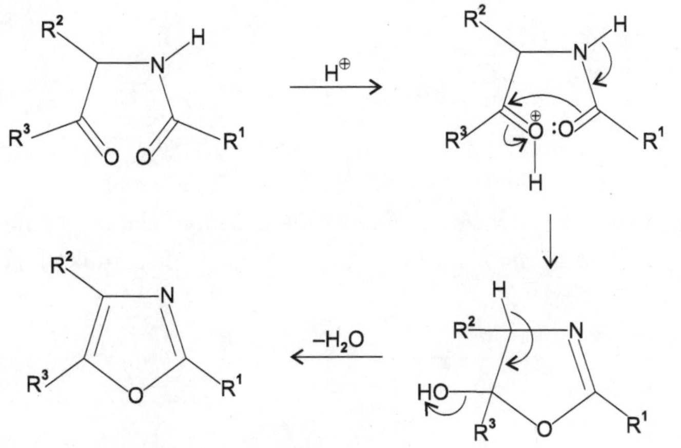

(*iii*) **Blümlein-Lewy synthesis of oxazoles**: Oxazoles can be synthesized by heating α-haloketone with amide or ammonium formate.

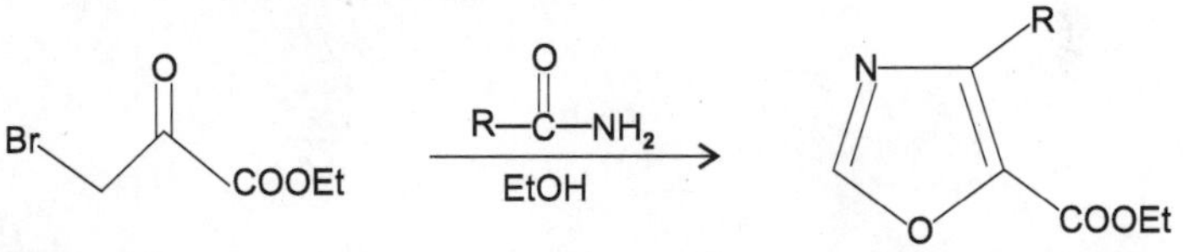

The reaction of a primary amide, with a bromopyruvate ester in alcoholic solvent leads to the corresponding oxazoline, before undergoing cyclodehydration to the oxazole. However, this method leads to low conversions and unwanted side product formation.

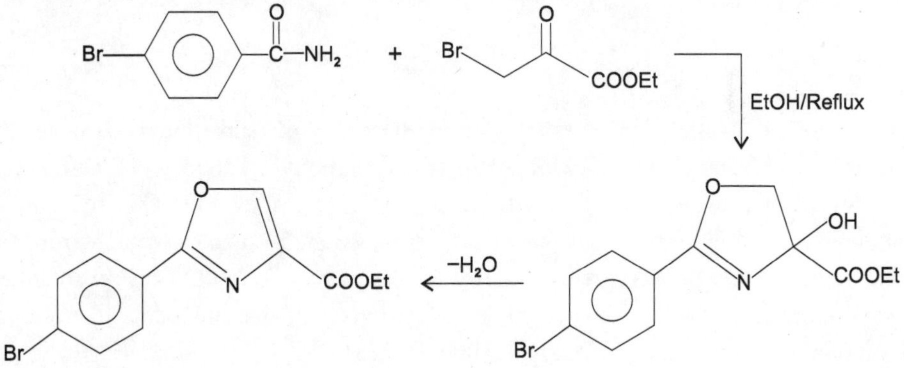

(*iv*) **Fischer oxazole synthesis**: Emil Fischer in 1896 reported the synthesis of 2,5-disubstituted oxazoles. In this reaction cyanohydrins is reacted with aldehyde under dry acidic conditions to form the corresponding oxazole. Normally aldehyde and cyanohydrins are aromatic in nature.

Fischer, E., Ber. **1896**, *29, 205.*

Mechanism: This reaction involves the addition of HCl to the cyanohydrins to form an iminochloride intermediate. Reaction of this intermediate with aldehyde, followed by loss of water gives chloro-oxazoline intermediate. Isomerization of chloro-oxazoline intermediate followed by the loss of an HCl molecule gives 2,5-diaryloxazole.

(v) Davidson oxazole synthesis: The preparation of substituted oxazole by condensation of O-acylacylion (acylated α-hydroxy carbonyl compound) with ammonia or ammonium acetate is known as Davidson oxazole cyclization. This reaction is found most suitable for the synthesis of 2,4,5-trisubstituted oxazoles with an aromatic substituent at C5, whereas it works poorly for the preparation of 2,4-disubstituted or monosubstituted oxazoles. The cyclization to form oxazole does not occur if ammonia or ammonium acetate is replaced by organic amines.

2-Acyloxyketone

Mechanism:

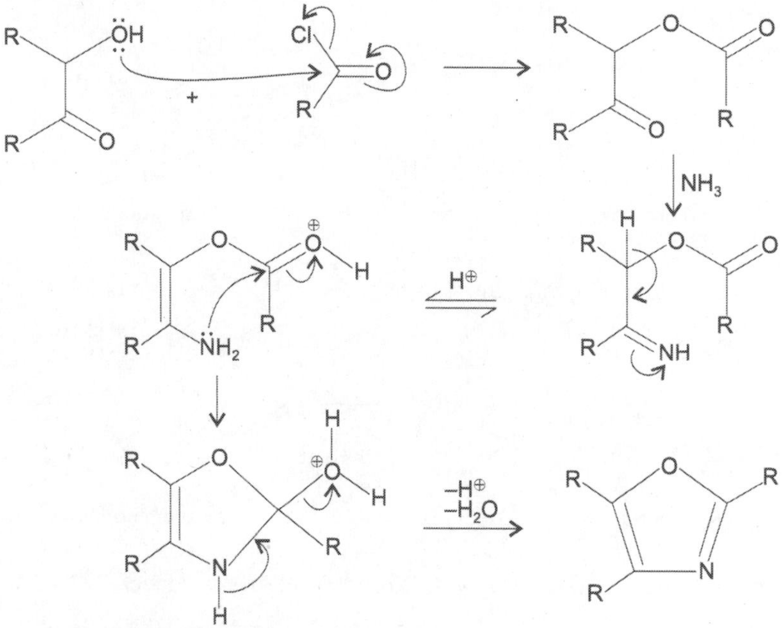

7.2.5.3 Methods of preparation of thiazole

(*i*) **Hantzsch synthesis of thiazoles**: The synthesis of thiazoles by the cyclic condensation of α-haloketone with thioamides is known as Hantzsch synthesis. For example the synthesis of 2,4-dimethylthiazole, by the reaction of thioacetamide with chloroacetone.

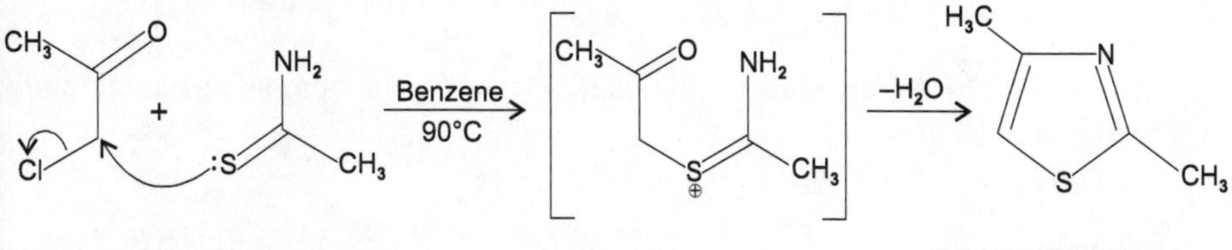

Chloroacetone Thioacetamide   **2,4-Dimethylthiazole**

Mechanism: It involves nucleophilic attack of sulphur (S-alkylation) on α-carbon of α-haloketone followed by cyclization and dehydration.

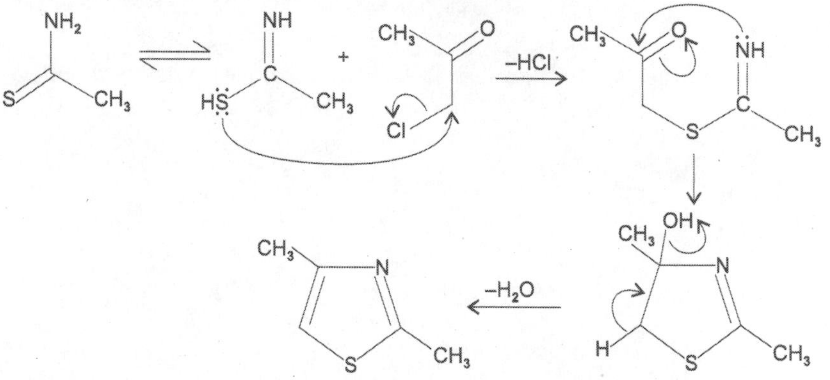

Similarly, synthesis of 2-aminothiazole is carried out by the reaction of 1,2-dichloroethyl ethyl ether and thiourea.

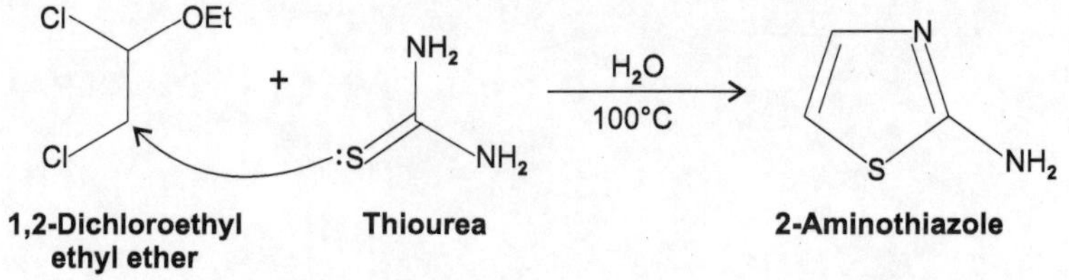

1,2-Dichloroethyl   **Thiourea**   **2-Aminothiazole**
ethyl ether

The reaction of thiourea with 2-chloroacetamides yields 2,4-diaminothiazole.

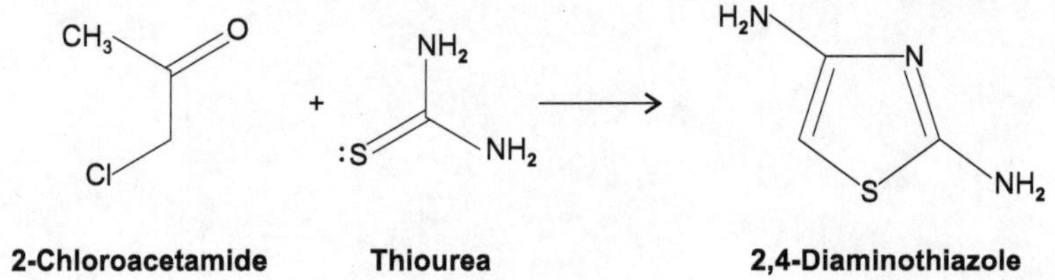

2-Chloroacetamide   **Thiourea**   **2,4-Diaminothiazole**

N-Acylthioureas on reaction with with 2-bromo ketones produces 2-(N-acylamino) thiazoles.

2-(N-acylamino)thiazoles

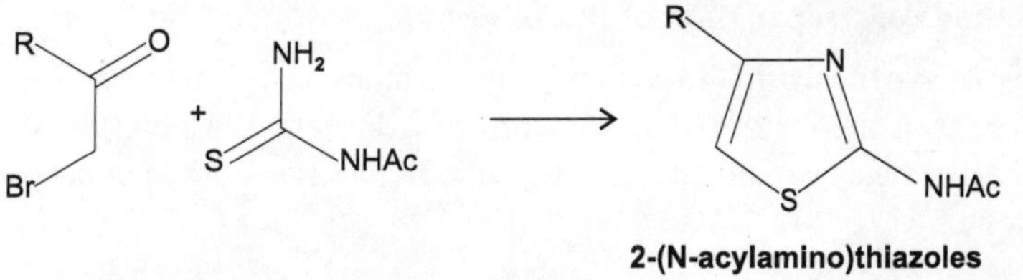

(*ii*) **From α-acylthioketones:** α-Acylthioketones react with ammonia or ammonium acetate to give thiazoles.

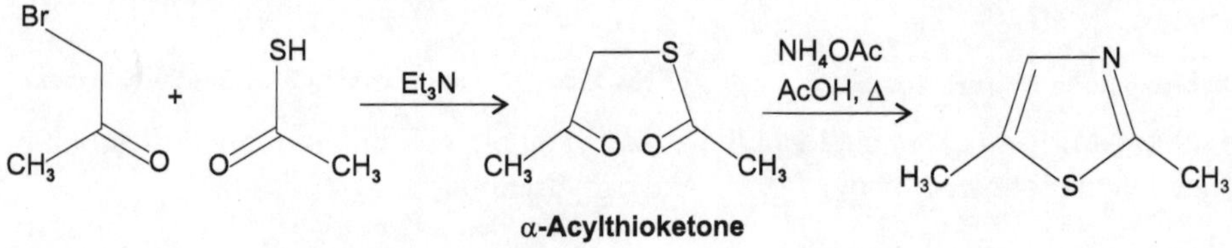

α-Acylthioketone

(*iii*) **Gabriel synthesis of thiazoles:** 2-Acylaminoketones react with phosphorus pentasulfide to form thiazoles.

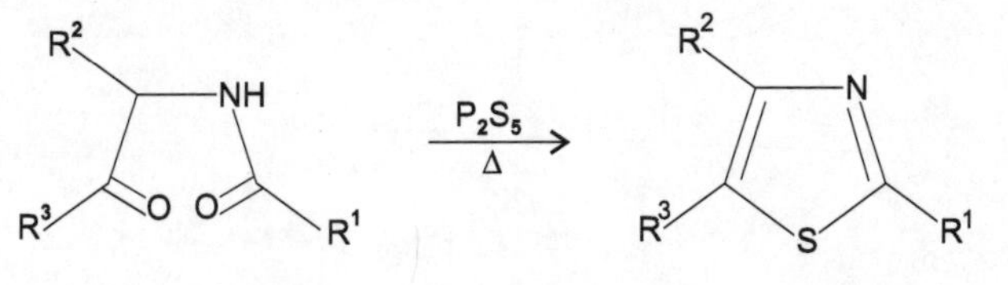

Mechanism:

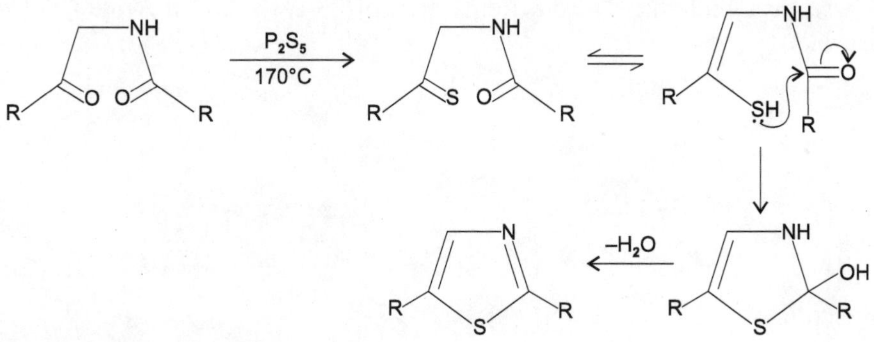

7.2.6 Reactions of 1,3-azoles

Basicity and acidity

Imidazole is amphoteric i.e., it can function both as an acid and as a base. As an acid, the pK_a of imidazole is 14.5, making it less acidic than carboxylic acids, phenols and imides, but slightly more acidic than alcohols and pyrrole (pK_a = 17.5). The acidic proton is located on N-1.

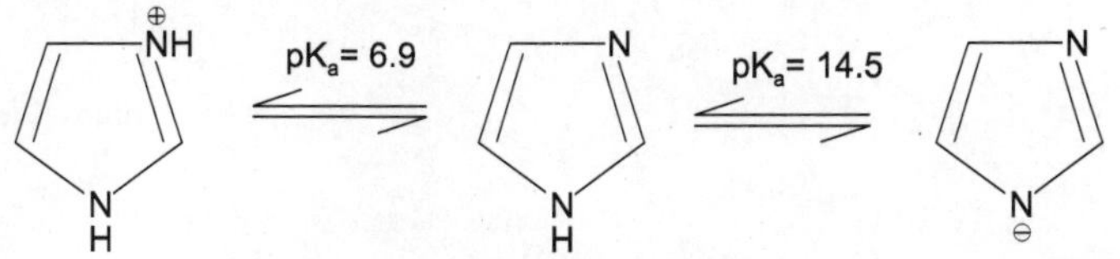

As a base, the pK_a of the conjugate acid of imidazole is 6.9, making imidazole approximately sixty times more basic than pyridine (pK_a of its conjugate acid is 5.2). The basic site is N-3. Protonation of imidazole gives the imidazolium cation, which is highly symmetric, and resonance stabilised.

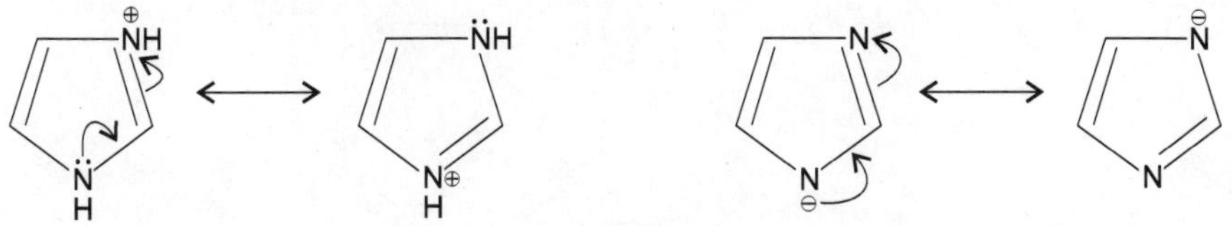

Equally contributing canonical forms of imidazole cation and anion

At physiological pH (pH = 7.3) imidazole exists both in protonated as well as in unprotonated form. This makes imidazole ring a very important component in several enzymatic reactions.

Oxazole is a weak base, its conjugate acid has a pK_a of 0.8. Basicity is due to presence of lone pair of electrons on pyridine type nitrogen which is available for protonation. However, oxazole is a weak base because the electron withdrawing inductive effect by the more electronegative oxygen atom is more predominant than electron releasing mesomeric effect.

Conjugate acid of thiazole has pK_a of 2.5. Thus it is more basic than oxazole but less basic than imidazole. The order of basicity of 1,3-azoles is **imidazole > thiazole > oxazole.**

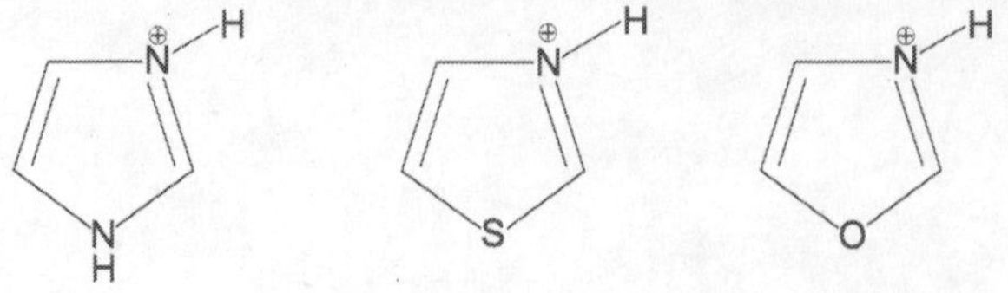

pK_a = 6.9   pK_a = 2.5   pK_a = 0.8

Protonation

Imidazole, alkyl oxazoles and thiazole form stable crystalline salts (imidazolium, oxazolium and thiazolium salts) with strong acids, by protonation of the imine nitrogen, N-3.

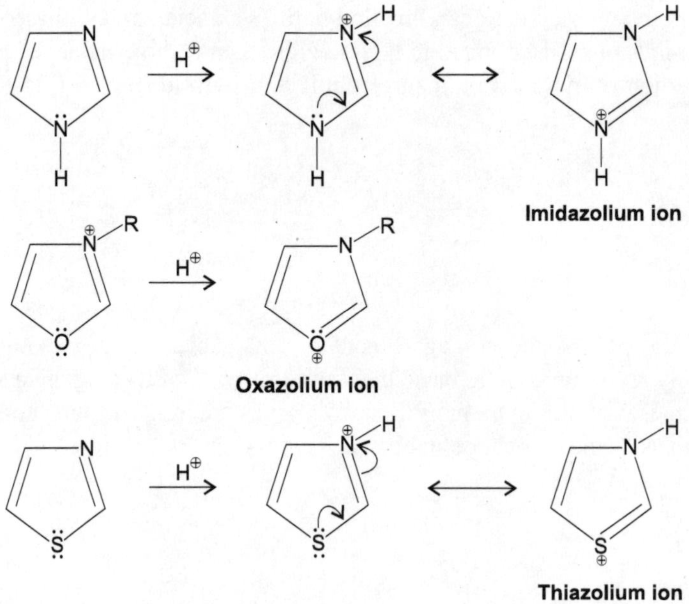

Imidazolium ion

Oxazolium ion

Thiazolium ion

Alkylation at nitrogen

The 1,3-azoles are quaternised easily at the imine nitrogen with alkyl halides. In the case of imidazoles the immediate product is a N-methylimidazolium salt, this can lose its N-hydrogen to form N-methyl imidazole, which further on methylation gives 1,3-dimethyl imidazolium salt.

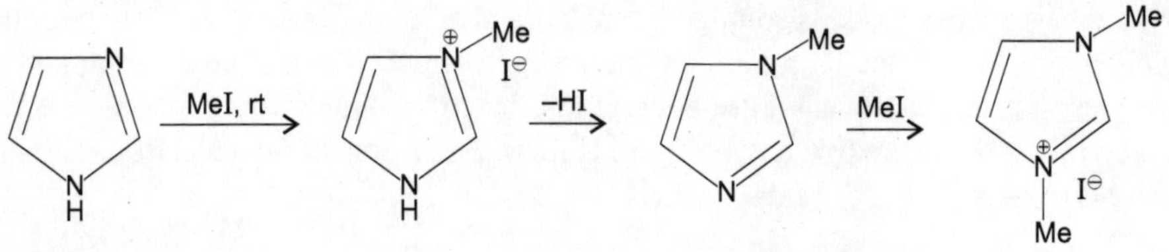

The 1,3-dialkyl imidazolium salts can also be prepared by reacting 1-trimethylsilylimidazole with an alkyl halide.

A 4(5)-substituted imidazole can give two isomeric 1-alkyl-derivatives: generally the main product results from alkylation of that tautomer which minimizes steric interactions, i.e. 4(5)-substituted imidazoles give mainly 1-alkyl-4-substituted imidazoles.

4-Substituted imidazole   **1-Alkyl-4-substituted imidazole**   **5-Subsituted imidazole**

N-Alkylation of imidazoles, carrying a phenylsulfonyl or acyl group on nitrogen, is difficult. However the reaction can be carried out using methyltriflate or a Meerwein salt (e.g. $Et_3O^+ BF_4^-$). Since acylation of 4(5)-substituted imidazoles gives the sterically less crowded 1-acyl-4-substituted imidazoles, subsequent alkylation, then hydrolytic removal of the phenyl sulfonyl or acyl group, produces 1,5-disubstituted imidazoles.

1,5-Disubstituted imidazole

Imidazole undergo 'normal' **Mannich reaction** to yield N-(dimethylaminomethyl) imidazole, by the attack at the imine nitrogen, followed by loss of proton from the other nitrogen.

Thiazoles react with alkyl halides to produce quaternary salts, as seen in the preparation of thiamine.

Thiamine

Acylation at nitrogen

Acylation of imidazole give N-acylimidazole *via* loss of proton from the initially formed N-acylimidazolium salt. N-Acylimidazoles are more easily hydrolysed than N-acylpyrroles even moist air is sufficient for the hydrolysis.

Deprotonation

Fast deprotonation of C2–H in 1,3-azoles at room temperature takes place in neutral or weakly basic solution, but not in acidic solution. The relative rates being in the order imidazole > oxazole > thiazole. The mechanism involves first formation of protonic salt, then C2–H deprotonation of the salt, producing a transient ylide. The ylide is resonance stabilised with its carbene form.

ylide   **carbene**

Electrophilic reaction

Electrophilic attack in imidazole is difficult under acidic conditions. Imidazole anion and neutral imidazole do undergo electrophilic substitution. Electrophilic attack occurs at C5 in imidazole. Also attack at C4 is equivalent to C5 attack in unsubstituted imidazole (due to tautomerism). But electrophilic attack at C2 in imidazole leads to formation of unstable intermediate.

(unstable)

In 1,3-azoles two types of structural effects are in operation one is mesomeric electron releasing effect of the heteroatom and other is the inductive electron withdrawing effect. The reactivity of these azoles depends on the predominance of one of these effect over the other.

Inductive effect
X = NH, O, S

Mesomeric effect

In case of oxazole electrophilic attack occurs at C4. This is because of greater inductive effect of oxygen at C5 and due to mesomeric electron releasing effect of oxygen atom the C4 has higher electron density.

In thiazole electrophilic reaction occurs at C5, thus it is regioselective at the 5 position.

The order of reactivity of 1,3-azoles towards electrophilic substitutions reactions is as follows: **Imidazole > Thiazole > Oxazole**. Oxazole is least reactive and generally undergo electrophilic substitution with great difficulty.

Nitration: Imidazole is much more reactive towards nitration than thiazole. Nitration of imidazole gives 4-nitroimidazole. Thiazole itself do not undergo nitration by nitric acid/oleum at 160°C, but 2-methylthiazole is sufficiently activated to undergo substitution to give mixture of 5-nitro and 4-nitro derivatives in 2:1 ratio. The 2-position is not attacked and 4,5-dimethylimidazole is resistant to nitration. The much less reactive oxazoles do not undergo nitration.

Sulfonation: Imidazole is sulfonated at 4-position with oleum at 160°C.

Imidazole-4(5)-sulfonic acid

Thiazoles are much less reactive than imidazoles, and generally require high temperatures and mercury(II) sulfate as catalyst for sulfonation. If the 5-position is blocked, sulfonation can occur at the 4-position.

Oxazole being least reactive among the 1,3-azoles does not undergo sulfonation at all.

Halogenation: Imidazole is brominated with remarkable ease at all free nuclear positions. Tribromoimidazole can be reduced with sodium sulfite to form 4(5)-bromoimidazole.

4(5)-Bromoimidazole

Thiazole does not undergo bromination easily, though 2-methylthiazole brominates at C5 if it is free.

Halogenation of simple oxazoles has not been reported.

Acylation: Azoles do not undergo Friedel-Craft acylation as the nitrogen interacts with the Lewis-acid catalyst used. However 1-alkylimidazoles undergo benzoylation with benzoyl chloride in the presence of triethylamine. The substitution proceeds *via* a N-acylimidazolium ylide.

Reactions with oxidising agents

The order of resistance to oxidation of 1,3-azoles is: thiazole > imidazole > oxazole. In the presence of peracids imidazole undergo degradation and while substituted thiazoles under similar conditions are converted to N-oxides.

Replacement of halogen

Thiazole, because of presence of nitrogen in the ring, is more susceptible to nucleophilic attack than thiophene.

A halogen at 2-position can undergo nucleophilic displacement for example, 2-halothiazoles react with sulfur nucleophiles to form 2-sulfur substituted thiazole.

Similarly, 2-bromothiazole reacts with sodium methoxide to give 2-methoxy thiazole.

Deprotonation of imidazole N-hydrogen and reactions of imidazolyl anions

The pK_a for loss of the N-hydrogen of imidazole is 14.5. Imidazole is a stronger acid than pyrrole (pK_a 17.5). This is because the anion formed after loss of proton has negative charge delocalized on both nitrogen atoms. The two canonical forms of imidazolyl anion are equivalent. Salts of imidazoles can be alkylated or acylated on nitrogen.

Metallation

Lithiation of oxazoles, thiazoles and N-methylimidazole takes place preferentially at C2, in the presence of strong base. If position 2 is substituted lithiation occurs at position 5. For example, N-substituted imidazole on treatment with n-BuLi forms the 2-lithioimidazole, which can further react with electrophilic moieties such as alkyl halide, trimethylsilylchloride, CO_2 and DMF giving the corresponding 2-substituted products.

Oxazoles are also lithiated at position 2. But the anion undergoes ring opening to form an enolate. However, electrophilic attack at carbon recloses the ring to form substituted oxazoles.

The electron withdrawing effects of the adjacent heteroatoms makes the 2-position of thiazole ring electron deficient, hence, it is deprotonated by strong base at C2 forming a nucleophilic carbanion. Such carbanion can react with electrophiles like acetaldehyde to produce substituted thiazoles.

Similarly, Grignard reagent prepared from 2-bromothiazole and magnesium in ethyl ether and ethyl bromide reacts with aldehydes to give 2-substituted thiazole.

Electrocyclic reactions

Oxazoles readily undergo **Diels-Alder** type cycloaddition reactions across the 2,5-positions as in case of furan. Thiazole and imidazole do not show this type of reactivity.

4-Phenyloxazole reacts with benzyne to form a cycloadduct.

EXERCISE

Q.1 Write the structure of oxazole and isoxazole.

Q.2 Draw the various resonating structures of imidazole.

Q.3 How will you synthesize lophine from benzil?

Q.4 Explain why:

(a) Imidazole is more basic than pyridine, but more acidic than pyrrole.

[Hint. More stable cation and anion of imidazole. Protonated imidazole has two resonance forms in which both nitrogens contribute equally in carrying the positive charge.]

(b) Pyrazole has higher boiling point as compared to isothiazole and isoxazole.

[Hint. Intermolecular H-bonding]

(c) 3-Methylpyrazole is a mixture of two inseparable isomers, 3-methyl and 5-methyl pyrazoles.

[Hint. Tautomerism]

(d) 1,2-Azoles (pyrazole, isoxazole and isothiazole) undergo electrophilic attack preferentially at 4-position.

(e) Pyrazole is appreciably stronger acid than imidazole, indole and pyrrole. Explain.

(f) Imidazole do not undergo Friedel-Craft acylation.

Q.5 Complete the following reaction:

(a) Imidazole $+$ H_2O_2 $\longrightarrow$

(b) Thiazole $+ H_2SO_4$-$SO_3 + HgSO_4$ $\xrightarrow{250°C}$

(c) Pyrazole $+$ $SOCl_2$ $+$ $CHCl_3$ $\longrightarrow$

(d) Isoxazole $+$ HNO_3/H_2SO_4 $\xrightarrow{40°C}$

(e) Propene $+ SO_2 + NH_3 + Al_2O_3$ $\xrightarrow{200°C}$

Ans. (a) Oxamide; (b) Thiazole-5-sulfonic acid; (c) 4-Chloropyrazole;

(d) 4-Nitroisoxazole; (e) Isothiazole

Q.6 Compare the basicity of imidazole with other azoles.

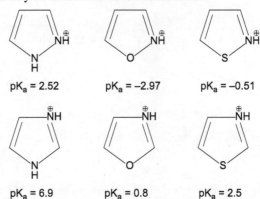

pKₐ = 2.52   pKₐ = -2.97   pKₐ = -0.51

pKₐ = 6.9   pKₐ = 0.8   pKₐ = 2.5

Ans. Azoles have a pyridine type nitrogen atom whose lone pair are not involved in delocalization. Thus all azoles are more basic than five membered aromatic heterocycles with one heteroatom (pyrrole, furan and thiophene). The basicity depends on the nature of the other heteroatom present, which has electron donating mesomeric effect as well as electron withdrawing inductive effect.

In case of 1,2-azoles the inductive effect is stronger than the mesomeric effect and reduce the basicity. In isoxazole and isothiazole the inductive effect is more pronounced and they are much weaker base than pyrazole.

For imidazole, the mesomeric effect is stronger than inductive effect, therefore it is most basic. But for oxazole and thiazole the oxygen and sulfur atom reduce basicity *via* electron withdrawing effect.

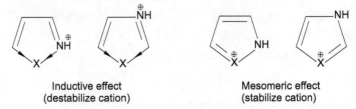

Inductive effect   Mesomeric effect
(destabilize cation)   (stabilize cation)

Q.7 Identify A and B in the following reaction

Q.8 In the following reaction two tautomeric five membered ring compounds A and B with molecular formula $C_6H_{10}N_2$ are obtained. Identify A and B.

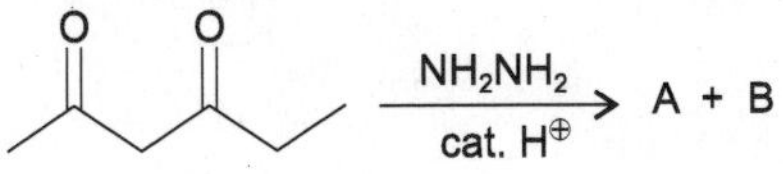

Ans.

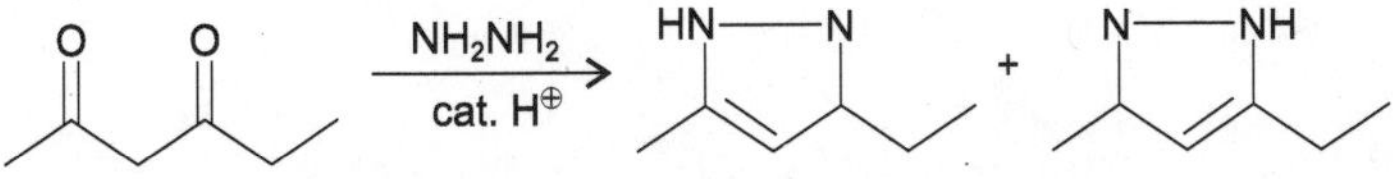

Q.9 Which of the two nitrogen atoms of imidazole is more basic?

Ans.

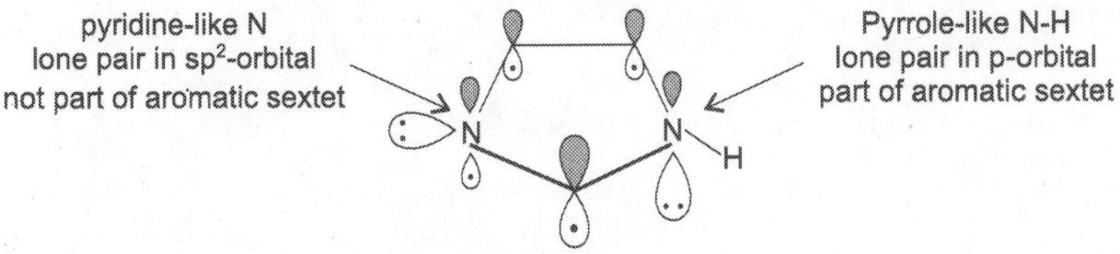

pyridine-like N
lone pair in sp^2-orbital
not part of aromatic sextet

Pyrrole-like N-H
lone pair in p-orbital
part of aromatic sextet

Imidazole is an aromatic heterocycle that has 6π-electrons. There are two nitrogen atoms in imidazole, one is pyrrole-like and the other is pyridine-like. The electron pair on the pyrrole-like nitrogen are involved in the aromatic sextet and thus unavailable for bonding to a proton. Therefore, the pyrrole-like nitrogen (NH) is not basic enough.

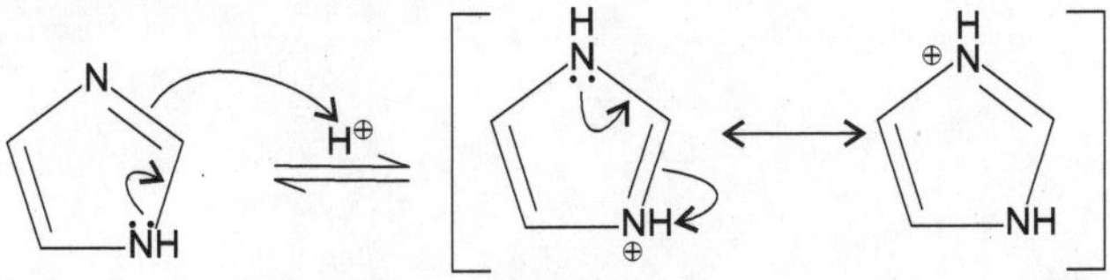

However, electron pair on the pyridine-like nitrogen are located in sp^2 hybrid orbital which is in the plane of the ring and thus not involved in the aromatic sextet. Therefore, they are available for bonding to a proton. It is for this reason that the pyridine-like nitrogen (N) is basic.

Q.10 Imidazole ring is present in the amino acid histidine, which acts as a base in many enzymatic reactions. Which nitrogen of the imidazole ring is responsible for its basic nature. Draw the structure of the conjugate acid of imidazole.

Ans. The lone pair of electrons on pyrrole-like nitrogen is delocalized over the imidazole ring to make it aromatic. These electrons are not available for protonation. The lone pair of electrons on the pyridine-like nitrogen is not delocalized over the ring and available for protonation. Thus, the pyridine-like nitrogen is more basic.

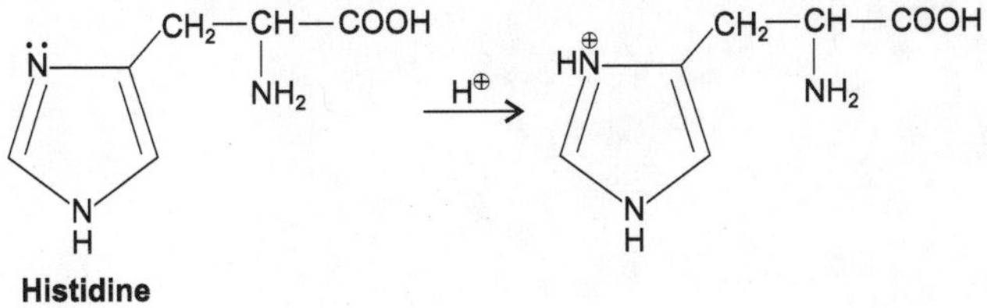

Histidine

Bicyclic Ring Systems Derived From 1,2-Azoles and 1,3-Azoles

Benzo-1,2-azoles: Indazole, Benzisoxazole and Benzisothiazole

Benzo-1,3-azoles: Benzimidazole, Benzoxazole and Benzothiazole

8.1 BENZO-1,2-AZOLES: INDAZOLE, BENZISOXAZOLE AND BENZISOTHIAZOLE

8.1.1 Introduction

Fusion of the 1,2-azole, pyrazole with benzene ring forms **1*H*-indazole**. However, oxazole can be fused to benzene ring giving rise to two different isomers, **1,2-benzisoxazole** (benz[d]isoxazole) and **2,1-benzisoxazole** (anthranil or benz[c] isoxazole). Similarly, two isomeric forms of benzisothiazole are known, depending upon the position of the ring fusion, **1,2-benzisothiazole** (benz[d]isothiazole) and **2,1-benzisothiazole** (benz[c]isothiazole or thioanthranil).

1*H*-Indazole

1,2-Benzisoxazole or Benz[d]isoxazole

2,1-Benzisoxazole or Benz[c]isoxazole or Anthranil

1,2-Benzisothiazole or Benz[d]isothiazole

2,1-Benzisothiazole or Benz[c]isothiazole or Thioanthranil

1*H*-Indazole is a colourless crystalline compound (mp 145-149°C) and soluble in hot water. 1,2-Benzisoxazole is a colourless liquid (mp 35°C). 2,1-Benzisoxazole has bp 94.5°C at 11 mmHg. 1,2-Benzisothiazole, which smells of almonds, is a pale yellow solid (mp 37°C, bp 220°C) and soluble in organic solvents, but insoluble in water. 2,1-Benzisothiazole (bp 242°C at 760 mmHg) is a pale yellow liquid.

8.1.2 Biological importance

Indazoles, e.g., **benzydamine** is analgesic, anti-inflammatory and antipyretic. The indazole derivative **APINACA** [N-(1-adamantyl)-1-pentyl-1*H*-indazole-3-carboxamide], acts as a potent agonist for the cannabinoid receptors. **Adjudin,** an indazole derivative has shown potent anti-spermatogenic activity in rats, rabbits and dogs and it may act as potential non-hormonal male contraceptive drug. 1,2-Benzisoxazoles ring system is present in **zonisamide**, a sulfonamide anticonvulsant and **risperidone** which is used to treat schizophrenia.

A 1,2-benzisothiazole derivative, **saccharin** [1,2-benzisothiazol-3(2*H*)-one-1,1-dioxide] is five hundred times sweeter than sugar in dilute solution. **Probenazole** is an agrochemical (plant activator) used against rice blast fungus.

Benzydamine

APINACA

Adjudin

Zonisamide

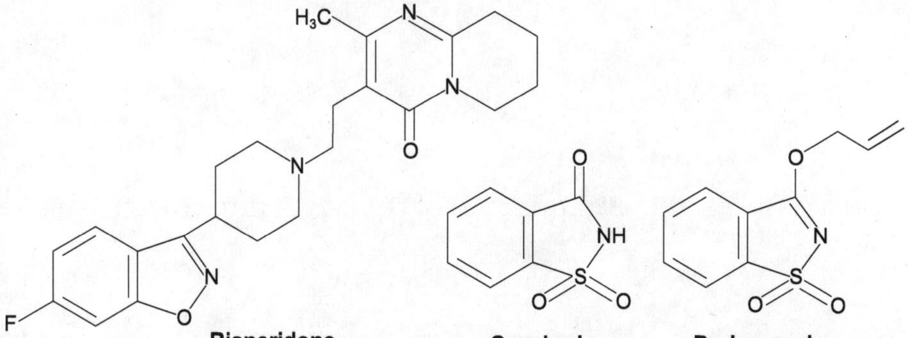

Risperidone   **Saccharin**   **Probenazole**

8.1.3 Structure

The order of aromaticity of 2,1-benzoisoxazole (anthranil), 1,2-benzoisoxazole and benzoxazole is as follows:

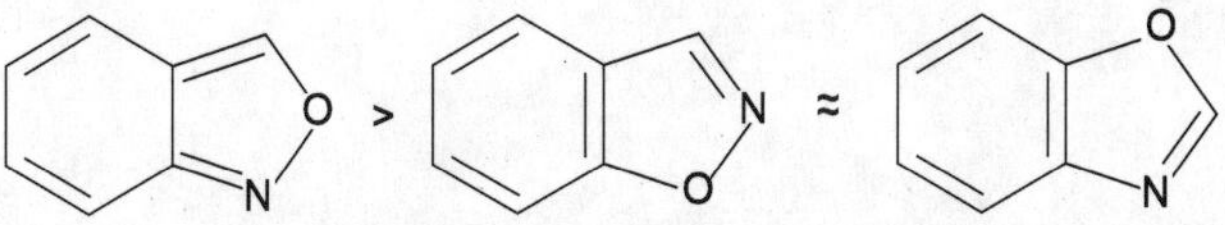

Anthranil   **1,2-Benzisoxazole**   **Benzisoxazole**

The resonance contributors of 2,1-benzisothiazole are shown below.

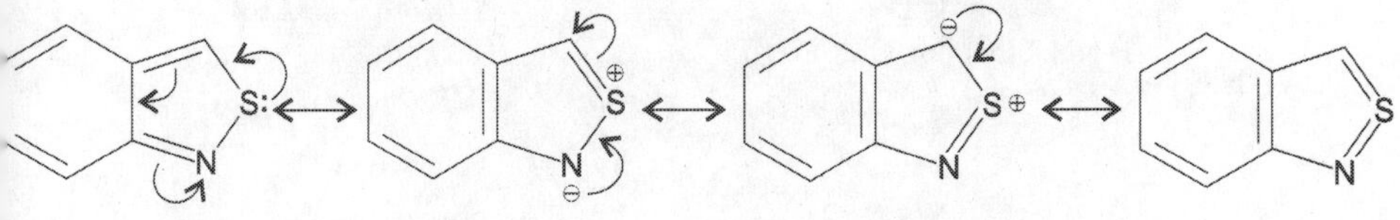

Resonance in 2,1-benzisothiazole

Indazole exists as a dimeric structure.

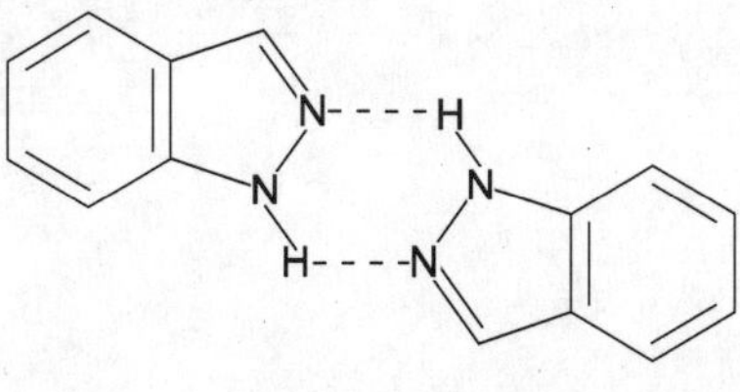

H-bonding in 1*H*-indazole

The 2*H*-indazole, which is tautomeric form of 1*H*-indazole, possesses an ortho-quinonoid structure. For this reason, the 2*H*-indazole is higher in energy by 15 kJ/mol in the gas phase.

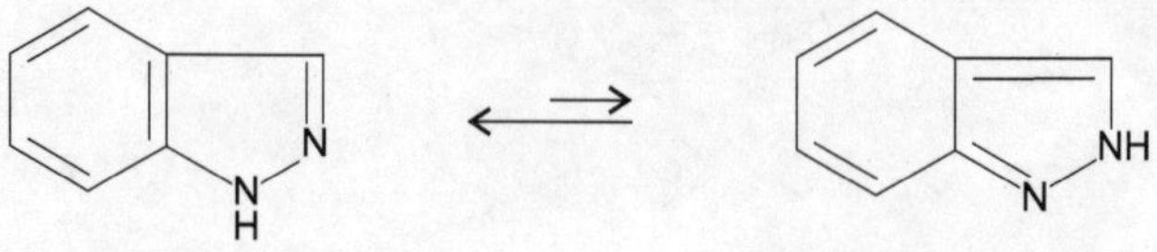

Indazole   **2*H*-Indazole**

Dipole moment (μ) of 1,2-Benzisothiazole is 3.03 D in benzene and of 1,2-benziso-thiazole is 2.44 D.

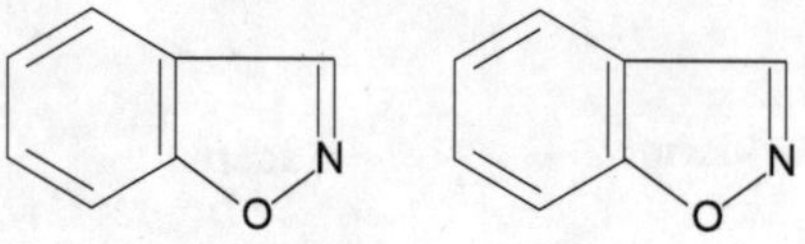

μ = 3.03 D μ = 2.44 D
(in Benzene)

8.1.4 Spectral data

The ^{1}H and ^{13}C NMR (given in brackets) values of benzo-1,2-azoles and benzo-2,1-azoles are shown below.

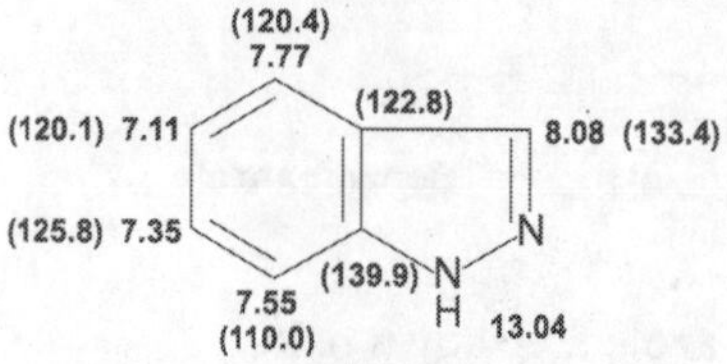

1*H*-Indazole
^{1}H NMR (δ, ppm, DMSO-d_6)
^{13}C NMR (δ, ppm, DMSO-d_6)

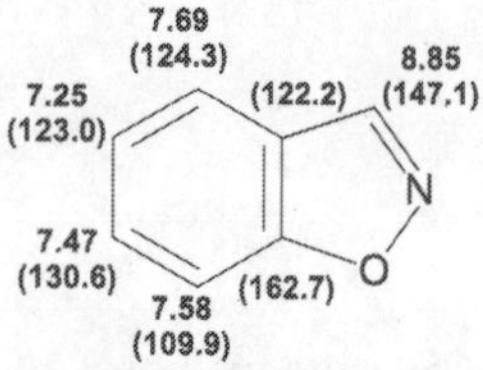

1,2-Benzisoxazole
^{1}H NMR (δ, ppm, neat)
^{13}C NMR (δ, ppm, DMSO-d_6)

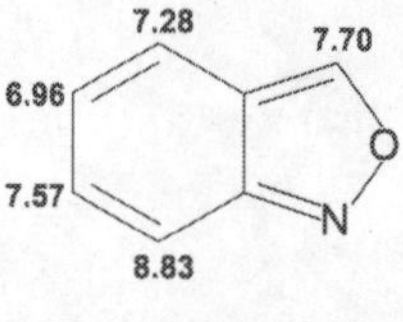

2,1-Benzisoxazole
^{1}H NMR (δ, ppm, neat)

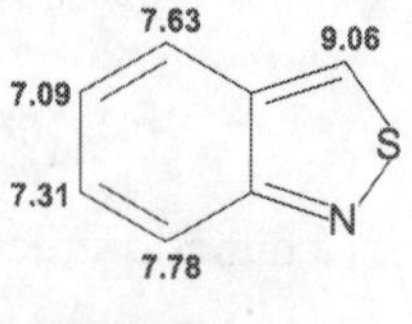

2,1-Benzisothiazole
^{1}H NMR (δ, ppm, neat)

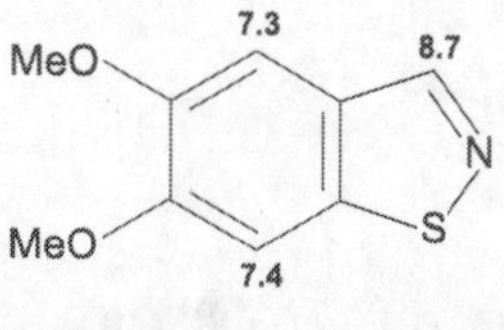

5,6-Dimethoxy-1,2-benzisothiazole
^{1}H NMR (δ, ppm)

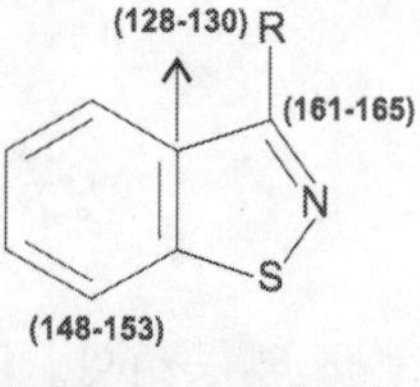

3-Alkyl-1,2-benzisothiazole
^{13}C NMR (δ, ppm, CDCl$_3$)

In the ^{1}H NMR spectra of 1,2-benzisothiazoles the H3 signal normally occurs at δ 8.2-8.8, thus, for 5,6-dimethoxy-1,2-benzisothiazole H3 resonates at δ 8.7. For the same compound the chemical shifts of H4 and H7 are δ 7.3 and 7.4, respectively.^{13}C NMR signals for C3 of 3-alkyl-1,2-benzisothiazoles occur at δ 161-165, the resonance of C3a is normally found at δ 128-130 and that of C7a is at δ 148-153 in CDCl$_3$.

8.1.5 Methods of preparation of benzo-1,2-azoles

8.1.5.1 Methods of preparation of indazole

(*i*) **From ortho-substituted anilines:** Indazoles can be synthesized from ortho-substituted anilines. For example ortho-toluidine on acetylation followed by nitrosation yields N-nitroso compound. The N-nitroso compound rearranges in benzene at 45-50°C giving the acetoxy compound which cyclizes to indazole.

Indazole can also be synthesized by diazotization of ortho-substituted anilines, in the presence of tetra-methylammonium acetate and chloroform in 92% yield.

(*ii*) **From ortho-fluorobenzaldehydes**: Indazoles can be synthesized on refluxing ortho-fluorobenzaldehydes with hydrazine.

8.1.5.2 Methods of preparation of 1,2-benzisoxazoles

(*i*) **From ortho-haloarylketone oxime:** 1,2-Benzisoxazoles are synthesized by the cyclization of ortho-haloarylketone oxime which involves an intramolecular nucleophilic substitution of the halide.

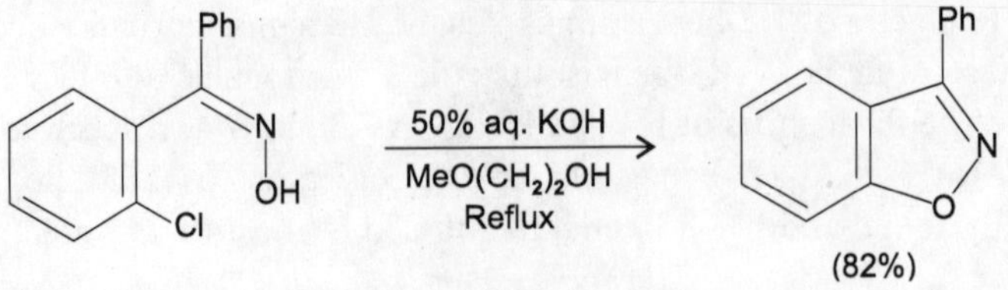

(82%)

(*ii*) **From 2-hydroxytolylbisazide:** 1,2-Benzisoxazoles can be prepared from 2-hydroxytolylbisazide on heating with acetic acid.

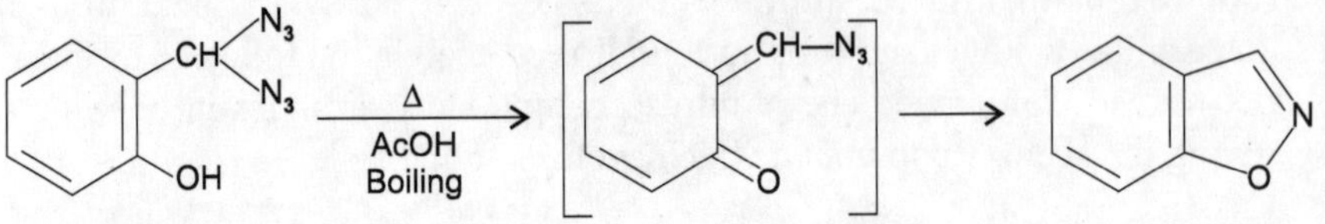

2-Hydroxytolylbisazide

(*iii*) **From salicylaldehyde:** Salicylaldehyde on reaction with urea in presence of base gives 1,2-benzisoxazoles **(Conduche reaction).**

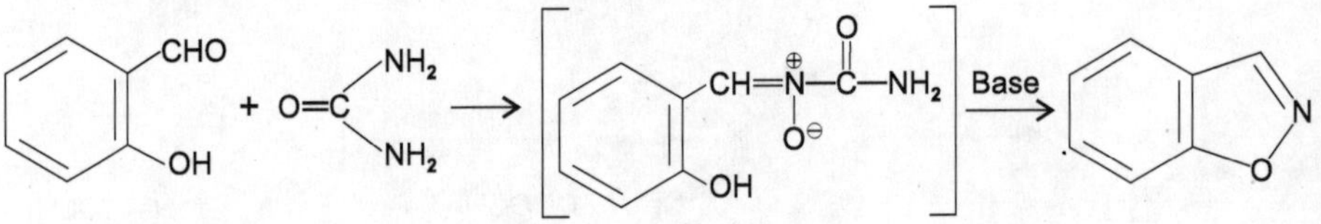

(*iv*) **From ortho-hydroxy carbonyls:** Carbonyl derivatives on boiling with hydroxylamine form 1,2-benzisoxazoles.

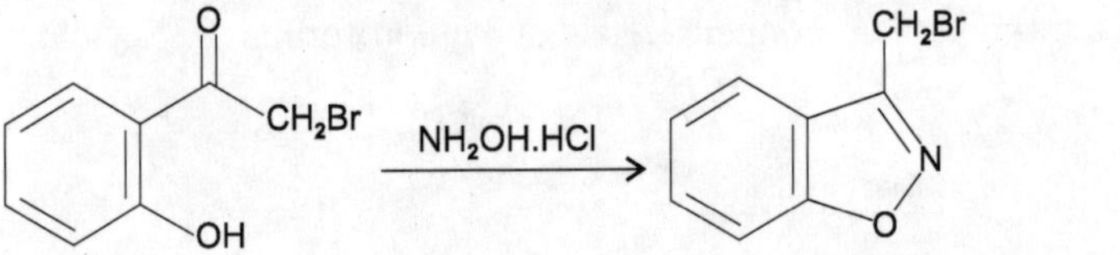

3-Bromomethyl-1,2-benzisoxazole

(*v*) **From benzyne:** Reaction of benzyne with nitrile oxide results into the formation of 1,2-benzisoxazole.

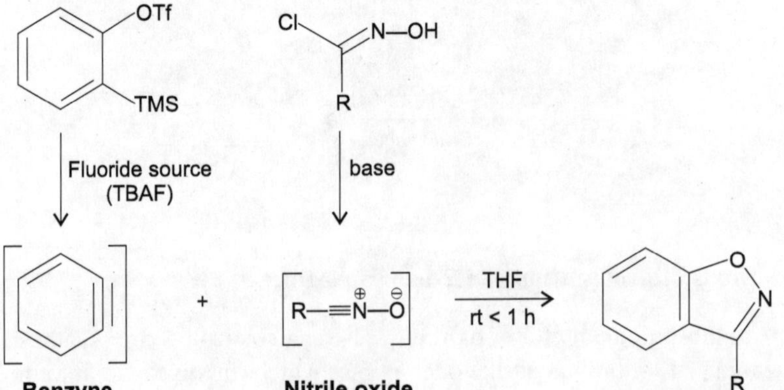

Benzyne   **Nitrile oxide**

1,2-Benzisoxazole

8.1.5.3 Methods of preparation of 2,1-benzisoxazole

(*i*) **From ortho-nitrobenzaldehydes**: 2,1-Benzisoxazoles can be synthesized from ortho-nitrobenzaldehydes on treatment with 2-bromo-2-nitropropane in presence of Zn dust.

3-Unsubstituted-2,1-benzisoxazoles

8.1.5.4 Methods of preparation of 1,2-benzisothiazole

(*i*) **From thiols:** Oxidative cyclization of 2-(aminomethyl)-4-methylbenzenethiol with iodine or bromine, or with alkaline potassium ferricyanide in alkaline solution gives 5-methyl-1,2-benzisothiazole in 89% yield, *via* a disulfide intermediate.

5-Methyl-1,2-benzisothiazole
(89%)

(*ii*) **From (aminosulfanyl)arenes:** The 1,2-benzisothiazoles are also prepared by the intramolecular cyclization of (aminosulfanyl)arenes bearing an unsaturated ortho substituent, such as a cyano or carbonyl group. For example, sodium 6-nitro-2-cyano-benzenethiolate reacts with chloroamine to form 6-nitro-2-(aminosulfanyl)benzonitriles *in situ*, which gets converted to 7-nitro-1,2-benzisothiazol-3-amine in 78% yield.

6-Nitro-2-(aminosulfanyl)   **7-Nitro-1,2-benzisothiazol-**
benzonitrile   **3-amine**
                                                      (78%)

(*iii*) **From oximes:** Cyclization of the oximes of 2-(alkylsulfanyl)benzaldehydes or 2-(alkylsulfanyl)phenyl ketones in the presence of acetic anhydride in pyridine, or polyphosphoric acid results into the formation of 1,2-benzisothiazoles.

$$\xrightarrow[\text{or}\;\text{PPA}]{\text{Ac}_2\text{O, Pyridine}}$$

(95%)

8.1.5.5 Methods of preparation of 2,1-benzisothiazole

(*i*) **From ortho-aminotoluenes:** 2,1-Benzisothiazoles are prepared from ortho-aminotoluenes by reaction with N-sulfinylmethanesulfonamide or thionyl chloride

$$\xleftarrow[\substack{\text{Pyridine, Benzene}\\\text{reflux}}]{\text{MeSO}_2\text{N}\!=\!\text{S}\!=\!\text{O}}$$

$$\xrightarrow[\substack{\text{p-xylene}\\\text{reflux}}]{\text{SOCl}_2}$$

2,1-Benzisothiazole   **2,1-Benzisothiazole**

(*ii*) **From 2-aminophenylmethanethiol**: Oxidation of 2-aminophenylmethanethiol with iodine in 2 M sodium hydroxide at pH 13.5 gives 2,1-benzisothiazole in 60% yield.

$$\xrightarrow[\text{pH = 13.5}]{\substack{\text{I}_2\text{, KI}\\\text{2M NaOH}}}$$

2-Aminophenylmethanethiol   **2,1-Benzisothiazole**
(60%)

8.1.6 Reactions of benzo-1,2-azoles

Basicity

Indazole (pK$_a$ of its conjugate acid is 1.25) is less basic than pyrazole (pK$_a$ of its conjugate acid is 2.5) but it is a stronger N-H acid. 1,2-Benzisoxazole is a weaker base (pK$_a$ of its conjugate acid is -4.2) than isoxazole (pK$_a$ of its conjugate acid is -3.0). 2,1-Benzisothiazole (pK$_a$ of its conjugate acid is -0.05) is a stronger base than isothiazole (pK$_a$ of its conjugate acid is -0.51).

pK$_a$ of conjugate acid:-   **1.25**   **2.5**

pK$_a$ of conjugate acid:-   **−4.2**   **−3.0**

pK$_a$ of conjugate acid:-   **−0.05**   **−0.51**

Electrophilic substitution reactions

Electrophilic attack takes place preferably at 5-position in indazole.

For example, halogenation of indazole gives 5-haloindazole. Nitration with sulfuric acid and nitric acid or with fuming nitric acid alone gives 5-nitroindazole. Sulfonation with oleum, however, yields indazole-7-sulfonic acid. Indazole couples with diazonium salts at the 3-position.

Iodination or bromination of indazole at 3-position can be carried by treating silver salt of indazole with bromine or iodine, respectively.

Nitration of 1,2-benzisoxazole affords almost exclusively corresponding 5-nitro-1,2-benzisoxazole. Further nitration of 3-alkyl-5-nitro-1,2-benzisoxazole gives 3-alkyl-5,7-dinitro-1,2-benzisoxazole.

1,2-Benzisoxazole   **5-Nitro-1,2-benzisoxazole**

3-Methyl-1,2-benzisoxazole   **3-Methyl-5,7-dinitro-1,2-benzisoxazole**

Similarly, if other substituent like hydroxyl, halo, ester are present at 3-position of 1,2-benzisoxazole the nitration occurs at 5-position.

X = OH, Cl, COOR

Nitration of 1,2-benzisothiazole gives a mixture of 5- and 7-nitro-1,2-benzisothiazoles.

5-Nitro-1,2-benzisothiazole   **7-Nitro-1,2-benzisothiazole**

Nitration of 2,1-benzisothiazoles with nitric acid and sulfuric acid gives a mixture of 4-nitro- (17%), 5-nitro- (57%), and 7-nitro-2,1-benzisothiazoles (26%).

(17%)   (57%)   (26%)

The bromination of 2,1-benzisothiazole with bromine in sulfuric acid and silver sulfate is a nonspecific reaction and a mixture of 5-bromo- (31%), 7-bromo (31%), and 4,7-dibromo-2,1-benzisothiazole (10%) is produced.

2,1-Benzisothiazole   (31%)   (31%)   (10%)

Sulfonation of 1,2-benzisoxazole either gives low yield or reaction does not takes place at all.

The acylation of 1,2-benzisoxazole is difficult and the yields are low. However, Friedel-Craft acylation of 3-alkyl-6-hydroxy-1,2-benzisoxazoles gives corresponding 7-acyl derivatives. Similarly the Riemer-Tiemann reaction gives the corresponding 7-formyl derivative.

Metallation

3-Bromoindazole can be converted into an N,C-di-lithio species, which reacts with acetone to give 3-substituted indazole.

1,2-Benzisothiazole and 1,2-benzisoxazole have not been lithiated in the heterocyclic ring, because it would lead to fragmentation of the heterocyclic ring.

Alkylation at nitrogen

Indazole forms quaternary salts by reaction at nitrogen, normally it reacts at the imine nitrogen, N-2.

N-Alkylation of 1,2-benzisoxazoles can also be achieved by treatment with triethyloxonium tetrafluoroborate, or with dimethyl sulfate in presence of perchloric acid. The salts formed are unstable when strongly heated.

N-Alkylation of 2,1-benzisothiazole occurs with alkyl halides (RX) giving 1-alkyl-2,1-benzisothiazolium salts in which the benzenoid ring is restored.

1-Alkyl-2,1-benzisothiazolium salt

Nucleophilic reactions

Nucleophilic reagents such as sodium methoxide, sodium cyanide, butyllithium, thiols, and some tertiary bases react with 1,2-benzisothiazoles, or 3-halo-1,2-benziso-thiazoles, leading to ring opening. For example, 1,2-benzisothiazoleis cleaved by treatment with sodium methoxide to sodium 2-cyanobenzenethiolate.

1,2-Benzisothiazole   **Sodium-2-cyanobenzenethiolate**

Cycloaddition

With benzyne, 2,1-benzisothiazole gives acridine in low yield (5%). Presumably here an adduct is formed as an intermediate, which loses sulfur to form acridine.

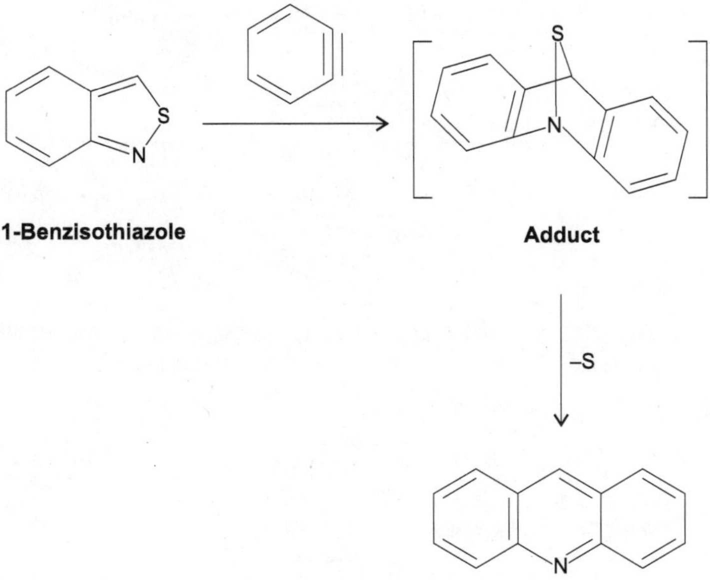

2,1-Benzisothiazole   **Adduct**

−S

Acridine
(5%)

Ring opening reaction

3-Alkyl and 3-aryl-1,2-benzisoxazoles are fairly stable towards bases, but 3-unsubstituted, 3-acyl and 3-carboxy derivatives are easily cleaved. The cleavage of ring involves a concerted E2 elimination process.

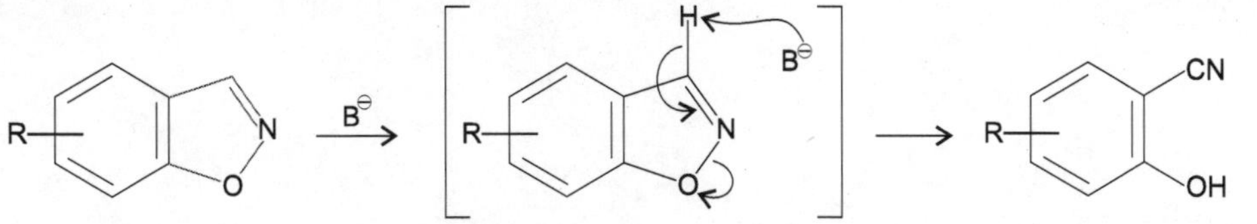

3-Unsubstituted 1,2-benzisoxazoles decompose to give corresponding salicylonitriles on heating.

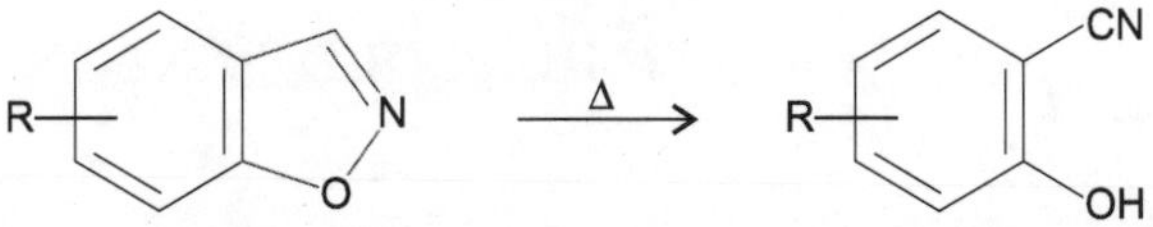

Treatment of 2,1-benzisothiazole with hydrazine (NH_2NH_2) causes the elimination of sulfur, but in this case the reagent is incorporated in the product so that 2-aminophenylhydrazone is formed in 23% yield.

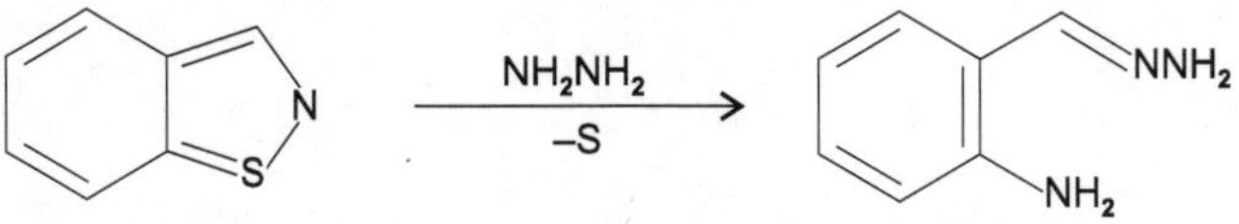

2-Aminophenylhydrazone
(23%)

2,1-Benzisothiazole on reaction with Raney Ni gives 2-alkylanilines.

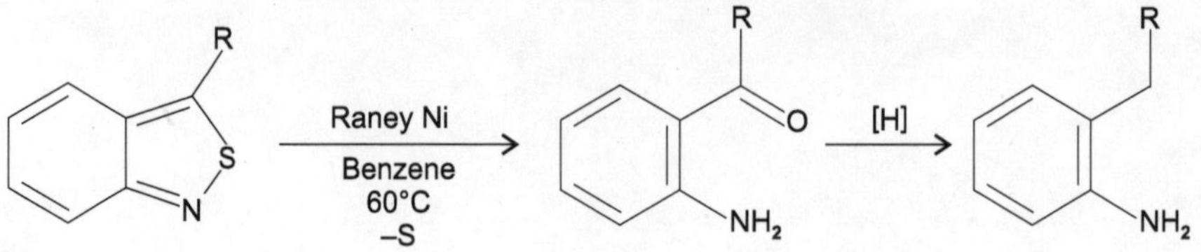

**3-Substituted-
2,1-benzisothiazoles**

1,2-Benzisoxazole gets converted to benzoxazole on irradiation. The conversion proceeds *via* a nitrile intermediate, which gets converted to isonitrile under reaction conditions.

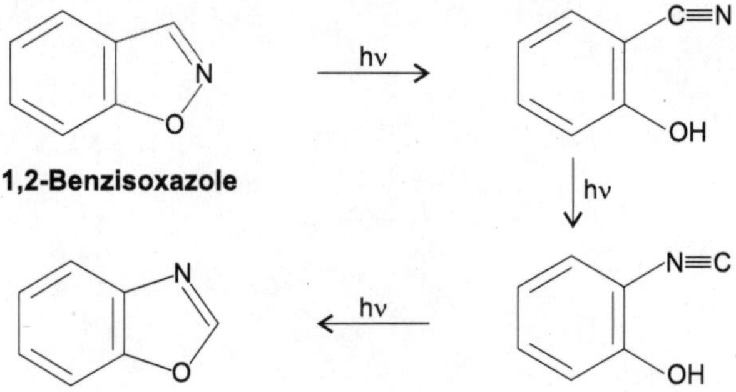

1,2-Benzisoxazole

Benzoxazole

3-Phenyl-1,2-benzisoxazole is converted to 2-phenylbenzoxazole on flash vacuum pyrolysis (FVP) at 800°C under low pressure (0.1-1 Torr) *via* a spiroazirine intermediate.

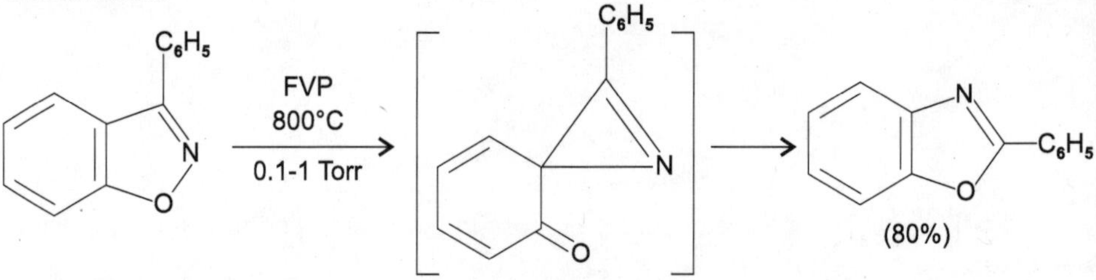

8.2 BENZO-1,3-AZOLES: BENZIMIDAZOLE, BENZOXAZOLE AND BENZOTHIAZOLE

8.2.1 Introduction

Benzene ring can fuse to imidazole, oxazole and thiazole to form corresponding heterocyclic aromatic compounds 1*H*-benzimidazole, benzoxazole and benzothiazole, respectively.

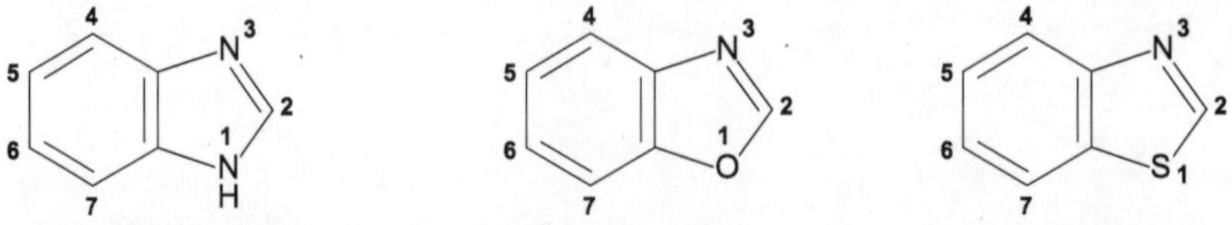

1*H*-Benzimidazole   **Benzoxazole**   **Benzothiazole**

Benzimidazole exists as colourless crystals (mp 169-171°C). It is moderately soluble in water and very soluble in ethanol. Benzoxazole also exist as colourless crystals (bp 182°C, mp 27-30°C). Benzothiazole is colourless, slightly viscous liquid (bp 227-228°C, mp 2°C).

8.2.2 Importance of benzo-1,3-azoles

Benzimidazole exhibits numerous biological activities and acts as antihelminthic, antifungal, anti-allergic, antimicrobial, antiviral and antineoplastic agent. **Thiabendazole** (antihelminthic), **diabazole** (vasodialator) and **omeprazole** (proton pump inhibitor) are some commercially available benzimidazole based drugs. Benzimidazole system is also present in **vitamin B12** (cyanocobalamin).

Thiabendazole **Diabazole**

Omeprazole

Benzoxazole is used in research as a starting material for the synthesis of large bioactive molecules. It is found within the chemical structures of pharmaceutical drugs such as **2-amino-5-chlorobenzoxazole** (sedative). **Fenoxaprop**, a biologically active compound containing the benzoxazole moiety is used as herbicide.

Fenoxaprop

2-Amino-5-chlorobenzoxazole

The light emitting component of **luciferin**, found in fireflies is a benzothiazole derivative. Some drugs contain this group, an example being **riluzole** (amyotrophic lateral sclerosis). Some dyes such as **thioflavin** have benzothiazolium group as a structural motif. The accelerators used for the vulcanization of rubber are based on **2-mercaptobenzothiazole**.

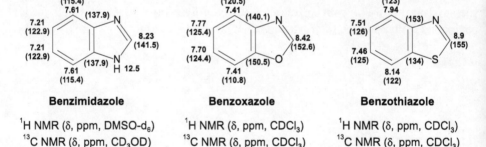

Luciferin

Riluzole

Thioflavin

2-Mercaptobenzothiazole

8.2.3 Structure

Benzimidazoles display annular tautomerism in solution. Two non-equivalent structures can be written, for 5(or 6)-methylbenzimidazole.

5-Methylbenzimidazole   **6-Methylbenzimidazole**

8.2.4 Spectral data

The ^{1}H and ^{13}C NMR (given in brackets) values of benzo-1,3-azoles are as follows:

Benzimidazole

^{1}H NMR (δ, ppm, DMSO-d_6)
^{13}C NMR (δ, ppm, CD$_3$OD)

Benzoxazole

^{1}H NMR (δ, ppm, CDCl$_3$)
^{13}C NMR (δ, ppm, CDCl$_3$)

Benzothiazole

^{1}H NMR (δ, ppm, CDCl$_3$)
^{13}C NMR (δ, ppm, CDCl$_3$)

8.2.5 Methods of preparation of benzo-1,3-azoles

(*i*) **From ortho-subsituted anilines:** The benzimidazoles, benzoxazoles and benzothiazoles are synthesized from ortho-substituted anilines. In this method a carboxylic acid is heated with ortho-phenylenediamine, ortho-aminophenol or ortho-aminothiophenol to form benzimidazole, benzoxazole and benzothiazole, respectively.

For example, ortho-phenylendiamine reacts with formic acid at 100°C to give benzimidazole in a yield of over 80%.

Mechanism: Benzimidazole synthesis using ortho-phenylendiamine involves nucleophilic attack of amine nitrogen on carbonyl carbon of formic acid, followed by a cyclocondensation.

Similarly, ortho-aminothiophenol reacts with formic acid in the presence of acetic anhydride to give benzothiazole .

ortho-Aminothiophenol   Formic acid

ortho-Esters can be used instead of acids along with KSF clay as catalyst for the synthesis of all the three unsubstituted benzo-1,3-azoles from ortho-substituted anilines.

X = NH, O, S   X = NH, O, S

(*ii*) **From indazoles**: Benzimidazole is formed in good yield by photolysis of indazoles.

Indazole   **Benzimidazole**

(*iii*) **From N-aryl amidines:** Barsche and Buchwald developed a simplified synthesis of benzimidazoles using N-aryl-amidines.

Benzimidazoles are also obtained from N-aryl amidines by reacting it with benzenesulfonyl chloride in triethylamine under anhydrous condition.

(*iv*) **From oximes:** The photolysis of oximes of 2-hydroxy benzaldehyde or acetophenone leads to benzoxazoles *via* Beckmann rearrangement and cyclization. The reaction is carried out in protic solvents such as water, methanol, or ethanol.

R = H, CH₃

(*v*) **From anilines:** 2-Aminobenzothiazoles are formed by the reaction of substituted anilines with potassium thiocyanate and aqueous solution of sodium dichloroiodate.

(*vi*) **From ortho-acylaminophenols:** 2-Methylbenzothiazoles are also formed by action of phosphorus pentasulfide on ortho-acylaminophenols.

ortho-Acylaminophenol

(*vii*) **From ortho-aminophenol:** 2-Mercaptobenzothiazole can be prepared by the reaction of carbon disulfide with ortho-aminophenol in presence of acetic anhydride.

2-Mercaptobenzothiazole

(*viii*) N-Arylthioamides can be cyclized oxidatively to give benzothiazoles.

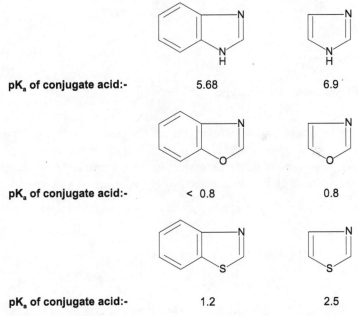

N-Arylthioamides   **2-Subsituted benzothiazoles**

8.2.6 Reactions of benzo-1,3-azoles

Basicity

The NH group present in benzimidazole is relatively strongly acidic (pK_a = 16.4) and also weakly basic. Another characteristic of benzimidazole is that they have the capacity to form salts. Benzimidazole (pK_a of its conjugate acid is 5.68) is less basic than imidazole (pK_a of its conjugate acid is 6.9). Benzoxazole (pK_a of its conjugate acid is < 0.8) is less basic than oxazole (pK_a of its conjugate acid is < 0.8). Benzothiazole (pK_a of its conjugate acid is 1.2) is a weaker base than thiazole (pK_a its conjugate acid is 2.5).

pK_a of conjugate acid:-	5.68	6.9
pK_a of conjugate acid:-	< 0.8	0.8
pK_a of conjugate acid:-	1.2	2.5

Some reactions of benzoxazoles such as salt formation and quaternization at the ring nitrogen are similar to oxazoles.

Alkylating agents such as iodomethane, trialkyloxonium tetrafluoroborates, dialkyl sulfates, methyl 4-toluenesulfonate, or methyl vinyl ketone are used to convert 2-alkylbenzoxazoles to their corresponding 2,3-dialkylbenzoxazolium salts. For example, alkylation of 2-methylbenzoxazole in presence of methyl iodide at 100°C yields 2,3-dimethylbenzoxazolium iodide in 89% yield.

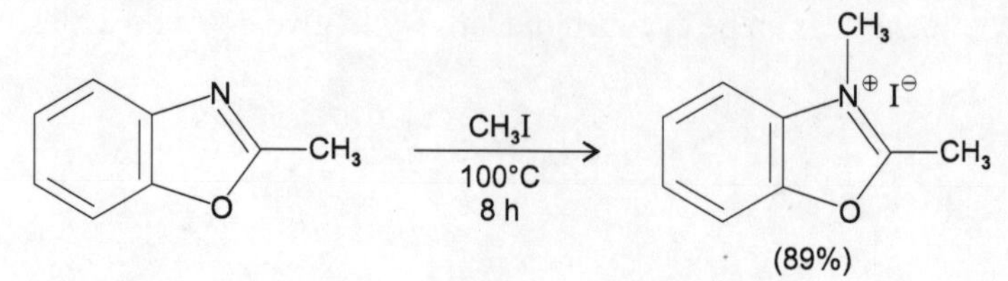

(89%)

Benzothiazole on heating with equimolar quantity of ethyl iodide forms N-ethylbenzothiazolium iodide which is used for the synthesis of thiocyanine dyes.

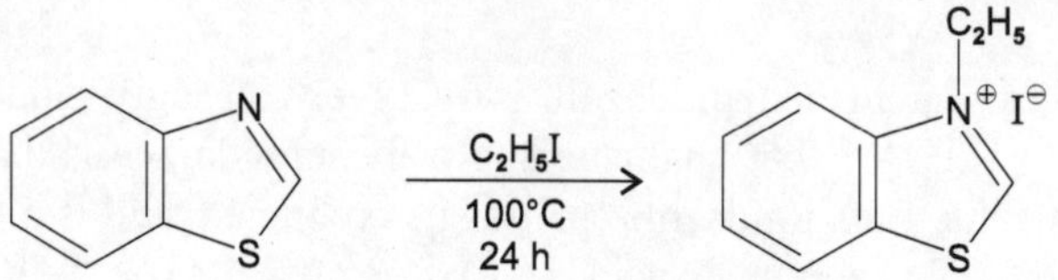

Benzimidazoles unsubstituted at 1-position undergo **Mannich reaction**.

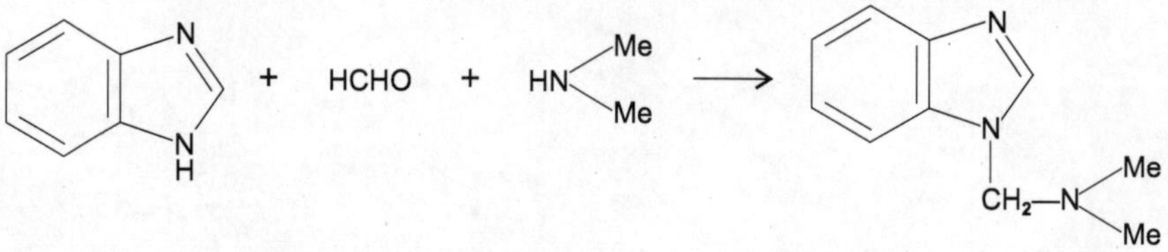

Electrophilic substitution reactions

Electrophilic substitutions occur only at the benzene ring. For example, in benzimidazole electrophilic substitutions on carbon take place first at 5-position and then at 7- or 6-position. Electrophilic substitution such as nitration of benzoxazole occurs at the 5- or 6-position of benzene ring with preference for the 6-position. While in benzothiazole electrophilic substitution may occur at 4-, 5-, 6- or 7- position.

The possible site of attack on benzo-1,3-azoles by electrophile is shown below:

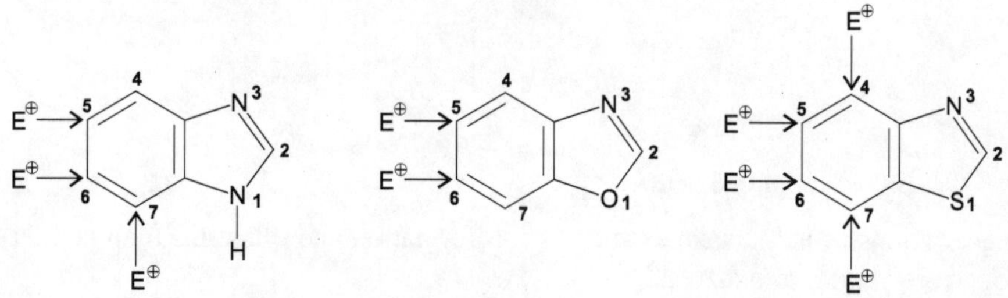

Nitration: Nitration of benzimidazole gives 5-nitrobenzimidazole.

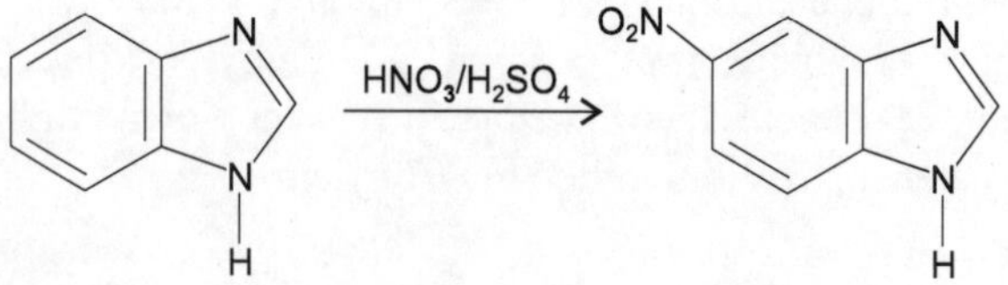

Nitration of 2-phenyl benzoxazole in presence of nitric acid at room temperature gives 6-nitro-2-phenylbenzoxazole in 83 % yield.

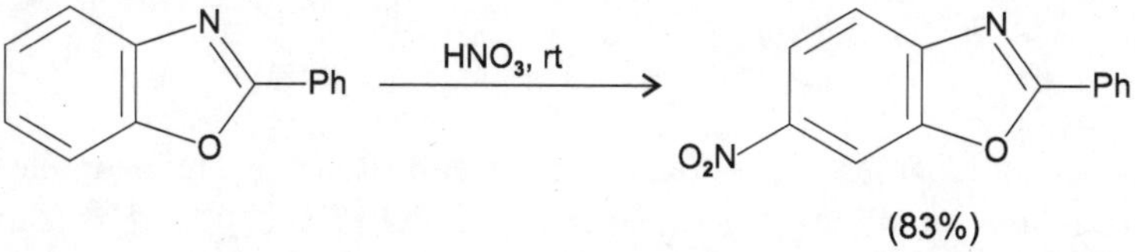

(83%)

Nitration of benzothiazole with nitrating acid at room temperature yields a mixture of 4-, 5-, 6- and 7-nitrobenzothiazole.

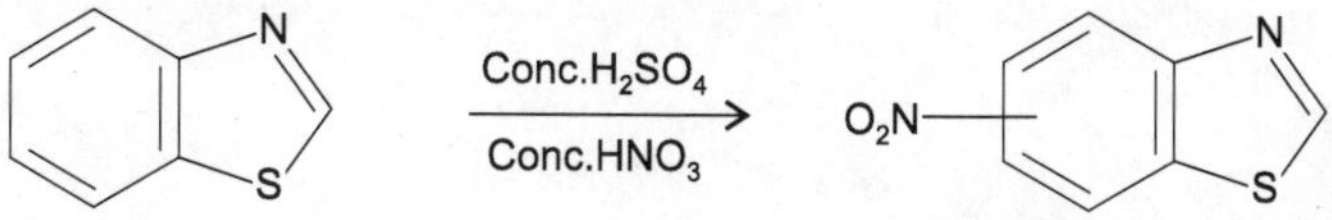

**Mixture of 4-, 5-, 6- and
7- Nitrobenzothiazole**

Halogenation: Benzimidazole undergoes 2-bromination with N-bromosuccinimide.

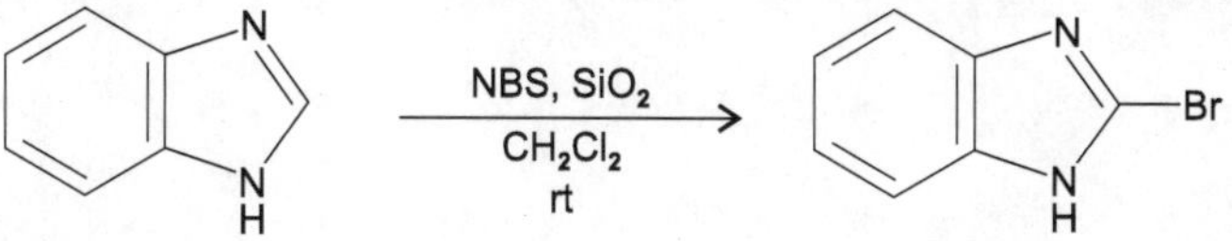

Nucleophilic reactions

Nucleophilic attack occurs at the 2-position in benzimidazole, benzoxazole and benzothiazole. The possible site of attack on benzo-1,3-azoles by nucleophile is shown below:

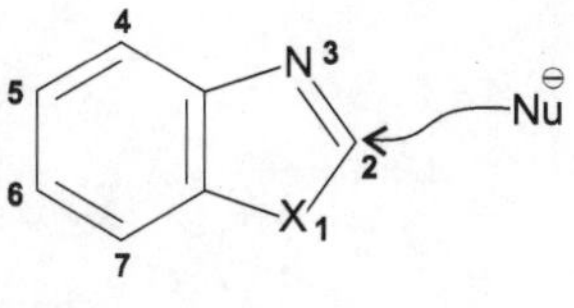

X = NH, O, S

Benzimidazoles are more reactive towards nucleophiles than imidazoles. Treatment of benzimidazole with sodium amide in xylene gives the corresponding 2-amino compound (**Chichibabin reaction**).

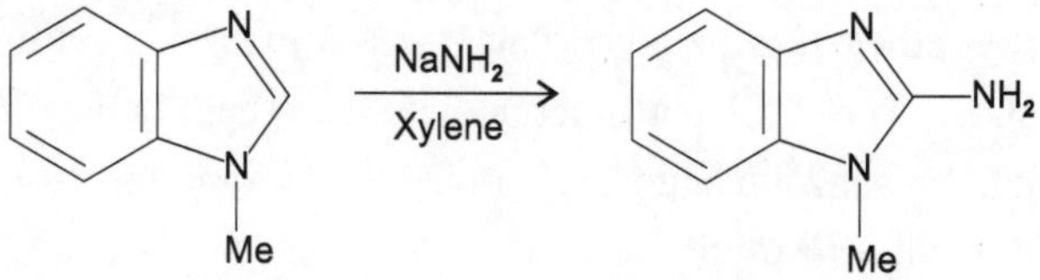

2-Aminobenzoxazole is obtained in low yield upon heating benzoxazole with hydroxylamine hydrochloride in sodium hydroxide solution.

In 2-chlorobenzimidazoles the halogen can be substituted by nucleophiles such as alkoxides, thiolates or amines to give the corresponding 2-substituted benzimidazoles.

$$Nu = \overset{\ominus}{O}R, \ \overset{\ominus}{S}R, \ \overset{..}{N}H_2\text{--}R$$

Similarly, 2-chlorobenzoxazoles reacts with various nucleophiles to give corresponding 2-substituted benzoxazoles.

Metallation

2-Lithiobenzoxazole is obtained from benzoxazole *via* lithiation with butyllithium at -110°C in diethyl ether. It is believed that it exists in an equilibrium with lithium 2-isocyanophenolate. The lithiated intermediate can be quenched with suitable electrophiles, such as benzaldehyde derivatives. For example, 2-lithiobenzoxazole on reaction with 2-chloro-6-fluorobenzaldehyde gives benzoxazol-2-yl(2-chloro-6-fluorophenyl)methanol in 54% yield.

Similarly, 2-lithiobenzothiazole can be obtained by treating benzothiazole with butyl lithium, which reacts with propanal or trimethylsilyl chloride.

2-Alkylbenzothiazoles, like 2-alkylthiazoles, are CH-acidic. They are deprotonated by butyllithium in THF at -78°C to form the lithium compound. This reacts with aldehydes or ketones giving alcohols.

Reaction with arynes

Benzimidazole reacts as a nucleophile with benzyne (obtained from iodobenzene in the reaction) to give 2-phenylbenzimidazole.

EXERCISE

Q.1 Carry out the following conversions:-

(a) Indazole to benzimidazole

(b) 1,2-Benzisoxazole to benzoxazole

Q.2 What happens when benzimidazole is treated with sodium amide in xylene?

Q.3 Write the reactions involved in the synthesis of benzimidazoles, benzoxazoles and benzothiazoles from ortho-substituted anilines.

Q.4 How will you synthesize 1,2-benzisoxazoles from salicylaldehyde?

Q.5 Write the product formed in the following reactions:

(a)

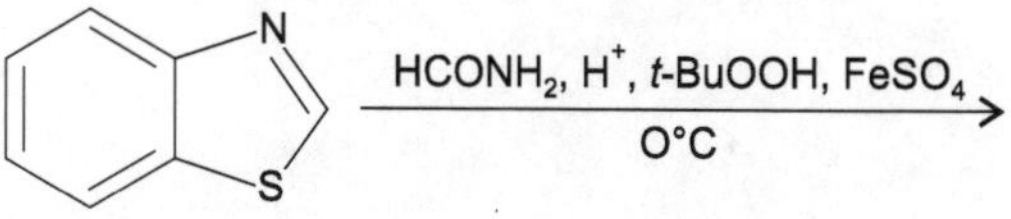

$$\xrightarrow[\text{0°C}]{\text{HCONH}_2, \text{ H}^+, \text{ }t\text{-BuOOH, FeSO}_4}$$

(b)

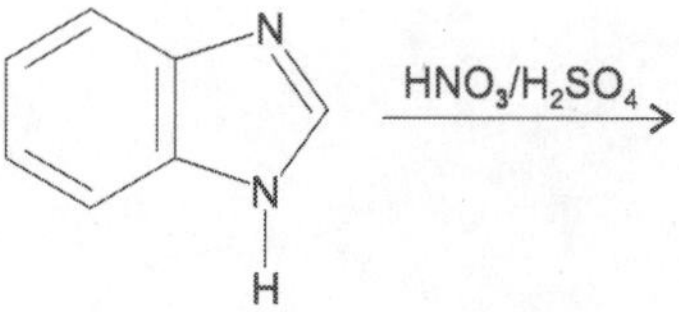

$$\xrightarrow{\text{HNO}_3/\text{H}_2\text{SO}_4}$$

(c)

+ HCHO + HN(Me)(Me) $\longrightarrow$

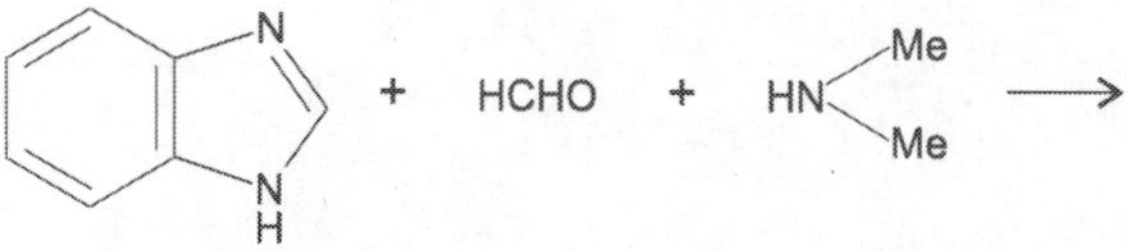

(d)

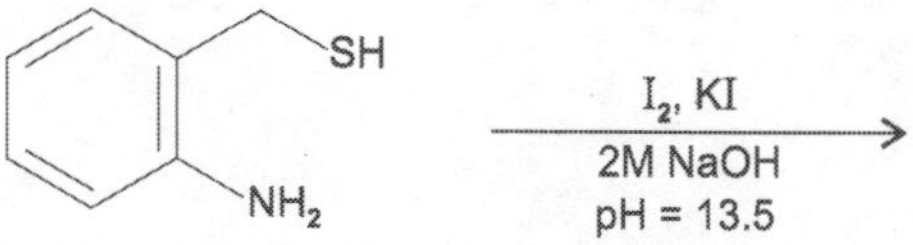

$$\xrightarrow[\substack{\text{2M NaOH}\\\text{pH = 13.5}}]{\text{I}_2,\text{ KI}}$$

Ans.

(a)

—CONH$_2$

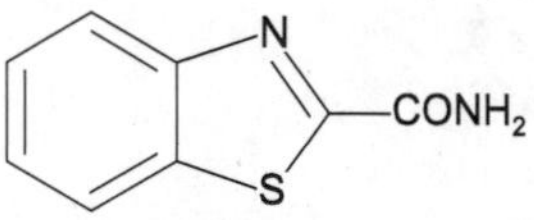

(b)

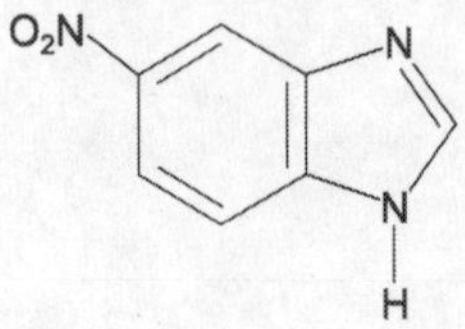

(c)

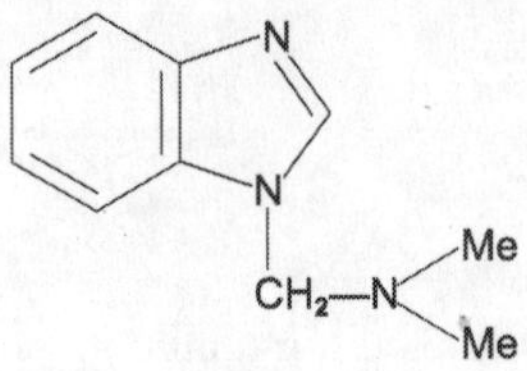

(d)

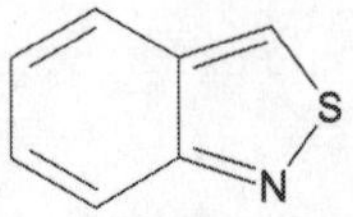

(e)

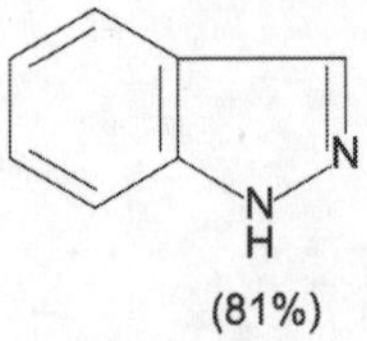

(81%)

9

Six Membered Heterocyclic Compounds with One Heteroatom

Pyridine, Pyridine N-oxide, Pyrilium salt and Thiopyrilium salt

The six membered heterocyclic compounds containing nitrogen, oxygen and sulfur are known. The introduction of nitrogen atom by the replacement of CH in benzene ring results into pyridine which is aromatic like benzene. However, replacement of CH in benzene ring with oxygen and sulfur alters the aromatic character of the corresponding heterocyclic rings. Therefore, pyrans (4*H*-pyran and 2*H*-pyran) and thiopyran (4*H*-thiopyran and 2*H*-thiopyran) are not aromatic. The pyrylium and thiopyrylium salts are aromatic, but are not very stable.

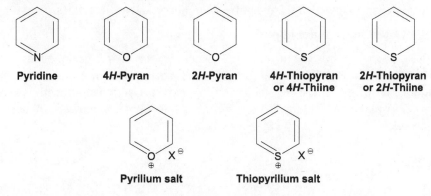

| Pyridine | 4*H*-Pyran | 2*H*-Pyran | 4*H*-Thiopyran or 4*H*-Thiine | 2*H*-Thiopyran or 2*H*-Thiine |

Pyrilium salt   Thiopyrilium salt

The reduced pyridines are dihydropyridines (1,4-DHP and 1,2-DHP), tetrahydro-pyridines (1,2,3,4-THP, 1,2,3,6-THP and 2,3,4,5-THP) and hexahydropyridine (piperidine).

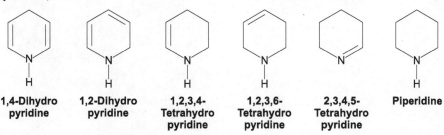

| 1,4-Dihydro pyridine | 1,2-Dihydro pyridine | 1,2,3,4-Tetrahydro pyridine | 1,2,3,6-Tetrahydro pyridine | 2,3,4,5-Tetrahydro pyridine | Piperidine |

Mono-, di-, and trimethylpyridines are known as picoline, lutidine and collidine, respectively, with position of the methyl groups denoted by symbols α, β, γ or arabic numerals.

α-Picoline   β-Picoline   γ-Picoline   2,6-Lutidine   2,4,6-Collidine

9.1 PYRIDINE

9.1.1 Introduction

Pyridine is an aromatic six membered heterocyclic compound with a nitrogen atom. Pyridine was first isolated from bone pyrolysates and can be isolated from coal tar. It was first synthesized from acetylene and hydrogen cyanide in 1876. The name pyridine is derived from the Greek word 'pyr' which means fire and the suffix 'idine'. It is stable and relatively unreactive liquid (mp −42°C and bp 115°C) with unpleasant smell. Pyridine is miscible with water and hydrocarbons. It is poisonous and inhalation of its vapour causes damage to the nervous system. **Pyridine** is aprotic but polar solvent. It is used as a solvent and base in many reactions such as acylation and tosylation. Pyridine has a disagreeable odour and a burning taste therefore, it is added to chemicals to make them unsuitable for human consumption. For example, it is used as a denaturant for ethyl alcohol, in antifreeze mixtures and as fungicides.

9.1.2 Biological importance of pyridine

Pyridine ring is present in several naturally occurring compounds like **pyridoxine**(vitamin B6), **Nicotinamide**(vitamin B3/niacin), **nicotine** and **nicotinic acid**. **Paraquat** one of the oldest herbicides, **sulfapyridine** a sulfonamide anti-bacterial agent and **isoniazid** an important drugs for the treatment of tuberculosis, all contain pyridine nucleus.

Pyridoxine
(Vit B6)

Nicotinamide
(Vit B3)

Nicotine

Nicotinic acid

Paraquat

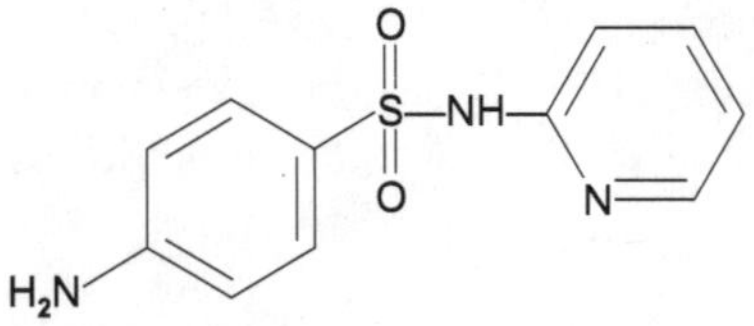

Sulfapyridine

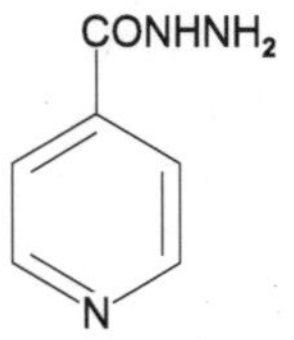

Isoniazid

9.1.3 Structure of pyridine

All the five carbon atoms and the nitrogen of pyridine are sp^2 hybridized. The two unpaired electrons of the sp^2 hybrid orbital of nitrogen are involved in the formation of two σ bonds with the adjacent carbons. There are also two paired electrons present in one of the sp^2 hybrid orbital of nitrogen atom as lone pair. The fifth valence electron of nitrogen occupies the unhybridized p orbital which is perpendicular to the ring plane. It is this electron present in p orbital which takes part in the π-electron system of pyridine. Hence, pyridine has six π-electrons from the three double bonds coming from the five carbon and one nitrogen atoms. Thus it is (4n+2) π-electron planar system and therefore aromatic. In the case of, the five membered nitrogen containing pyrrole ring, the lone pair of electrons on nitrogen is delocalized over the ring and responsible for its aromaticity. However, this is not in the case of pyridine. In pyridine the lone pair present in sp^2 hybrid orbital of nitrogen is not involved in delocalization but it is responsible for the basic character of pyridine.

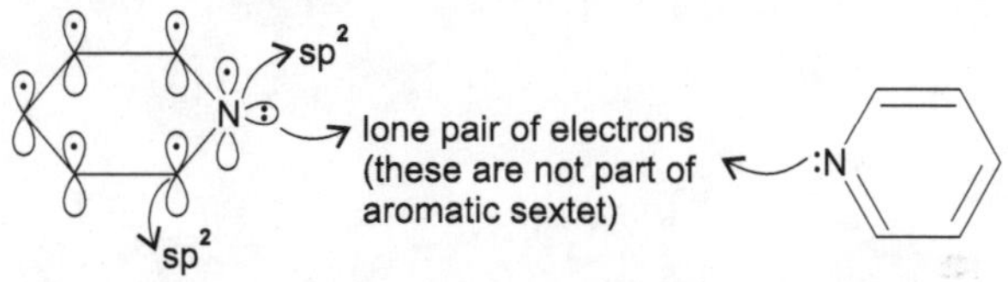

The resonance energy of pyridine is about 117 kJ/mol or 28 kcal/mol and it is less aromatic than benzene (resonance energy 36 kcal/mol) but more aromatic than pyrrole (resonance energy 21.6 kcal/mol. Resonating structures of pyridine are shown below. The electronegative nitrogen atom pulls electron towards itself and positive charge appears at position 2, 4 and 6, as seen in the resonating structures of pyridine. Therefore, in pyridine positions 2, 4 and 6 are the possible sites of attack by a nucleophile.

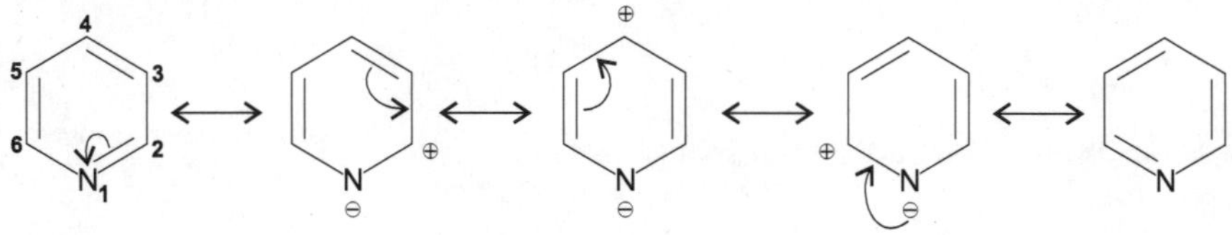

Resonance in pyridine

9.1.4 Dipole moment

Pyridine is aprotic and polar compound with a non-delocalized lone pair of electron on nitrogen atom. This structure results in a very strong moisture affinity caused by the H-bonding. The dipole moment of pyridine is 2.22 D. The inductive (caused by electronegative nitrogen) and mesomeric (as seen in the resonating structures) effects polarize it in the same direction therefore, the permanent dipole is towards the nitrogen atom. In the saturated analogue, piperidine the dipole moment (μ = 1.17 D) is due to the induced polarization of the σ-skeleton. The higher dipole moment of pyridine in comparison to piperidine is because of the additional polarization due to distortion of the π-electron system in pyridine.

Pyridine
μ = 2.22 D

Piperidine
μ = 1.17 D

9.1.5 Basicity

Pyridine behaves as a tertiary base and can easily be protonated, yielding a pyridinium cation. The pK_a value of conjugate acid of pyridine i.e., the pyridinium cation is 5.2 in H_2O. It is a weaker base than similar cyclic secondary amines such as piperidine (pK_a of its conjugate acid is 11.2). In pyridine the lone pair on N are present in sp^2 hybrid orbital whereas in piperidine sp^3. The lone pair of electrons are available for bond formation unlike in pyrrole which are delocalised over the ring. Pyridine is more basic than aniline (pK_a = 4.6) because in aniline the lone pair are delocalised towards the ring. Electron donating substituents (alkyl, amino, alkoxy) increase basicity while electron withdrawing substituents (nitro, halo) decrease basicity of substituted pyridines in comparison to pyridine.

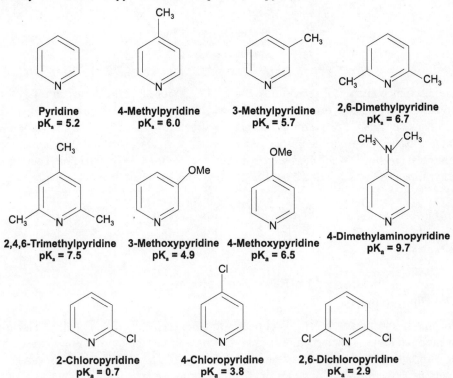

Pyridine pK_a = 5.2	**4-Methylpyridine** pK_a = 6.0	**3-Methylpyridine** pK_a = 5.7	**2,6-Dimethylpyridine** pK_a = 6.7
2,4,6-Trimethylpyridine pK_a = 7.5	**3-Methoxypyridine** pK_a = 4.9	**4-Methoxypyridine** pK_a = 6.5	**4-Dimethylaminopyridine** pK_a = 9.7
2-Chloropyridine pK_a = 0.7	**4-Chloropyridine** pK_a = 3.8	**2,6-Dichloropyridine** pK_a = 2.9	

(pK$_a$ values given here are for the conjugate acids)

9.1.6 Spectral data

The ring carbons of pyridine are electron deficient therefore the five ring protons appear in a deshielded region in the ^{1}H NMR. The protons of pyridine display chemical shifts in the ^{1}H NMR spectrum that are typical of aromatic protons. The ^{1}H and ^{13}C NMR chemical shift values (given in brackets) for pyridine ring atoms are shown below.

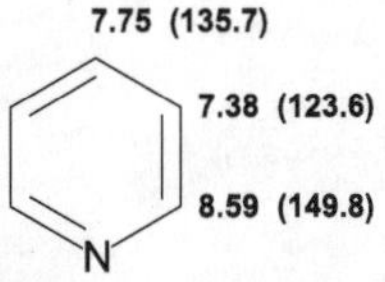

7.75 (135.7)
7.38 (123.6)
8.59 (149.8)

^{1}H NMR (δ, ppm, CDCl$_3$)
^{13}C NMR (δ, ppm, CDCl$_3$)

9.1.7 Methods of preparation of pyridine

(*i*) **Hantzsch pyridine synthesis**: In this method firstly dihydropyridine derivatives are prepared by a four component reaction of an aldehyde with two equivalents of a β-ketoester in the presence of ammonia. This is followed by oxidation (or dehydrogenation) to give pyridine-2,6-dimethyl-4-substituted-3,5-dicarboxylates. Subsequent hydrolysis of the ester groups and decarboxylation yields the corresponding pyridines.

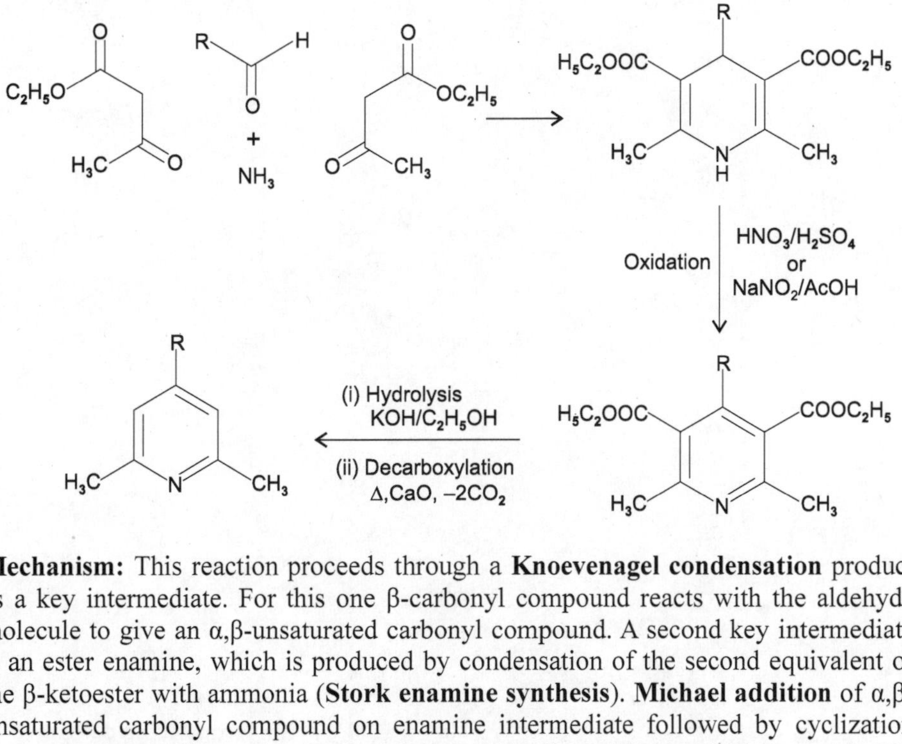

Mechanism: This reaction proceeds through a **Knoevenagel condensation** product as a key intermediate. For this one β-carbonyl compound reacts with the aldehyde molecule to give an α,β-unsaturated carbonyl compound. A second key intermediate is an ester enamine, which is produced by condensation of the second equivalent of the β-ketoester with ammonia (**Stork enamine synthesis**). **Michael addition** of α,β-unsaturated carbonyl compound on enamine intermediate followed by cyclization gives dihydropyridine derivative. Oxidation of dihydropyridine with HNO$_3$ and H$_2$SO$_4$ leads to the formation of pyridine derivative.

(I) Knoevenagel reaction

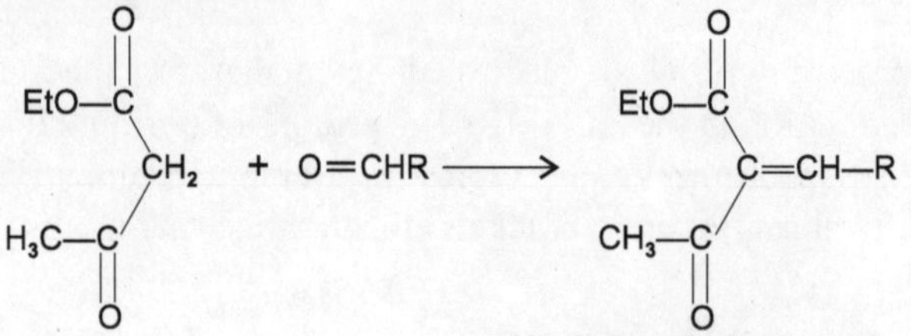

Alkylidine derivative of acetoacetic ester

(II) Stork enamine synthesis

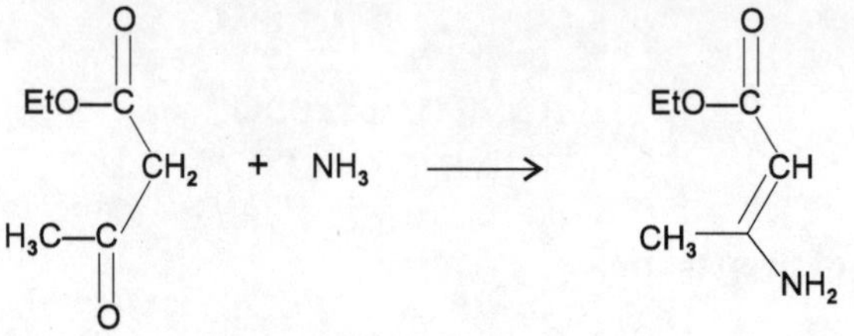

(III) Michael addition followed by cyclisation

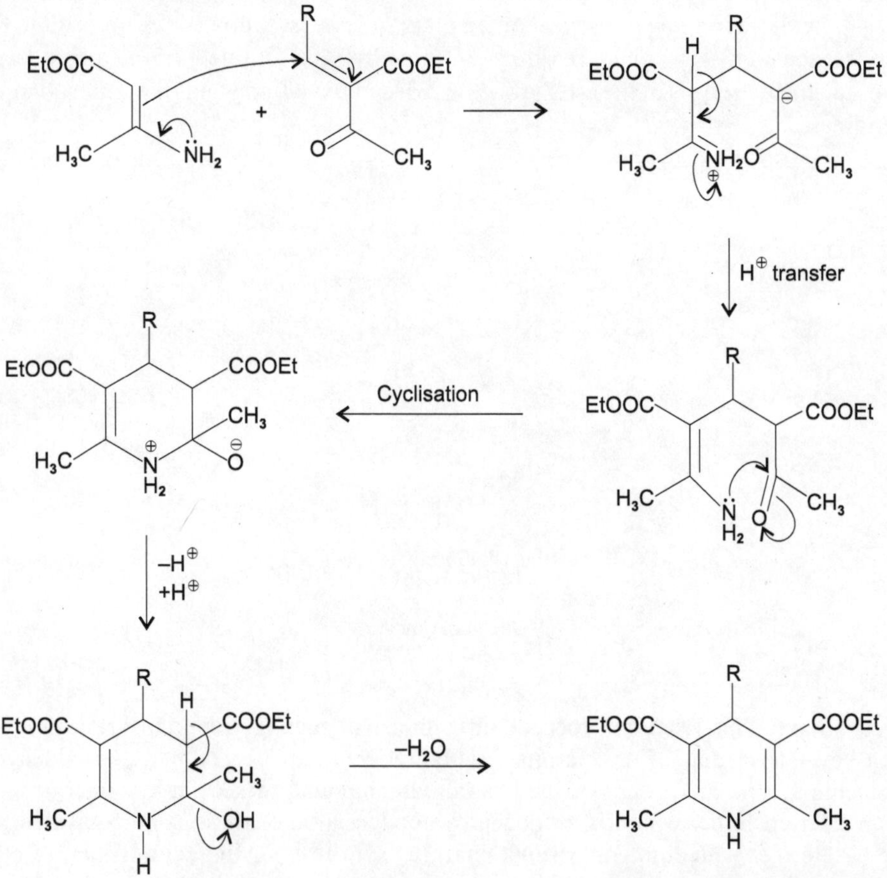

$H^\oplus$ transfer

Cyclisation

$-H^\oplus$
$+H^\oplus$

$-H_2O$

Dihydropyridine derivative

(*ii*) **Chichibabin pyridine synthesis:** Another important method for the synthesis of pyridine is Chichibabin pyridine synthesis. This method involves the condensation of carbonyl compounds with ammonia or amines under pressure to form pyridine derivatives. The reaction is reversible and produces different derivatives. The ammonia or ammonia derivatives are used as a catalyst in this reaction. This reaction can also be carried out in vapour phase by passing the mixture of aldehyde and ammonia over an aluminium oxide catalyst at 300°C. But in vapour phase other products are also formed.

*Chichibabin, A.E., J. Russ. Phys. Chem. Soc. **1906**, 37, 1229.*

$$3 \, RCH_2CHO \ + \ NH_3 \ \underset{\xrightarrow{\hspace{1cm}}}{\overset{NaNH_2}{\rightleftharpoons}}$$

Mechanism: The first step involves formation of imine. The second step is an aldol condensation between two molecules of the carbonyl compound. The aldol product undergoes Michael addition to the imine obtained from the previous step to form the pyridine ring.

(I) Imine formation

Enamine-Imine tautomerism

(II) Aldol condensation

(III) Michael addition

(*iii*) **From 1,5-dicarbonyl compound:** Another method for the preparation of pyridine and its derivative, is by the cyclization of 1,5-dicarbonyl compound in the presence of ammonia, followed by the oxidation of dihydropyridine to pyridines. Unsaturated 1,5-dicarbonyl compounds directly give pyridine without the need of oxidation step. Pyridine itself is prepared from a 1,5-dicarbonyl compound known as glutaconic aldehyde. If hydroxylamine is used instead of ammonia, the oxidation step can be avoided. The final product pyridine is obtained by loss of water molecule.

(*iv*) **From pyrones:** The pyrones (α or γ) are converted into corresponding pyridones (α or γ) by heating with ammonia. The mechanism of this reaction probably involves the opening of pyrone ring to 1,5-dicarbonyl compound which is then cyclized to give pyridone.

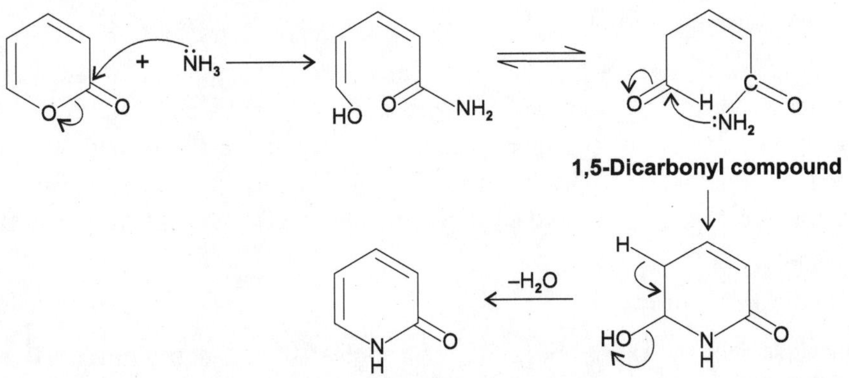

1,5-Dicarbonyl compound

(*v*) **By reaction of alkynes with nitrile:** The Co(I)-catalyzed co-oligomerization of nitrile and acetylenes results in the formation of 2-substituted pyridines. The side reaction leading to cyclotrimerization of acetylene to benzene can be suppressed by using an excess of nitrile.

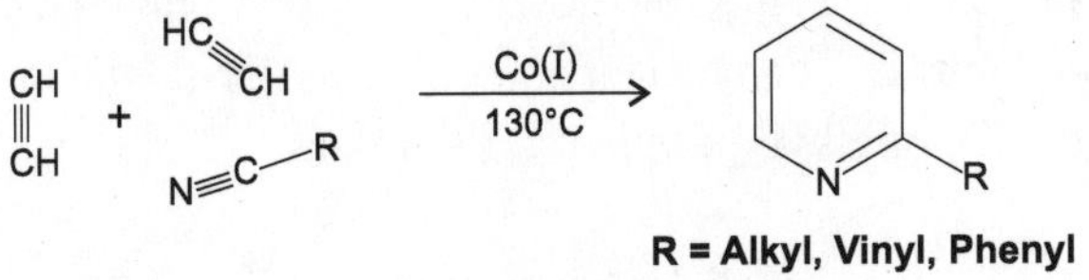

R = Alkyl, Vinyl, Phenyl

(*vi*) **Guareschi synthesis:** Cyanoacetamide and a β-diketone reacts in the presence of base to form pyridine. This is called the Guareschi synthesis. Pyridoxine (vitamin B6), has been synthesised by Guareschi ring synthesis.

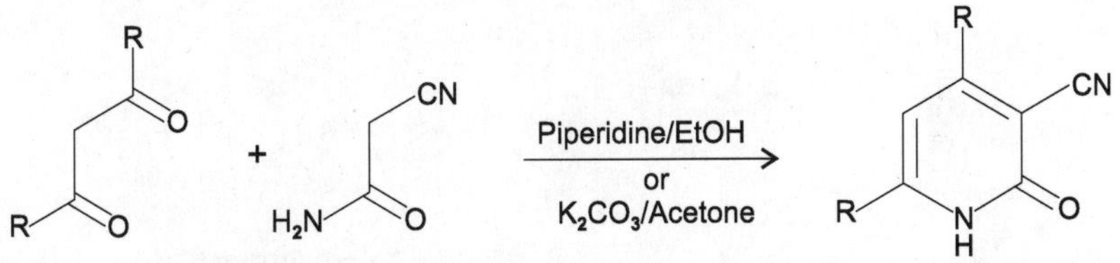

Mechanism: The β-diketone undergoes nucleophilic attack by the carbanion generated from cyanoacetamide in the presence of a base.

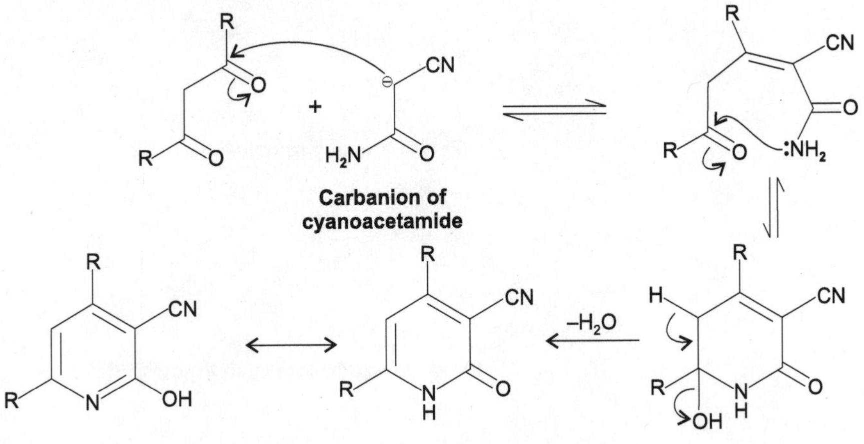

Carbanion of cyanoacetamide

9.1.8 Reactions of pyridine

Pyridine is stable, not easily oxidised at carbon and undergoes substitution rather than addition reactions. Nucleophilic substitution (S_NAr) in pyridine occurs more readily than benzene. Attack of nucleophile takes place at the α- and γ-C-atoms (2- and 4- positions) of pyridine. However, it is reluctant towards electrophilic substitution reactions (S_EAr). Attack of electrophiles takes place at N-atom and at the β-C-atoms (3-position).

Reactions at ring nitrogen

Lewis acids such as $AlCl_3$, $SbCl_3$, SO_3, etc. form stable N-adducts. Bronsted acids form salts with pyridine e.g., the pyridinium chlorochromate (PCC) and pyridinium dichromate (PDC) salts obtained from pyridine and CrO_3/HCl and $H_2Cr_2O_7$, respectively. These salts (PCC and PDC) are used as oxidizing agents for the conversion of alcohols to carbonyls. The pyridinium poly (hydrogen fluoride) or PPHF known as Olah's reagent is obtained by the reaction of pyridine with hydrogen fluoride (in acetone-dry ice bath) is a convenient source of HF for the addition of fluorine to alkenes and preparation of alkyl fluoride from alcohols.

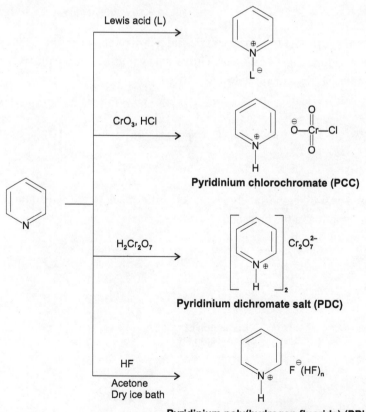

Pyridinium chlorochromate (PCC)

Pyridinium dichromate salt (PDC)

Pyridinium poly(hydrogen fluoride) (PPHF)

Pyridine undergoes N-quaternization in the presence of alkyl halides, alkyl tosylates or dialkyl sulfates to give N-alkylpyridinium salts. The N-arylpyridinium salts are also obtained with activated haloarenes, e.g. 1-chloro-2,4-dinitrobenzene. N-Alkylation with alkyl acrylate or vinyl nitrile leads to Michael addition of pyridine in the presence of HX (acid catalyzed).

The N-acyl salts obtained by the reaction of pyridine and its derivatives with acid chlorides may serve as acyl transfer agents for the synthesis of amides and esters.

The quarternary salts of pyridine such as N-nitropyridinium tetrafluoroborate, N-pyridinium sulfonate and N-fluoropyridinium triflate are used as nitrating, sulfonating, and fluorinating agents, respectively.

N-Nitropyridinum tetrafluoroborate

(Mild nitrating agent)

N-Pyridinium sulfonate

(Sulfonating agent)

N-Fluoropyridinium triflate

(Fluorinating agent)

N-Nitration occurs by nitrating agent such as nitronium tetrafluoroborate.

Sulphonation at nitrogen forms pyridine-sulfur trioxide complex that can serve as a mild sulfonating agent and as an activating agent for DMSO in Moffatt oxidation (oxidation of 1° and 2° alcohols).

Pyridinium sulfonate

Pyridine reacts with bromine in the presence of CCl_4 as solvent to form crystalline salt, containing at N – Br bond.

Reaction of pyridine hydromide with bromine forms pyridinium bromide perbromide (a trihalide anion)which can be used as a convenient source of molecular bromine for electrophilic aromatic substitution reactions.

Pyridine hydrobromide

Pyridinium bromide perbromide

Electrophilic substitution reactions in pyridine

Being an aromatic compound, pyridine participates in certain electrophilic substitution reactions. The presence of an electronegative nitrogen atom in the pyridine ring decreases the availability of electrons at the ring carbons. Thus the attack of the electrophiles at the carbon atoms of the ring becomes more difficult. Therefore, the nucleophilicity of the π-electron system is decreased. The nitrogen atom deactivates the 2- and 4- positions more than 3-position. This is indicated by the three resonating structures of pyridine (see sec. 9.1.3) which contain positively charged carbons, (at C2 and C4). The rate of electrophilic substitution reaction in pyridine is much lower than that of electrophilic substitution in benzene. It requires drastic conditions. Thus the reactivity of pyridine is comparable to that of nitrobenzene.

Electrophilic substitution is more favoured at 3-position in pyridine as compared to positions 2 and 4. This can also be explained by the stability of the σ-complexes formed from the addition of electrophiles at 2-, 3- and 4-positions in pyridine. It is clearly seen that when electrophile attacks 3-position the energy-rich nitrenium canonical form is not formed.

Attack of electrophile at 2- or 4-position: When the attack takes place at 2- or 4-position the positive charge gets delocalized and the intermediate has positive charge on the nitrogen atom.

Attack at 2-position

High energy contributing structure

Attack at 4-position

High energy contributing structure

Attack of electrophile at 3-position: The intermediate formed by attack of electrophile at 3-position is more stable than those formed by attack at 2- or 4-position. Hence, substitution at 3-position is favoured. During electrophilic attack at 3-position none of the contributing structures bears a positive charge on nitrogen atom.

Attack at 3-position (Favoured)

The pyridine and pyrrole nitrogen atoms influence electrophilic substitution reactions in different manner. In pyridine the nitrogen atom slows down electrophilic substitution reactions similar to presence of a strong meta-directing substituent in benzene. In pyrrole the nitrogen atom accelerates electrophilic substitution just as a strongly activating or ortho/para directing substituent does in benzene. Electrophilic substitution is favoured at 2-position in pyrrole.

Nitration: Nitration of pyridine with HNO_3/H_2SO_4 occurs under drastic conditions (300°C), but yields only 15% of 3-nitropyridine. However nitration with N_2O_5 gives rise to good yields of 3-nitropyridine (70%) in the presence of nitromethane or sulfur dioxide.

Sulfonation: Sulfonation of pyridine with oleum at 250°C with Hg(II) catalyst yields the pyridine-3-sulfonic acid in 70% yield. Further on heating at 360°C pyridine-3-sulfonic acid gets converted into pyridine-4-sulfonic acid.

The electrophilic substitution on pyridine occurs easily if the ring is activated by electron donating substituents such as NH_2 or OCH_3. For example 2-aminopyridines can be nitrated or sulfonated under much milder condition than pyridine itself.

Halogenation: Halogenation of pyridines is easier than nitration or sulfonation. This is because the complex formed by attack of halogen at the pyridine nitrogen is less stable and thus the free base is available for halogenation.

Halogenation of pyridine occurs with elemental chlorine or bromine at high temperature. At 300°C 3-halo- and 3,5-dihalopyridines are formed as a result of an ionic S_EAr process. Bromination with Br_2 in presence of conc. H_2SO_4 gives 3-bromopyridine. Chlorination with Cl_2 in presence of $AlCl_3$ gives 3-chloropyridine.

Pyridine does not undergo Friedel-Crafts alkylation and acylation. These reactions require Lewis acids which form complex with nitrogen and makes it completely unreactive towards electrophiles. Pyridine thus acts as a good ligand for metals such as Al(III) or Sn(IV).

Stable Complex

Nucleophilic substitution reactions

Reactions of pyridine with nucleophiles take place preferentially at the 2-, 4- and 6- positions. Attack of nucleophile at 2 or 4 position gives intermediates with negative charge on more electronegative nitrogen atom, while attack at 3-position does not form any such intermediate.

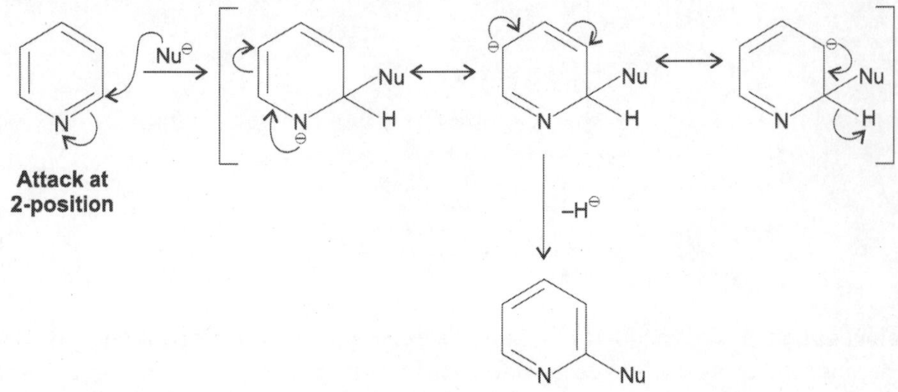

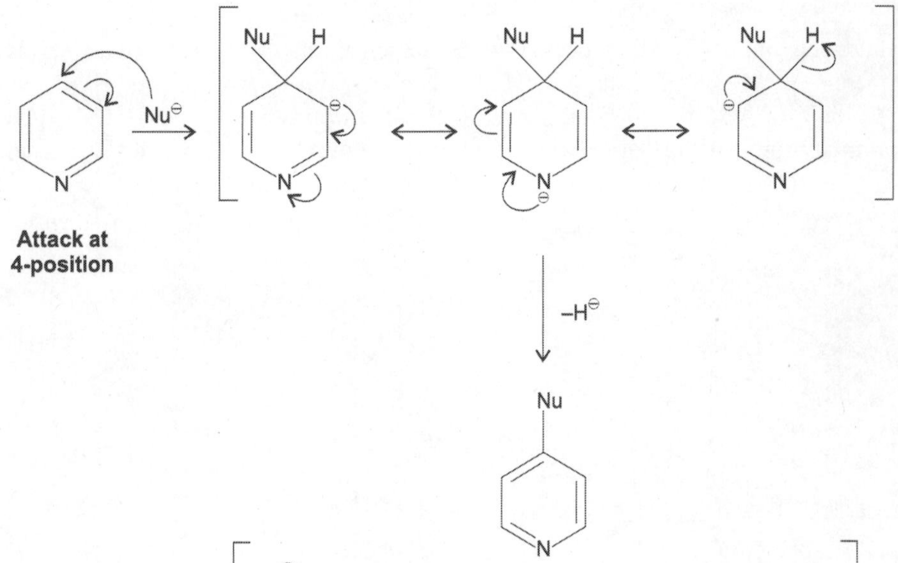

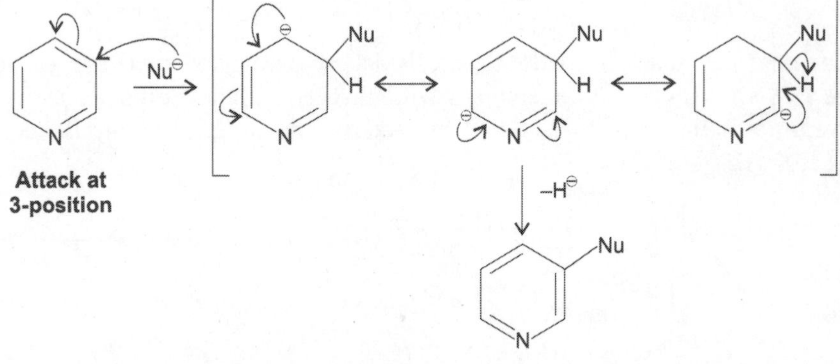

Attack at 2-position

Attack at 4-position

Attack at 3-position

Chichibabin reaction: The Chichibabin reaction is the first S_NAr reaction known for pyridine. It involves reaction with sodium amide (in toluene or dimethylaniline) and produces 2-aminopyridine regioselectively.

Mechanism: The mechanism of the Chichibabin reaction involves loss of hydrogen at 2-position as a hydride ion. The reaction starts with coordination of the pyridine to the $NaNH_2$ surface. The regioselectivity is controlled by Na coordination in the addition complex and formation of the intermediate amide.

Addition complex (Meisenheimer σ adduct)

Alkylation or arylation with PhLi or BuLi also occurs at 2-position of pyridine.

When a good leaving group is already present at 2-position or at 4-position of pyridine ring it can be displaced with a nucleophile.

The first step is the attack of nucleophile on the carbon bearing the leaving group, known as ipso attack. The intermediate anion loses the best leaving group to regenerate the stable aromatic system. Thus a leaving group at the 2- or 4-position is displaced through an addition-elimination process.

X = Halogen

Nu = $\overset{\ominus}{N}H_2$, $\overset{\ominus}{N}H_3$, RNH_2, $\overset{\ominus}{O}H$, $R\overset{\ominus}{O}$, $R\overset{\ominus}{S}$, RLi, $Al\overset{\ominus}{H}_4$

The nitrogen, oxygen, sulfur, carbon and hydride nucleophiles (such as NH_2^-, NH_3, RNH_2, OH^-, RO^-, RS^-, RLi, AlH_4^-) attack the ring carbon atoms of pyridine. Addition of the nucleophile and elimination of a pyridine substituent as leaving group occur in a two-step process.

However, if the leaving group is present at the 3-position of pyridine ipso attack of nucleophile does not place. Because in this case the negative charge is not placed on nitrogen atom in the intermediate anion.

In the case of 3-halopyridines, the nucleophilic substitution takes place by an aryne mechanism. For instance, reaction of 3-chloropyridine with KNH_2 in liquid NH_3 yields a mixture of 3- and 4-aminopyridines. This clearly indicates that reaction proceeds *via* a 3,4-dehydropyridine (hetaryne) intermediate. The 3,4-pyridyne intermediate can be treated with furan to yield a cycloadduct.

3-Chloropyridine   **Hetaryne**   **Cycloadduct**

Reaction of 4-substituted-3-halopyridines with butyl lithium results into the formation of 2,3-pyridyne which on treatment with furan forms cycloadduct.

2,3-Pyridyne   **Cycloadduct**
                          5,8-Dihydroquinoline endoxide

Reduction

Reduction of pyridine with Na and NH_3 gives 1,4-dihydropyridine. Hydrogenation in presence of Pt catalyst in acetic acid gives piperidine

Piperidine   **1,4-Dihydropyridine**

9.2 PYRIDINE N-OXIDE

9.2.1 Introduction

Pyridine N-oxide (PNO) was first described by Meisenheimer in 1926. Pyridine N-oxides are useful synthetic intermediates, protecting groups, auxiliary agents, oxidants, ligands in metal complexes and catalysts. Pyridine N-oxide has much higher dipole-moment ($\mu = 4.25$ D) than pyridine ($\mu = 2.22$ D). Pyridine N-oxide is much weaker base (pK$_a$ of its conjugate acid is 0.79) than pyridine (pK$_a$ of its conjugate acid 5.2).

9.2.2 Methods of preparation of pyridine N-oxide

Pyridine is oxidized to pyridine N-oxide on reaction with sulfuric acid/hydrogen peroxide (Caro's acid) or peracids. Pyridine can be obtained back from pyridine N-oxide on treatment with zinc in dilute acids.

9.2.3 Reactions of pyridine N-oxide

The N-O moiety of pyridine N-oxides can act as a push electron donor and as a pull electron acceptor group. The pyridine N-oxide is reactive towards both electrophiles and nucleophiles. However, electrophilic substitution occurs more readily in pyridine N-oxides than in pyridine. The resonating structures of the pyridine N-oxide have both positive and negative charges at 2- and 4-position. Thus, both electrophile and nucleophile can attack at 2- or 4-positions. Thus, Pyridine N-oxide is more nucleophilic and more electrophilic than pyridine.

Pyridine N-oxide is especially active in nucleophilic aromatic substitution because the attack of nucleophile at 2- or 4- positions annihilates the positive charge on nitrogen.

For example, the reaction of pyridine N-oxide with PCl_3 or $SOCl_2$ gives corresponding chloropyridine. Pyridine N-oxide reacts with $SOCl_2$ or PCl_3 through oxygen and the chloride ion released in this reaction attacks at 2- or 4- position. The loss of good leaving group and proton regenerates the aromaticity.

Nitration of pyridine N-oxide takes place in presence of fuming HNO_3 and conc. H_2SO_4 at 100°C.

The N-oxides easily gets alkylated at the oxygen atom and the entire N-substituent can be removed with base.

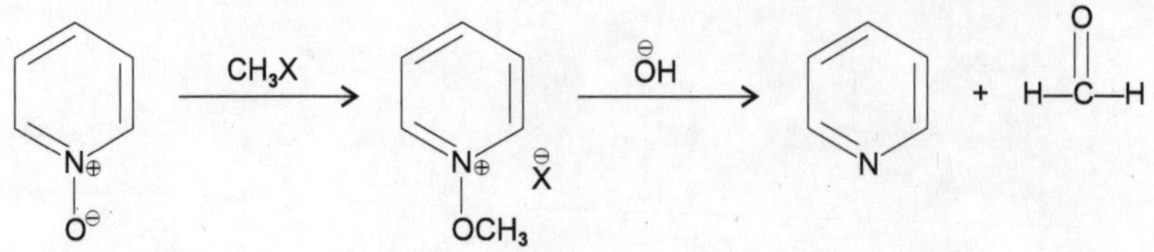

Pyridine N-oxide on reaction with acetic anhydride results into the formation of 2-pyridone.

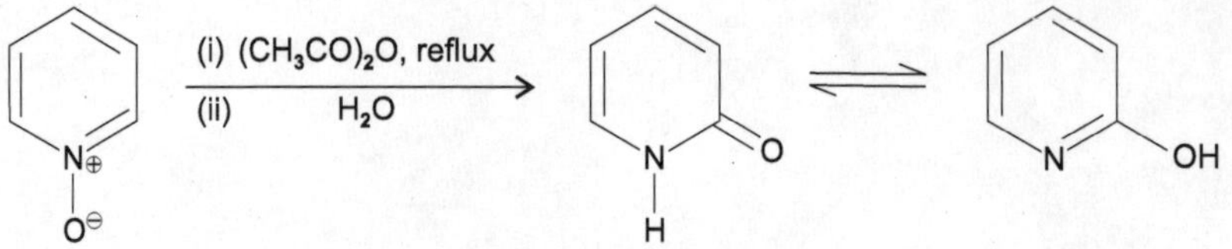

Reduction of pyridine N-oxides with ammonium formate/palladium on carbon in methanolic solution overnight at room temperature results into the formation of piperidine.

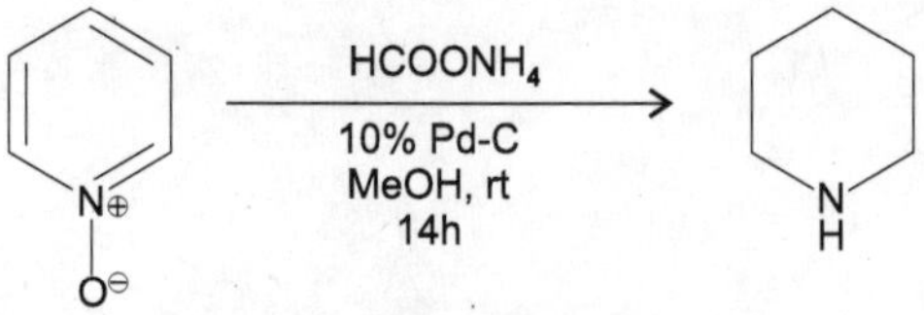

9.3 PYRYLIUM AND THIOPYRYLIUM SALTS

9.3.1 Introduction

Both pyrylium and thiopyrylium salts have 6π-electrons. The aromaticity of pyrylium salts is decreased by the strong electronegativity of the oxygen heteroatom. The resonance energy is appreciably lower than that of benzene, but the aromatic character allows pyrylium salts to be stable in aqueous media at pH $\leq$ 7. The chemical behaviour of pyrylium salts differs from benzene or pyridine. The thiopyrylium salts are less reactive than pyrylium salts due to the lower electronegativity of the sulfur atom.

The resonating structures of pyrylium and thiopyrylium are shown below.

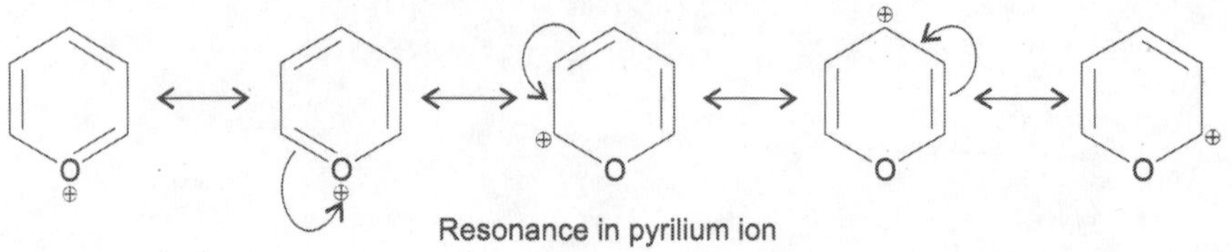

Resonance in pyrilium ion

Resonance in thiopyrilium ion

9.3.2 Methods of preparation of pyrylium and thiopyrylium salts

(*i*) **Pyrylium salts from alkenes:** Alkenes with more than three carbon atoms undergo diacylation in the presence of strong acids to form pyrylium salts with same substituents at both the α-positions. This method was discovered independently by Balaban and Nenitzescu and by Praill. Acylation is carried out with acetyl chloride in presence of Lewis acids (FeCl$_3$, AlCl$_3$, SnCl$_4$, ZnCl$_2$) or anhydride in presence of Bronsted acids (HClO$_4$, HBF$_4$, HPF$_6$, CF$_3$COSO$_3$H).

(*ii*) **Thiopyrylium salts from 1,5-dicarbonyls:** Thiopyrylium salts can be prepared from 1,5-dicarbonyl compounds on reaction with phosphorous pentasulphide in dioxane as H$_2$PS$_2$O$_2$ salt.

9.3.3 Reactions of pyrylium and thiopyrylium salts

Pyrylium salts react with nucleophiles having hydrogen atom such as ammonia, primary amines, hydroxylamine, hydrazine derivatives, phosphine, hydrogen sulfide, nitromethane, acetonitrile) to form a variety of six-membered aromatic carbocyclic or heterocyclic compounds. In these reactions attack by nucleophile proceeds with ring opening followed by ring closure.

When 2,4,6-triphenyl thiopyrylium salt is treated with active methylene compounds in presence of potassium tert-butoxide in tert-butanol, the corresponding 1-substituted 2,4,6-triphenylbenzene derivative is obtained.

X = Y = CN
X = Y = COMe

The reactions of 2,4,6-triphenylthiopyrylium perchlorate with primary and secondary amines in Me_2SO-d_6 and/or CD_3CN leads to the complete disappearance of starting cation, and 2H-thiopyran is formed.

EXERCISE

Q.1 Discuss the basicity of pyridine in comparison to aliphatic and aromatic amines

Q.2 Write any two reactions of pyridine with nucleophiles.

Q.3 Discuss the mechanism of Hantzch pyridine synthesis

Q.4 Give any two methods for the synthesis of pyridine.

Q.5 Mention the name of two vitamins which contain pyridine nucleus.

Ans. Vitamin B3 and B6.

Q.6 Give an account on the aromaticity and basicity of pyridine.

Q.7 How is pyridine N-oxide prepared? Why does it undergoes electrophilic substitution reactions more readily than pyridine.

Ans. Pyridine N-oxide can be prepared by the oxidation of pyridine with hydrogen peroxide.

In the resonating structures of pyridine N-oxide there is negative charge at α-and γ-positions, hence it is more reactive towards electrophilic substitution reactions.

Q.8 Draw the resonating structures of thiopyrylium salt.

Q.9 Convert pyridine to:

(a) 2-Amino pyridine

(b) 3-Nitropyridine

(c) 4-Amino pyridine

(d) Pyridine-N-oxide

Q.10 How will you obtain pyridine from pyrrole?

Q.11 Explain why:-

(a) Pyridine is a weaker base than piperidine. (Refer sec. 9.1.5)

(b) Pyridine is a much stronger base than pyrrole. (Refer sec. 9.1.5)

(c) Pyridine undergoes nucleophilic substitution at 2-position

(d) Electrophilic substitution in 2-aminopyridine is much easier than pyridine

(e) Pyridine undergoes electrophilic substitution only under vigrous conditions

(f) pK_a of conjugate acid of piperidine ($pK_a = 11.2$) is more than that of morpholine ($pK_a = 9.28$).

(g) 2- and 4-Chloropyridines react with nucleophiles such as RNH_2 but 3-chloropyridine does not.

(h) Pyridine acts like a deactivated benzene ring and electrophilic substitution occurs at 3-position.

Q.12 Write the product of reaction of :-

(a) Pyridine-1-oxide with acetic anhydride

(b) Pyridine with SO_2Cl_2

Q.13 Complete the following reactions:

(a) Pyridine + $NaNH_2$ + NH_3 + heat ———→

(b) Pyridine + CH_3I ———→

Q.14 Arrange the following compounds in decreasing order of their basicity giving reason.

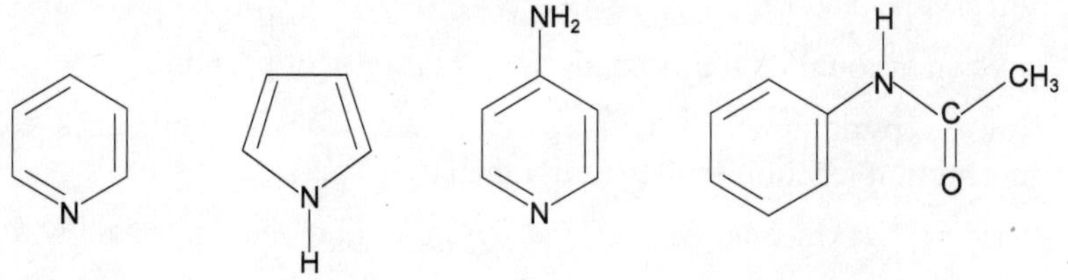

Pyridine   Pyrrole   Aniline   Acetanilide

Ans. Pyridine (pK_a = 5.2) > aniline (pK_a = 4.6) > acetanilide (pK_a = 0.61) > pyrrole (pK_a = –3.8). (pK_a are for conjugate acids.)

Pyridine and aniline are basic, as both have lone pair of electron on nitrogen which may involve in protonation. Pyridine is more basic than aniline because the lone pair of pyridine nitrogen are not involved in delocalization, whereas in aniline the lone pair on nitrogen are delocalized over the ring. In case of acetanilide the lone pair electron take part in resonance with the carbonyl group, thus not free. But, in pyrrole the nitrogen lone pair electron are completely delocalized and involved in the aromaticity of the ring, thus it is least basic.

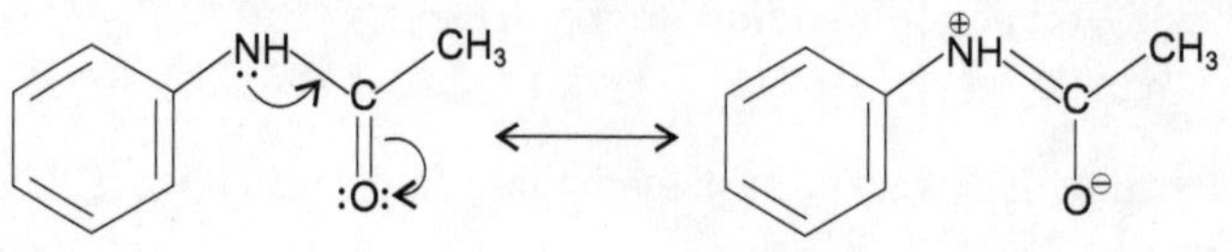

Q.15 Under what conditions pyridine can be converted into methyl pyridinium iodide.

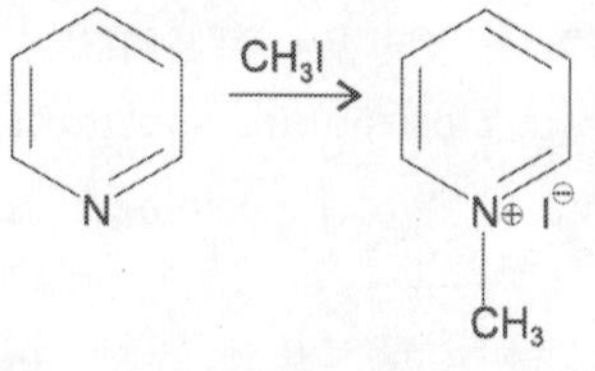

Ans. NO_2BF_4 or CH_3COCl or SO_3

Q.17 Write the product formed in the reactions given below:

(a) [pyridine] $\xrightarrow[\text{SO}_2]{\text{N}_2\text{O}_5}$

(b) [pyridine] $\xrightarrow[\substack{\text{HgSO}_4 \\ 250°\text{C}}]{\text{Oleum}}$

(c) [pyridine] $\xrightarrow{\text{AlCl}_3}$

(d) [pyridine] $\xrightarrow[70\text{-}100°\text{C}]{\text{H}_2\text{SO}_4/\text{H}_2\text{O}_2}$

Ans. (a) [3-nitropyridine with NO$_2$ substituent]

(b) [pyridine with SO$_3$H substituent]

(c) [pyridine with N$^{\oplus}$ and $^{\ominus}$AlCl$_3$]

(d) [pyridine N-oxide with N$^{\oplus}$ and O$^{\ominus}$]

Bicyclic and Tricyclic Ring Systems Derived from Pyridine

Quinoline, Isoquinoline and Acridine

10.1 QUINOLINE AND ISOQUINOLINE

10.1.1 Introduction

There are three different types of benzofused pyridines namely **benzo[b]pyridine** (or **quinoline**), **benzo[c]pyridine** (or **isoquinoline**) and **benzo[a]pyridinium salt** (or **quinolizinium salt**).

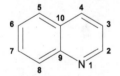

Benzo[b]pyridine or
Quinoline or
1-Azanapthalene

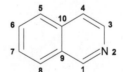

Benzo[c]pyridine or
Isoquinoline or
2-Azanapthalene

Benzo[a]pyridinium salt
Quinolizinium salt

Quinoline is a high boiling liquid (mp –15°C, bp 237-238°C) while isoquinoline is a low melting solid (mp 26-28°C, bp 242-243°C). Quinoline and isoquinoline were first isolated from coal tar in 1834 and 1885, respectively. Quinoline is colourless hygroscopic liquid but when exposed to light becomes yellow and later brown. Isoquinoline is also colourless, however impure samples can appear brownish.

10.1.2 Significance of quinoline and isoquinoline

Quinoline system is found in anti-malarial drugs such as **quinine**, a natural product isolated from the bark of the *Cinchona* tree and **chloroquine**, a completely synthetic compound. **Quinaldine** (2-methylquinoline) is also anti-malarial and used to prepare other drugs.

Quinine

Chloroquine

**2-Methylquinoline
(Quinaldine)**

Isoquinoline system is found in **emetine**, which is the principle alkaloid of ipecac (ground roots of *Uragoga ipecacuanha*). It is a drug used as an anti-protozoal (amoebicide) and to induce vomiting. **Lycorine** present in bulbs of *Narcissus* and *Amaryllis* also contains isoquinoline system. **Papaverine,** an alkaloid isolated from the opium poppy, contains isoquinoline ring system. It is a smooth muscle relaxant and a coronary vasodilator.

Lycorine

Emetine

Papaverine

10.1.3 Structure

Both quinoline and isoquinoline are planar 10 π-electron aromatic systems. All atoms of quinoline and isoquinoline are sp^2 hybridized and contribute one electron each in their orthogonal p-orbitals for delocalization over the rings. The resonance energies of quinoline and isoquinoline are 198 and 143 kJ/mol, respectively. Both are more aromatic as compared to pyridine (resonance energy 117 kJ/mol or 28 kcal/mol). The resonating structures of quinoline and isoquinoline are given below.

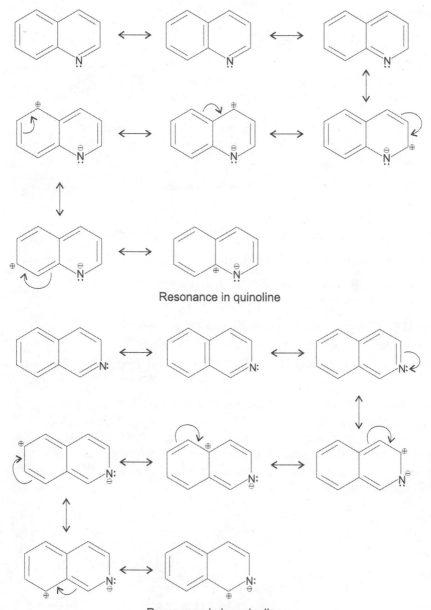

Resonance in quinoline

Resonance in isoquinoline

10.1.4 Dipole moment

The dipole moment of isoquinoline (2.60 D) is more than quinoline (2.10 D).

μ = 2.10 D   μ = 2.60 D

10.1.5 Spectral data

The ¹H and ¹³C NMR (given in brackets) of quinoline and isoquinoline are shown below:

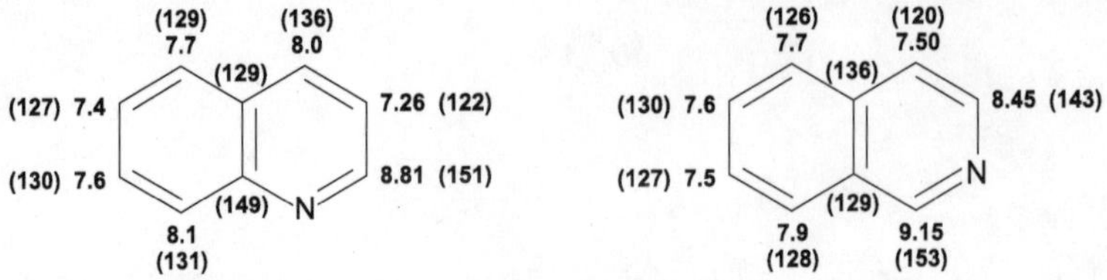

Quinoline   Isoquinoline

¹H NMR (δ, ppm, CDCl₃)

¹³C NMR (δ, ppm, CDCl₃)

10.1.6 Methods of preparation of quinoline and isoquinoline

10.1.6.1 Methods of preparation of quinoline

Quinoline can be obtained by the distillation of coal tar. Other methods for the synthesis of quinolines are as follows.

(*i*) **Skraup synthesis:** Skraup described the synthesis of quinoline in 1880 by heating a mixture of aniline, glycerol and nitroethane with concentrated sulfuric acid.

Skraup, Z.H., Monatsh. Chem. 1880, 1, 316-318.; Skraup, Z.H., Monatsh.Chem. 1881, 2,139-170.; Skraup, Z.H., Monatsh. Chem. 1881, 2, 587-609.; Skraup, Z.H., Ber. Dtsh. Chem. Ges. 1882, 15, 897.

It has been reported that initial Skraup protocol usually gives a very low yield of quinolines. Therefore, extensive modifications have been made based on the initial protocol. Skraup reaction can be carried out in protic acid or in the presence of a Lewis acid. In the typical **Skraup reaction**, aniline is heated with glycerol, sulfuric acid and an oxidizing agent such as nitrobenzene to yield quinoline. Here, nitrobenzene serves both as a solvent and an oxidizing agent.

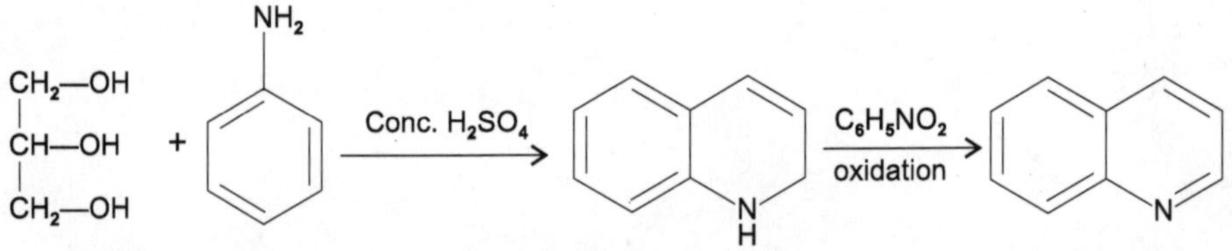

Glycerol   Aniline   1,2-Dihydroquinoline   Quinoline

Mechanism: The mechanism involves dehydration of glycerol *in situ* to give acrolein. The acrolein undergoes conjugate addition with aniline. Subsequent tautomerism, acid catalyzed cyclization and dehydration leads to the formation of 1,2-dihydroquinoline. Which is oxidized *in situ* using nitrobenzene as co-solvent or by using an oxidant such as iodine or an iron(III) salt to quinoline.

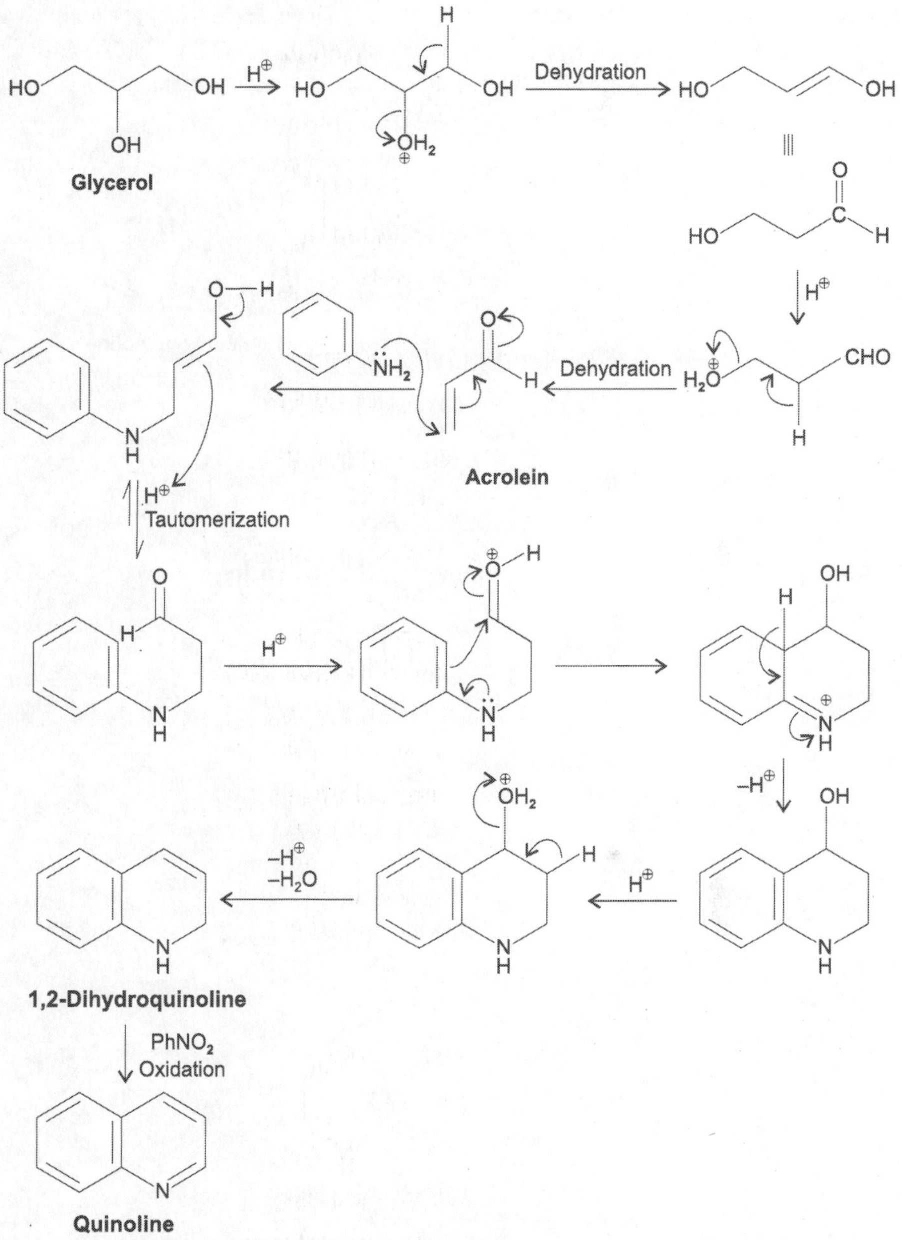

Glycerol

Acrolein

1,2-Dihydroquinoline

Quinoline

Later on several modifications have been reported to improve the yield and reproducibility of the Skraup quinoline synthesis. Various moderators such as acetic acid, boric acid, ferrous sulfate, thorium oxide, vanadium oxide, iron oxide used to accelerate the reaction and make it higher yielding.

(*ii*) **Doebner-von Miller synthesis:** The formation of quinoline nucleus by heating primary aromatic amine with α,β-unsaturated carbonyl compound (instead of glycerol used in Skraup synthesis) in the presence of conc. hydrochloric acid or sulfuric acid or iodine is called **Doebner-von Miller synthesis**. The Skraup-Doebner-von Miller synthesis of quinolines can also be catalyzed by a number of Lewis ($SnCl_4$, $Yb(OTf)_3$, $Sc(OTf)_3$, $ZnCl_2$, $InCl_3$) and Bronsted acids (TsOH, $HClO_4$, Amberlite) in addition to iodine.

Doebner, O.; Miller, W.V., Ber. Dtsh. Chem. Ges. 1881, 14, 2812-2817.

4-Methylquinoline

Mechanism: The mechanism is quite similar to **Skraup synthesis**.

(*iii*) Knorr quinoline synthesis: The **Knorr quinoline synthesis** (first described by Ludwig Knorr in 1886) is an intramolecular organic reaction converting a β-ketoanilide to 2-hydroxyquinoline using sulfuric acid. The β-ketoanilide can be prepared by reaction of aniline with β-ketoester at high temperature.

β-Ketoester

β-Ketoanilide   2-Hydroxyquinoline

Mechanism: In the Knorr quinoline synthesis the β-ketoanilide undergoes cyclization followed by dehydration to give quinoline derivative.

(iv) Friedlander synthesis: Amongst various methodologies reported for the preparation of quinolines, the Friedlander reaction is still one of the simplest and most straightforward protocols. **Friedlander synthesis** of substituted quinolines involves cyclodehydration of ortho-amino substituted aromatic aldehydes or ketones with a ketone possessing α-methylene group. Quinoline itself was synthesized by Friedlander in 1882 by the condensation of ortho-amino benzaldehyde with acetaldehyde in the presence of sodium hydroxide. While different catalysts have been proposed for Friedlander annulations, it has been shown that acidic catalysts are superior to basic ones. Other than different acidic catalysts, such as Bronsted acids and Lewis acids, ionic liquids and other catalysts have also been employed to promote this reaction. Some quinolines can also be prepared simply by heating a mixture of the reactants with or without solvent.

Mechanism: The first step of the Friedlander quinoline synthesis is a mixed aldol condensation. This is followed by base- or acid-catalyzed cyclocondensation to produce a quinoline derivative. With this method, structurally varied substrates can give the corresponding quinoline products in moderate to high yields.

*Friedländer, P., Chem. Ber. **1882**, 15, 2572; Marco-Contelles, J.; Perez-Mayoral, E.; Samadi, A.; do Carmo Carreiras, M.; Soriano, E., Chem. Rev. **2009**, 109, 2652.*

(*v*) **From indole:** Indole on reaction with methyllithium in dichloromethane solution leads to the formation of quinoline. The reaction proceeds *via* formation of carbene and its addition to the double bond to form a cyclopropyl intermediate. It undergoes rearrangment to generate quinoline.

(*vi*) **Combes synthesis:** The synthesis of quinolines by the condensation of primary aromatic amines with acetoacetone or other β-diketones followed by cyclization in the presence of sulfuric acid is referred to as **Combes quinoline synthesis**. This method provides a rapid access to the 2,4-substituted quinoline skeleton. The polyphosphoric acid can also be used as catalyst for this cyclization.

*Combes, A. Bull. Soc. Chim. France **1888**, 49, 89.*

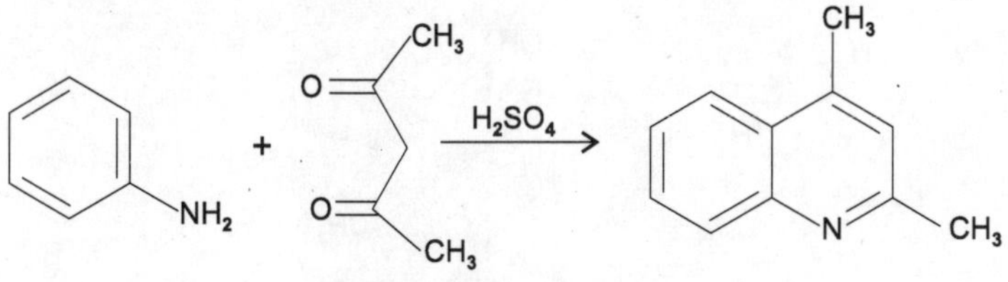

**Acetylacetone
(Acetoacetone)**

Mechanism: The Combes synthesis involves nucleophilic addition of aniline to the carbonyl group of the 1,3-diketone to form enamine intermediate followed by electrophilic aromatic annulations to yield 2,4-dimethylquinoline.

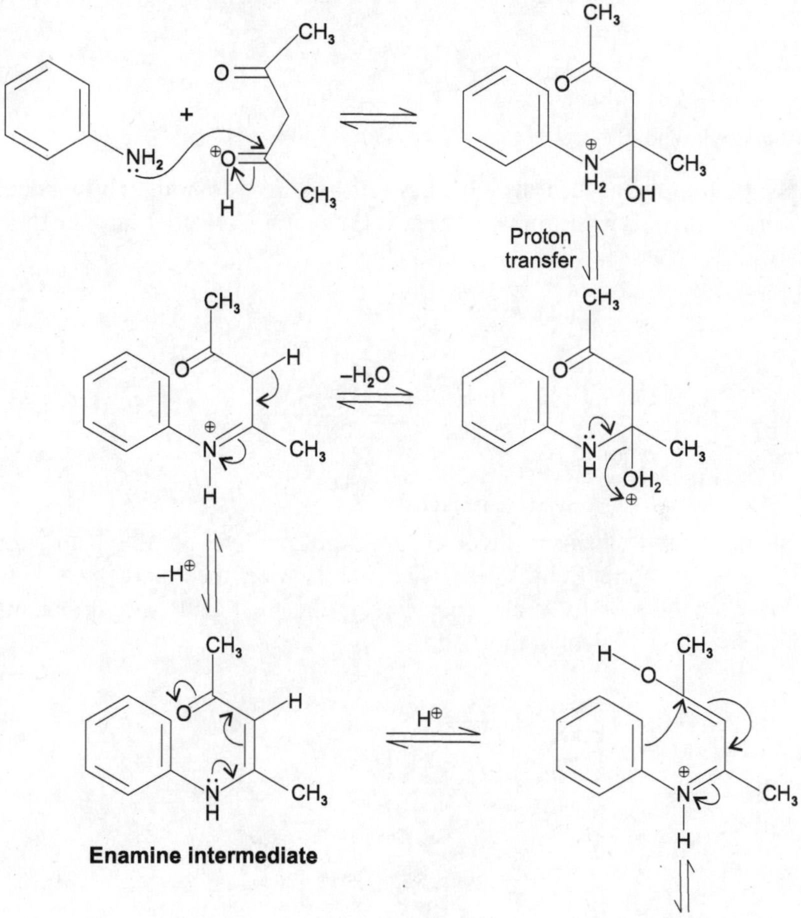

Proton transfer

−H_2O

−$H^\oplus$

Enamine intermediate

$H^\oplus$

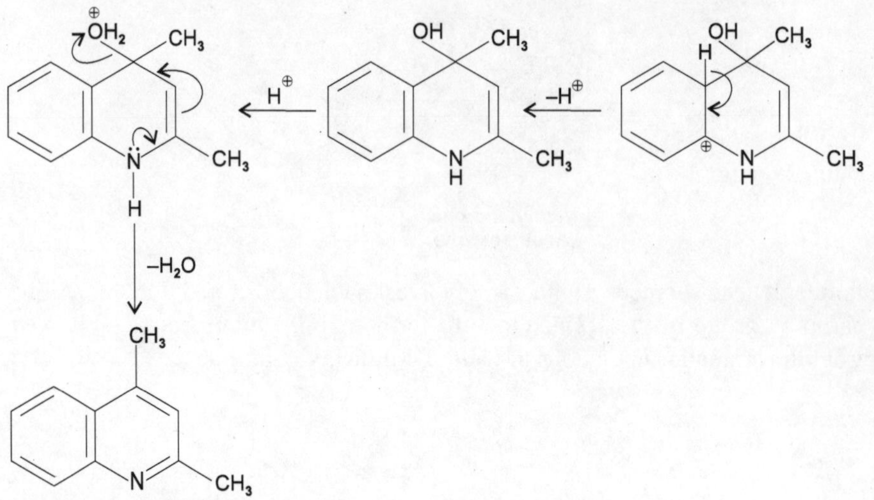

2,4-Dimethylquinoline

(*vii*) **Conrad-Limpach synthesis**: Synthesis of 4-hydroxyquinolines by condensation of esters of β-keto acids with aromatic amines is known as **Conrad-Limpach quinoline synthesis**.

Conrad, M.; Limpach L., Ber. 1887, 20, 944.

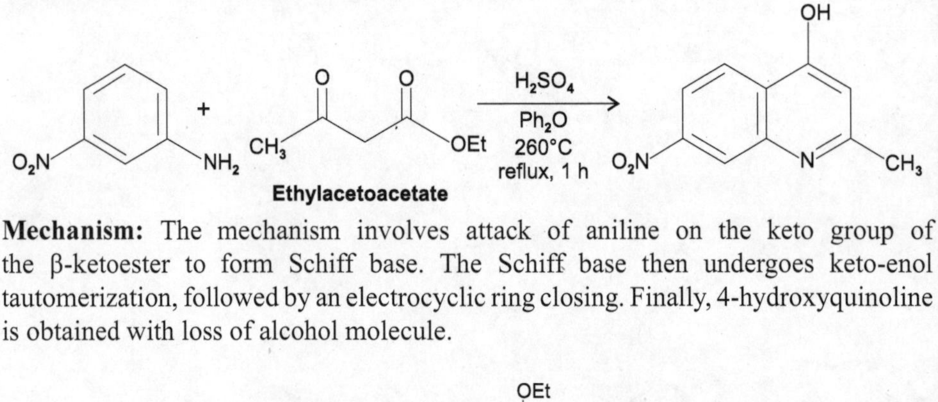

Ethylacetoacetate

Mechanism: The mechanism involves attack of aniline on the keto group of the β-ketoester to form Schiff base. The Schiff base then undergoes keto-enol tautomerization, followed by an electrocyclic ring closing. Finally, 4-hydroxyquinoline is obtained with loss of alcohol molecule.

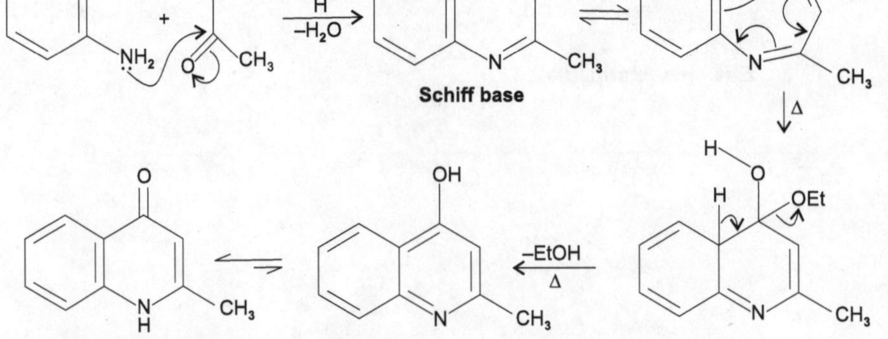

2-Methyl-4-quinolone

Synthesis of chloroquine:
Chloroquine can be synthesized using Conrad-Limpach method as shown below:

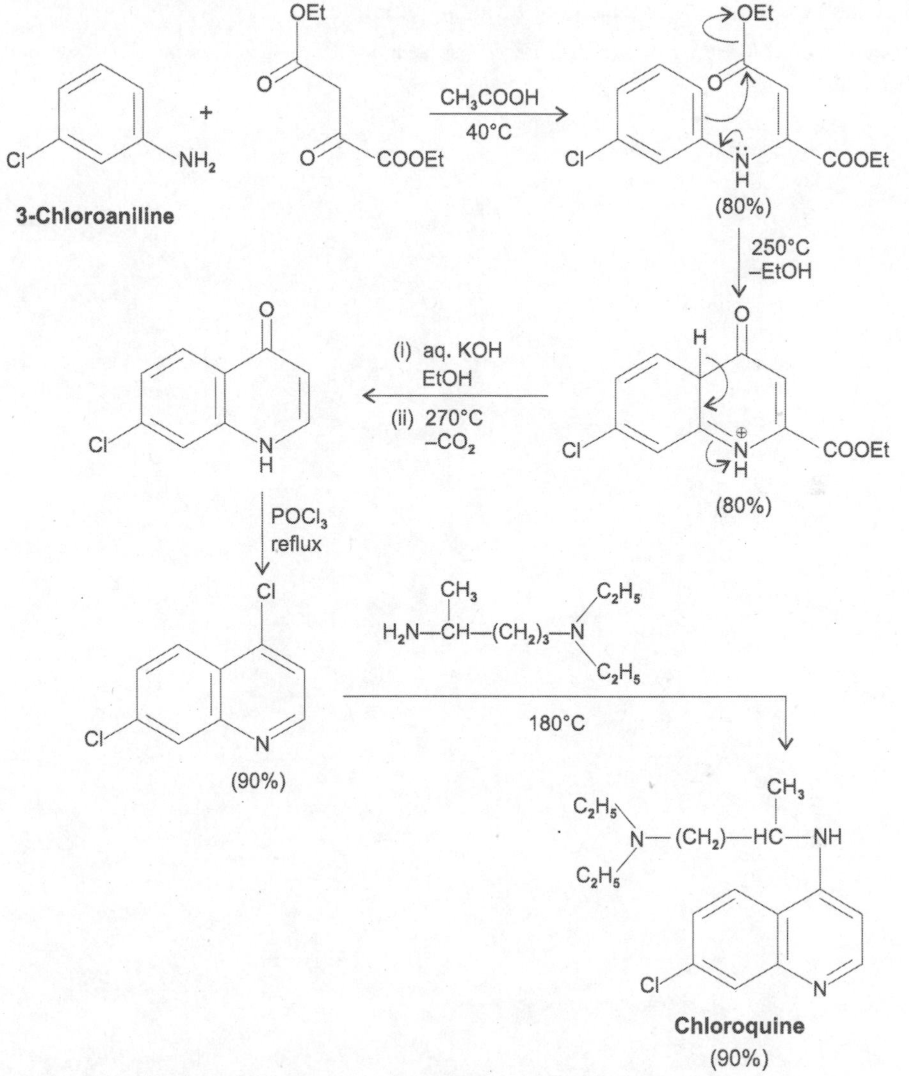

3-Chloroaniline

(80%)

250°C
–EtOH

(i) aq. KOH
EtOH

(ii) 270°C
–CO₂

(80%)

POCl₃
reflux

$$H_2N—CH—(CH_2)_3—N\begin{matrix}C_2H_5\\C_2H_5\end{matrix}$$

with CH₃

180°C

(90%)

Chloroquine
(90%)

10.1.6.2 Methods of preparation of isoquinoline

(*i*) **Pomeranz-Fritsch synthesis:** In the **Pomeranz-Fritsch synthesis** of isoquinolines aromatic aldehyde or ketone is treated with amino acetaldehyde diethylacetal in acidic medium.

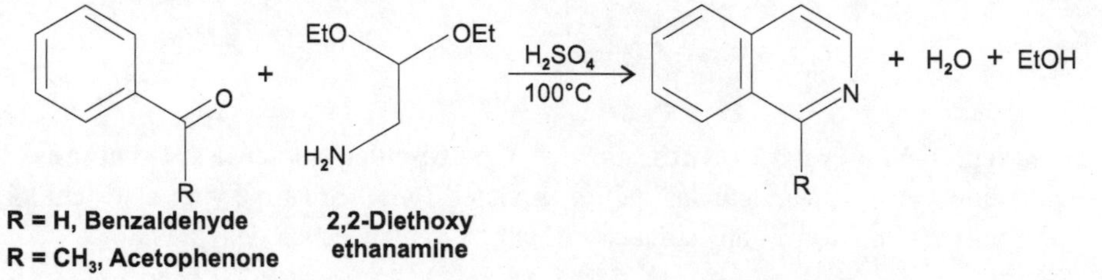

R = H, Benzaldehyde
R = CH₃, Acetophenone

2,2-Diethoxy
ethanamine

Mechanism: The mechanism of this reaction involves the reaction of aromatic aldehyde or ketone with aminoacetal (or 2,2-diethoxyethanamine) to generate imine that can be cyclized with an acid to form isoquinoline.

*Pomeranz, C., Monatsh **1893**, 14 116; Fritsch, P., Ber. **1893**, 26, 419; Gensler, W.J., Org. React.*
***1951**, 6, 191-206.*

(*ii*) **Bischler-Napieralski synthesis**: In the **Bischler-Napieralski synthesis** of isoquinoline   the   2-arylethylamine is acylated with a carboxylic acid chloride or   anhydride   to   form   an   amide,   which   is   then   cyclodehydrated   to   form

3,4-dihydroisoquinoline. The cyclization takes place in the presence of phosphorus pentoxide, phosphoryl chloride or phosphorus pentachloride. Dehydrogenation of 3,4-dihydroisoquinoline using palladium, sulfur or diphenyl disulfide leads to isoquinoline.

Synthesis of papaverine: Papaverine is an opium alkaloid found in the opium poppy. It is a smooth muscle relaxant. It can be synthesized using the **Pictet-Gams** modification of Bischler-Napieralski method.

Papaverine

Whaley, W.M.; Govindachari, T.R., Org. React. **1951**, *6, 74.*

(*iii*) **Pictet-Spengler synthesis:** The β-arylethylamine on reaction with an aldehyde in the presence of acid forms an imine which on protonation and cyclization yields 1,2,3,4-tetrahydroisoquinoline. Dehydrogenation of which in the presence of Pd/C yields isoquinoline.

Pictet, A; Spengler, T., Ber. **1951**, *44, 2030; Whaley, W.M.; Govindachari, T.R., Org. React.* **1951**, *6, 151.*

10.1.7 Reactions of quinoline and isoquinoline

Basicity

Both quinoline and isoquinoline are weak bases. Their basicity is similar to pyridine. Quinoline is slightly less basic than pyridine. The pK_a values of conjugate acids of aniline, pyridine, quinoline and isoquinoline are given in Table 10.1.

Table 10.1

Compound	Structure	Hybridization of nitrogen	pK_a of conjugate acid
Aniline	![NH₂ structure]	sp^3	4.6
Pyridine	![pyridine structure]	sp^2	5.2
Quinoline	![quinoline structure]	sp^2	4.9
Isoquinoline	![isoquinoline structure]	sp^2	5.4

Though the nitrogen atom of aniline is sp^3 hybridized it is least basic amongst these amines because the lone pair of electron present on nitrogen is involved in delocalization towards the ring.

Reactions with electrophiles at ring nitrogen

Quinolines form corresponding salts on reaction with HX, RX and RCOX (X= Cl, Br, I).

Electrophilic substitution reactions

The nitrogen of the quinoline and isoquinoline has deactivating effect on the ring towards electrophilic substitution therefore, attack of electophile occurs at the benzo-rather than hetero-ring. However, reactions are faster than those of pyridine but slower than those of naphthalene. Under strongly acidic conditions, reaction occurs *via* the formation of ammonium salt. The electrophilic attack occurs preferentially in the 5- and 8-positions to give mixture of two products in equal amounts. The reactions of isoquinoline closely parallel those of quinoline and pyridine.

The order of reactivity towards electrophiles is as follows:
benzene >> isoquinoline ~ quinoline > pyridine, with relative rates 1, ~10^{-5}, ~10^{-6} and ~10^{-12}, respectively.

Nitration: Nitration of quinoline with HNO_3/H_2SO_4 at 0°C affords equal amounts of both 5- and 8-nitroquinolines. Whereas, isoquinoline exclusively yields 5-nitroisoquinoline under similar conditions.

5-Nitroquinoline   **8-Nitroquinoline**
(50%)   (42%)

5-Nitroisoquinoline
(72%)

Nitration can be carried out in the hetero-ring using nitric acid and acetic anhydride at 3-position in case of quinoline and at 4-position in the case of isoquinoline.

3-Nitroquinoline

4-Nitroisoquinoline

Halogenation: Halogenation is also possible in quinoline and isoquinoline but products formed depend upon the conditions used.

Bromination of quinoline and isoquinoline in the presence of strong protic acids or $AlCl_3$ leads to the 5-bromo compound. In concentrated sulfuric acid, bromination of quinoline gives a mixture of 5- and 8-bromo derivatives.

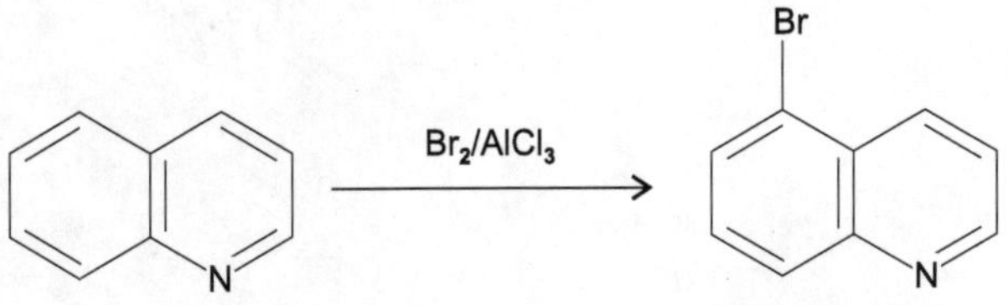

5-Bromoquinoline

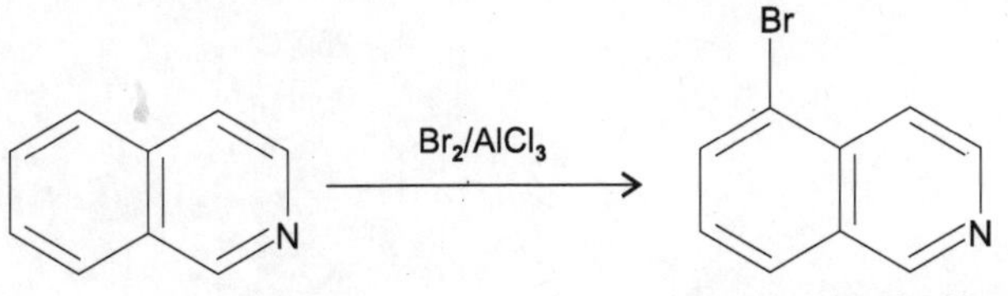

5-Bromoisoquinoline

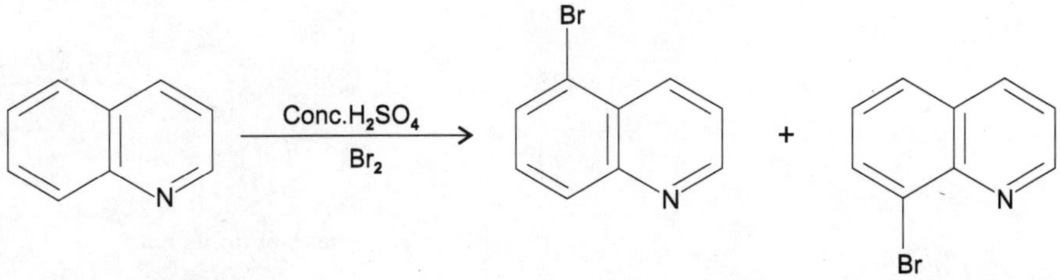

5-Bromoquinoline   **8-Bromoquinoline**

It is also possible to introduce halogen into the hetero ring under mild conditions. Here, the addition of halide to a salt initiates the sequence. The quinoline or isoquinoline hydrochlorides are treated with bromine to give 3-bromoquinoline and 4-bromoisoquinoline, respectively.

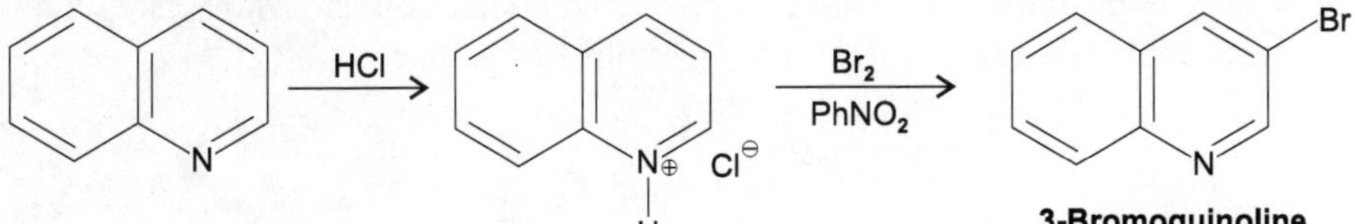

3-Bromoquinoline

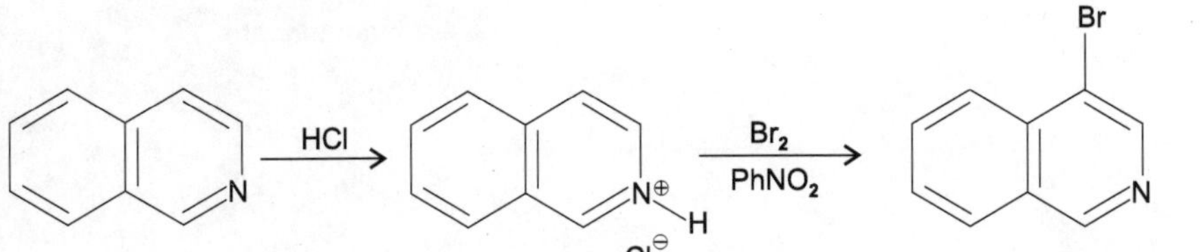

4-Bromoisoquinoline

Sulfonation: Sulfonation with H_2SO_4/SO_3 yields largely 8-sulfonic acid and small amount of 5-sulfonic acid. At higher temperature i.e. under thermodynamic controlled conditions both quinoline-8-sulfonic acid and quinoline-5-sulfonic acid isomerizes to the quinoline-6-sulfonic acid. Sulfonation of isoquinoline affords the isoquinoline-5-sulfonic acid.

Quinoline H_2SO_4, SO_3

(Major)   **(Minor)**

$> 250°C$

Quinoline-6-sulfonic acid

H_2SO_4, SO_3

Isoquinoline-5-sulfonic acid

Friedel-Crafts alkylation or acylation is not usually possible, with quinoline and isoquinoline.

Nucleophilic substitution

Isoquinoline undergoes nucleophilic aromatic substitution faster than quinoline. The reaction proceeds in a manner analogues to pyridine For example, quinoline undergoes **Chichibabin reaction** to give 2-aminoquinoline, while isoquinoline undergoes Chichibabin reaction to give 1-aminoisoquinoline.

$NaNH_2$, liq. NH_3

2-Aminoquinoline

$NaNH_2$, liq. NH_3

1-Aminoisoquinoline

Oxidation

Quinoline undergoes oxidative cleavage with alkaline potassium permanganate to give unstable pyridine-2,3-dicarboxylic acid that undergoes decarboxylation to give nicotinic acid. Similarly, isoquinoline on oxidation with alk. $KMnO_4$ gives pyridine-3,4-dicarboxylic acid and phthalic acid.

Pyridine-2,3-dicarboxylic acid   **Nicotinic acid**

Pyridine-3,4-dicarboxylic acid   **Phthalic acid**
(Cinchomeronic acid)

Quinoline and isoquinoline both form N-oxides when treated with hydrogen peroxide in acetic acid or with organic peracids.

Quinoline-N-Oxide

Isoquinoline-N-Oxide

Quinoline on reductive ozonolysis yields o-xylene.

o-xylene

Reduction

The pyridine ring in quinoline and isoquinoline is more easily reduced. Quinoline is converted to 1,2,3,4-tetrahydroquinoline by catalytic hydrogenation or with tin and hydrochloric acid. Quinoline can be selectively reduced at 1,2-bond by reaction with lithium aluminium hydride. However, 1,2-dihydroquinolines are unstable and disproportionate easily to give 1,2,3,4-tetrahydroquinoline and quinoline.

1,2,3,4-Tetrahydroquinoline

1,2-Dihydroquinoline   **1,2,3,4-Tetrahydroquinoline**   **Quinoline**

Isoquinoline can be converted to 1,2,3,4-tetrahydroisoquinoline with sodium/ethanol and to 1,2-dihydroisoquinoline by diethyl aluminium hydride or LiAlH$_4$, respectively.

1,2-Dihydroisoquinoline   **1,2,3,4-Tetrahydroisoquinoline**

10.2 ACRIDINE

10.2.1 Introduction

Acridine is an aza derivative of anthracene. It is also known as dibenzopyridine or 2,3,5,6-dibenzopyridine or 10-azaanthracene. Acridine is a known human carcinogen. It has lachrymatory effect. This effect is minimized when methyl substituents are present in the positions 4 and 5. It is colourless to light yellow solid (mp 110°C). Solutions of acridine salts give blue fluorescence.

Acridine

10.2.2 Significance of acridine derivatives

Acridine and its derivatives, such as **9-aminoacridines**, are widely used in medicine and biochemistry as drugs, dyes and labeling agents. The important alkaloids such as **melicopicine** contain acridine ring system. A number of therapeutic agents are based on acridine nucleus such as **quinacrine** (antimalarial), **acriflavine** (antiseptic), **proflavine** (antiseptic), **ethacridine** (abortifacient)**, amsacrine** (antineoplastic) and **tacrine** (anticholinesterase). Some amino acridines, particularly **acridine orange L** is used as dyestuff.

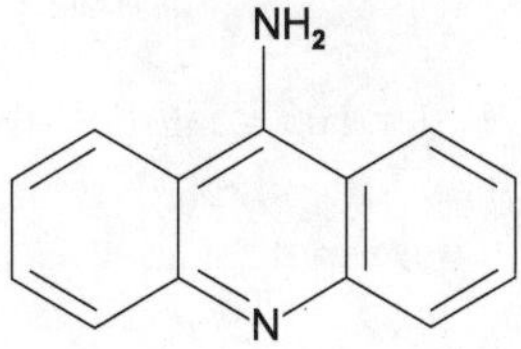

9-Aminoacridine

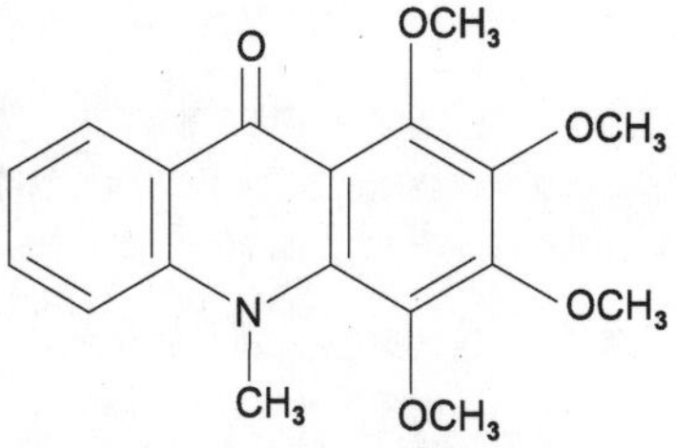

Melicopicine

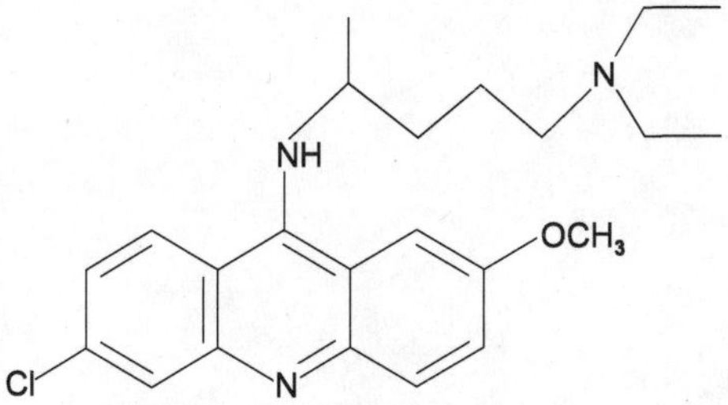

Quinacrine (Mepacrine)

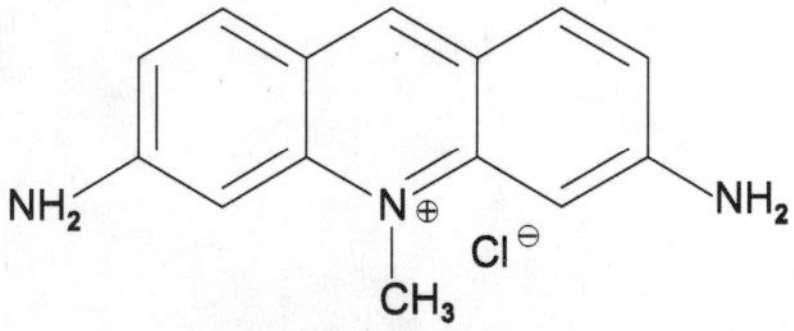

**Acriflavinium chloride
(Acriflavine)**

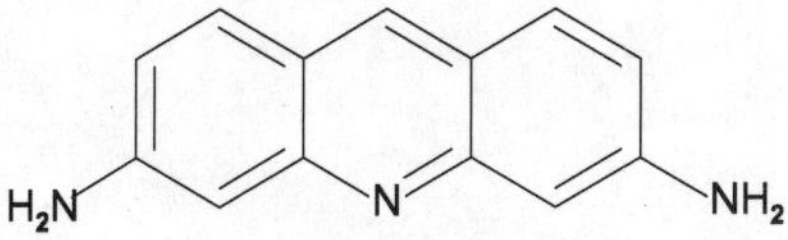

Proflavine

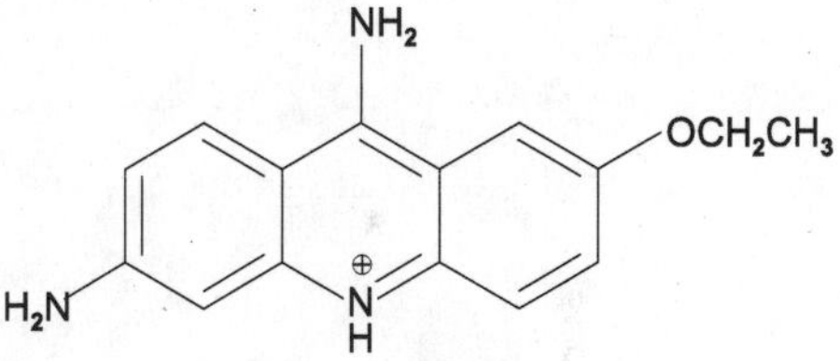

Ethacridine

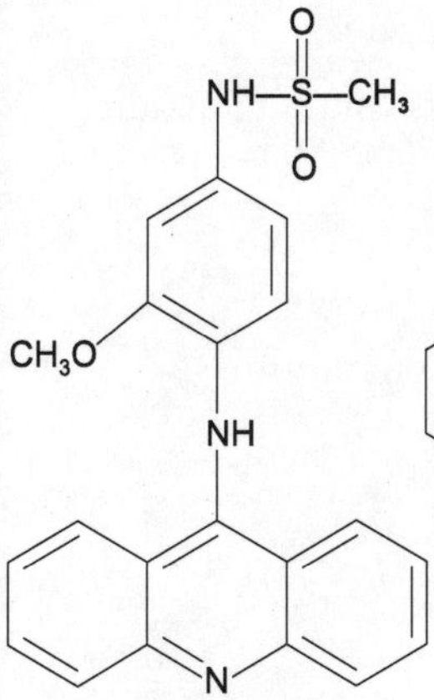

Amascrine

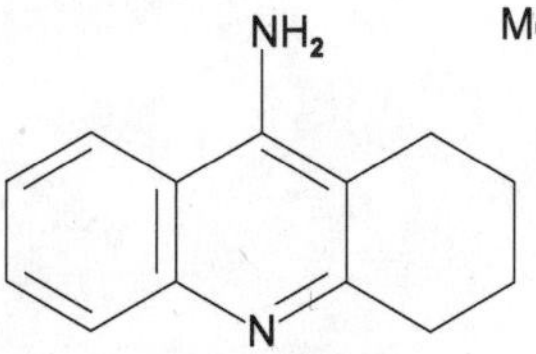

Tacrine

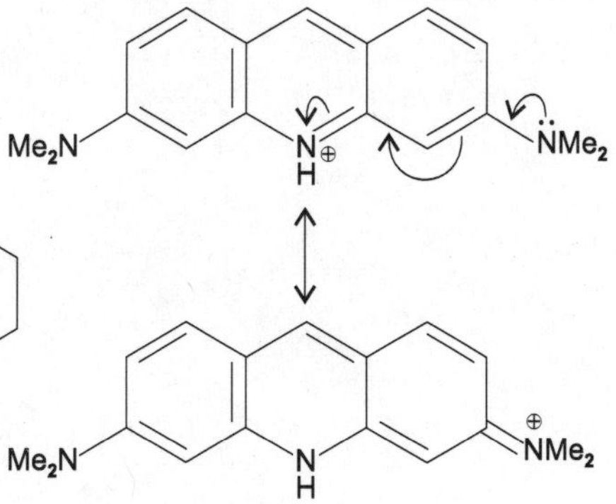

Acridine orange L

10.2.3 Structure

Acridine is highly aromatic with resonance energy of the order 105 kcal/mol. It is a resonance hybrid of all possible Kekule structures, as well as few others. One of the resonating structures with unpaired electrons participates in homolytic reactions.

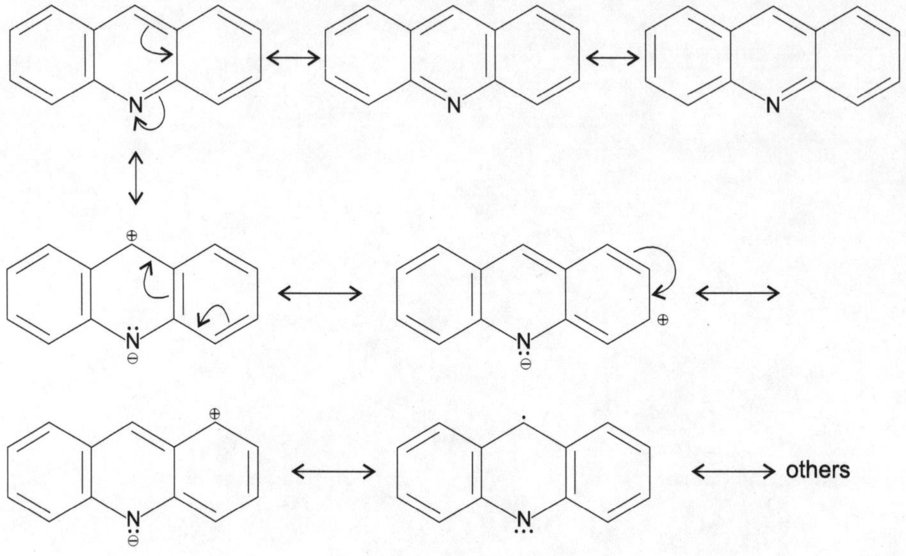

Resonance in acridine

10.2.4 Dipole moment

A dipole moment of 2.09 D has been reported for acridine.

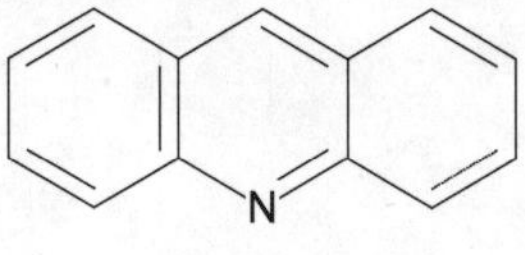

$\mu = 2.09$

10.2.5 Spectral data

The ^1H NMR and ^{13}C NMR (given in brackets) chemical shift values of acridine are shown below:

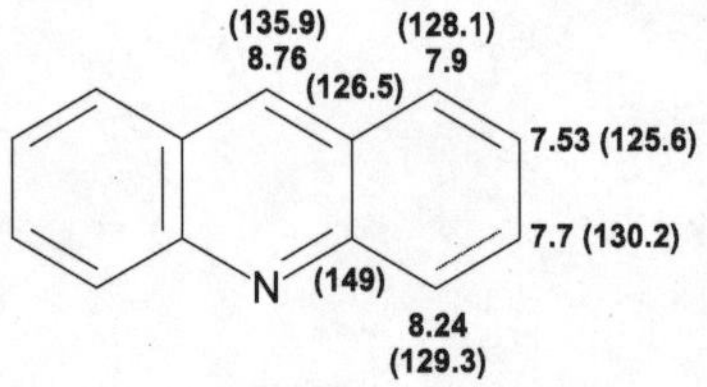

(135.9)   (128.1)
8.76   7.9
(126.5)

7.53 (125.6)

7.7 (130.2)

N (149)

8.24
(129.3)

^1H NMR (δ, ppm, CDCl$_3$)
^{13}C NMR (δ, ppm, CDCl$_3$)

10.2.6 Methods of preparation of acridine

Acridine is separated from coal tar by extracting with dilute sulfuric acid. Addition of potassium dichromate to this solution precipitates acridine dichromate. The acridine dichromate is decomposed by ammonia to get acridine. Other important methods for the synthesis of acridine and its derivatives are discussed below.

(*i*) **Ullmann synthesis:** The preparation of acridine derivatives by the treatment of N-arylanthranilic acid with an acylating agent, such as sulfuric acid, trifluroacetic acid, polyphosphoric acid etc, followed by reduction and dehydrogenation is generally referred to as the **Ullmann acridine synthesis**.

In this reaction, cyclization of diphenylamine-2-carboxylic acid in the presence of sulphuric acid or phosphoric acid gives acridone, which on reduction with sodium/ amyl alcohol gives 9,10-dihydroacridine. Further oxidation of 9,10-dihydroacridine by air or ferric chloride yields acridine. The diphenylamine-2-carboxylic acid (N-phenyl anthranilic acid) can be prepared from ortho-chlorobenzoic acid and aniline in the presence of copper powder and potassium carbonate under reflux conditions. Alternatively, it can also be prepared from o-aminobenzoic acid and chlorobenzene

**N-Phenylanthranilic acid
(Diphenylamine-2-carboxylic acid)**

9,10-Dihydroacridine

9-Acridone

Acridine

(*ii*) **Goldberg synthesis:** Diphenylamine-2-carboxylic acid on reaction with POCl₃ gives 9-chloroacridine which on reduction with Raney nickel gives 9,10-dihydroacridine. Further oxidation of 9,10-dihydroacridine with chromic acid yields acridine.

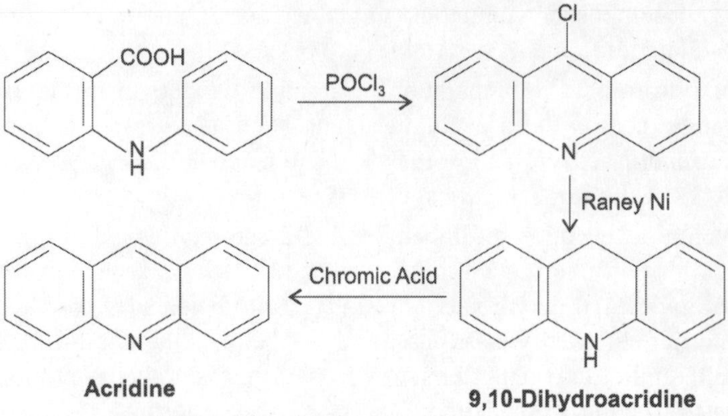

Acridine   **9,10-Dihydroacridine**

Mechanism: The mechanism of Goldberg synthesis is shown below.

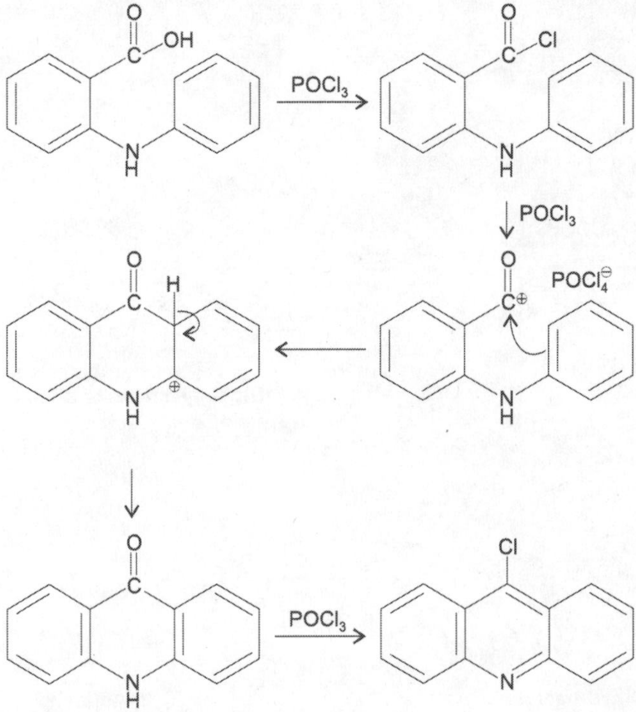

9-Substituted acridines can be prepared from 9-chloroacridine by nucleophilic displacement reaction. For example, 4-methoxy-9-chloroacridine on reaction with hydrazine hydrate yields 9-hydrazino-4-methoxyacridine derivative.

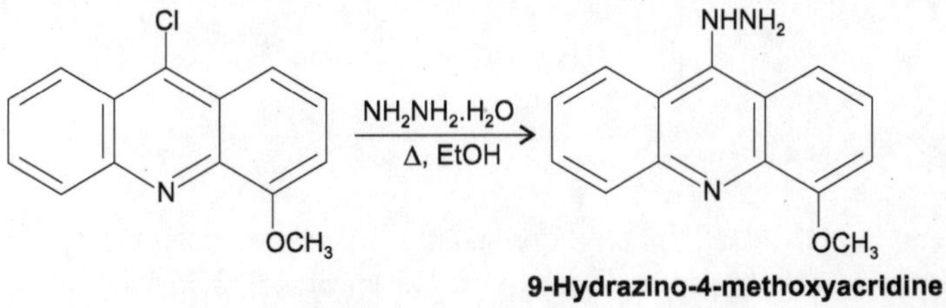

9-Hydrazino-4-methoxyacridine

Synthesis of quinacrine: The antimalarial compound quinacrine can also be obtained by nucleophilic reaction of 9-chloroacridine derivative with a suitable aliphatic amine.

Qunacrine (Antimalarial)

(*iii*) **Friedlander synthesis:** In the Friedlander synthesis of 9-methylacridine, ortho-aminoacetophenone is reacted with cyclohex-2-enone at 120°C.

o-Aminoacetophenone Cyclohex-2-enone

(*iv*) **Bernthsen synthesis:** The Bernthsen synthesis of acridines involves the reaction of diphenylamine with carboxylic acid or acid anhydride in the presence of anhydrous zinc chloride.

Bernthsen, A., Ann, 1878, 1, 192

Mechanism: The mechanism of Bernthsen acridine synthesis is shown below.

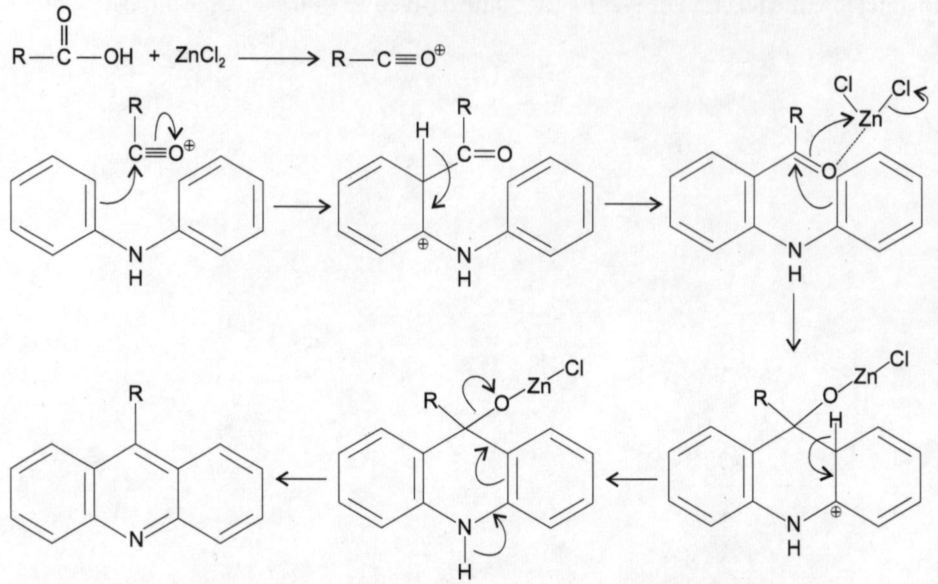

(v) From 2-methyldiphenylamine: Oxidation of 2-methyldiphenylamine with lead dioxide gives acridine in 20-30% yield. The 2-methyldiphenylamine can be synthesized by the reaction of o-toluidine with aniline in presence of iodine at high temperature.

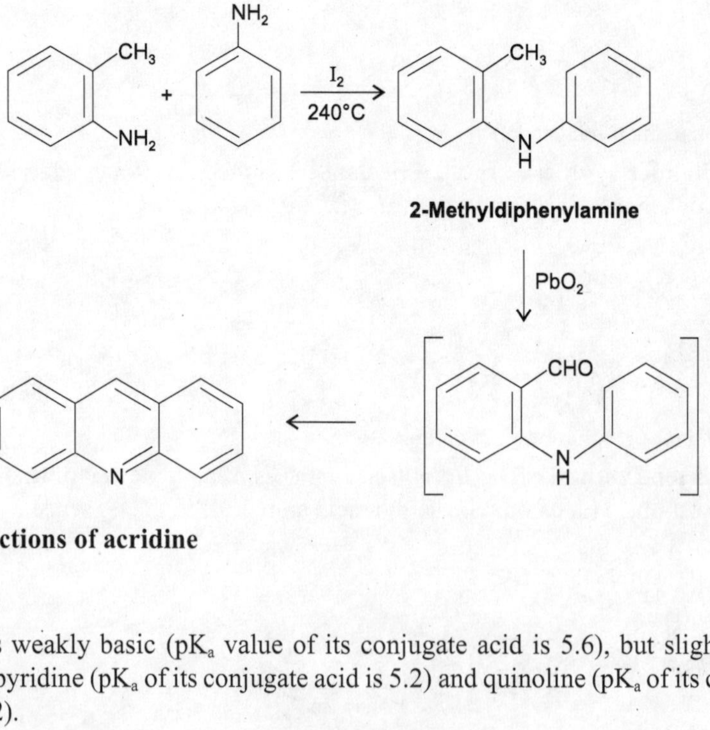

2-Methyldiphenylamine

10.2.7 Reactions of acridine

Basicity

Acridine is weakly basic (pK_a value of its conjugate acid is 5.6), but slightly more basic than pyridine (pK_a of its conjugate acid is 5.2) and quinoline (pK_a of its conjugate acid is 4.92).

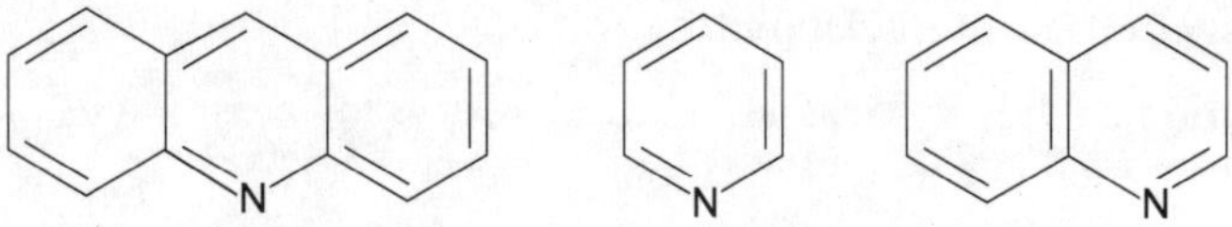

pK$_a$ of conjugate acid:-	5.6	5.2	4.92

Electrophilic substitution reactions of acridine

Electrophilic substitution reactions of acridine as in the case of quinoline or isoquinoline occur in the benzenoid ring and often results in disubstitution at the 2- and 7-positions.

Nitration: The nitration of acridine with HNO$_3$/H$_2$SO$_4$ yields 2,7-dinitroacridine.

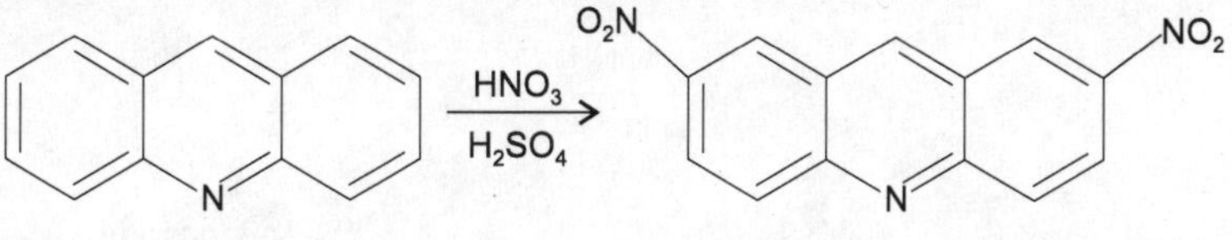

2,7-Dinitroacridine

Halogenation: Chlorination of acridine with thionyl chloride gives 9-chloroacridine, this is due to lower steric hindrance compared to bromination.

Bromination of acridine in acetic acid gives mixture of 2-bromoacridine and 2,7-dibromoacridine. Bromination of acridine with NBS, however yields 9-bromoacridine.

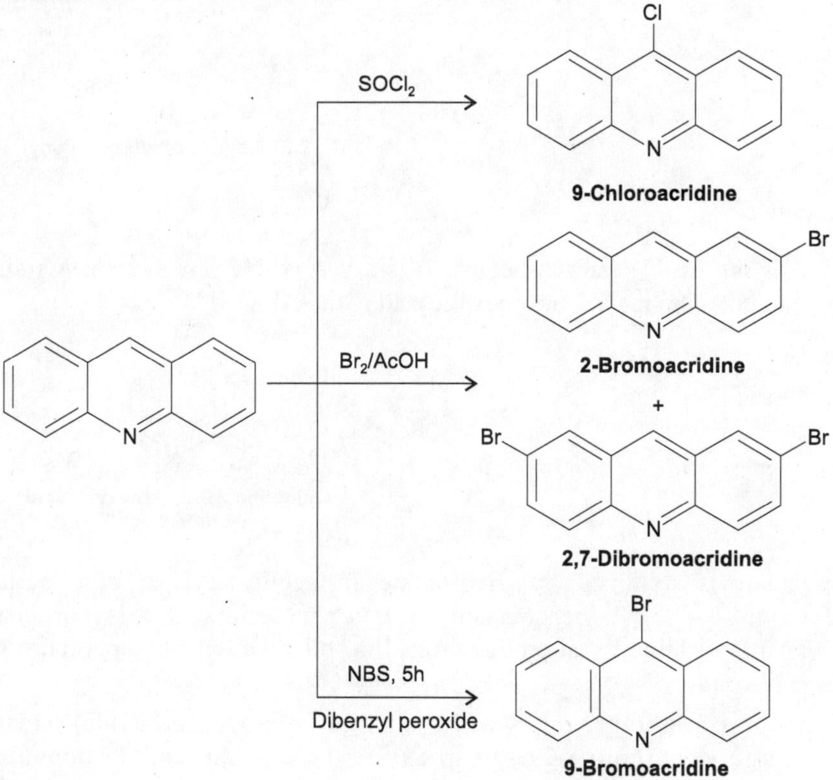

9-Chloroacridine

2-Bromoacridine

+

2,7-Dibromoacridine

9-Bromoacridine

Reactions with nucleophiles

Acridine shows variable regiochemistry towards nucleophiles. The nucleophilic substitution occurs at 9-position.

For example, reaction of acridine with $NaNH_2$ in liquid ammonia leads to formation of 9-aminoacridine (**Chichibabin amination**). However, reaction of acridine with N,N-dimethylaniline and $NaNH_2$ gives 9,9'-bis-9,10-dihydroacridinyl as a major product. A radical SET mechanism was proposed for dimerization instead of S_NAr mechanism.

9-Aminoacridine

9,9'-Bis(9,10-dihydroacridiny)

Oxidation

Acridine is not easily oxidized because it is a very stable ring system. Acridine is converted into quinoline-2,3-dicarboxylic acid with $KMnO_4$.

**Quinoline-2,3-dicarboxylic acid
(Acridinic acid)**

However, it is oxidized by dichromate in acetic acid to give acridone (10H-acridin-9-one). Perbenzoic acid oxidation of acridine occurs more readily than that of pyridine or quinoline. Acridine on reaction with peracids yield acridine N-oxide.

Ozonation of acridine, followed by alkaline hydrogen peroxide oxidation, yields largely quinoline-2,3-dicarboxylic acid with small amounts of 9(10H)-acridanone or acridone.

Reductive alkylation

Acridine on reaction with n-pentanoic acid, in the presence of ultraviolet light, gives 9-n-butylacridine.

9-n-Butylacridine

Reduction

The 9-position is activated by ring nitrogen thus addition reaction to 9- and 10-positions takes place readily. For example, reduction of acridine with Zn/HCl gives 9,10-dihydroacridine (pyridine ring reduced). However, reduction of acridine with H_2/Pt/HCl results in the formation of 1,2,3,4,5,6,7,8-octahydroacridine (benzene rings reduced). Reduction of acridine with lithium-liq. ammonia in ethanol forms 1,4,5,8-tetrahydroacridine.

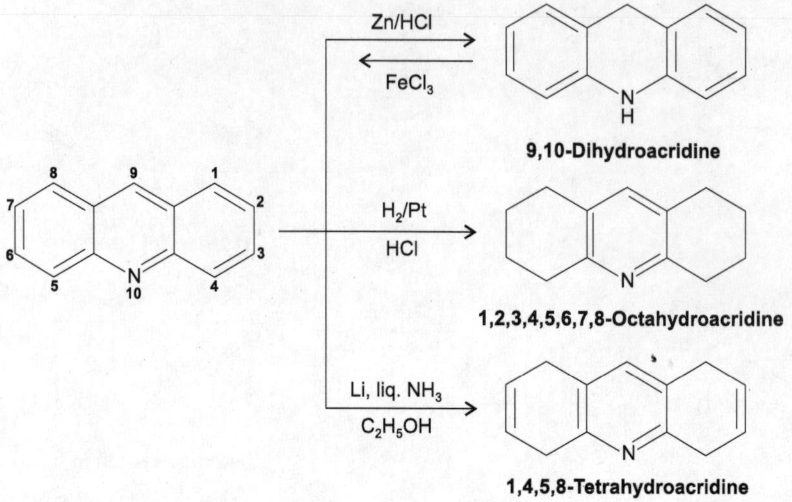

9,10-Dihydroacridine

1,2,3,4,5,6,7,8-Octahydroacridine

1,4,5,8-Tetrahydroacridine

Reactions of acridine N-oxide

Nitration and bromination of acridine N-oxide occurs only at the pyridine ring. It undergoes nitration and bromination at position-9. Reaction of acridine N-oxide with potassium cyanide and potassium ferrocyanide gives 9-cyanoacridine N-oxide. Acridine N-oxide can be converted to acridine on reaction with $NiCl_2.2H_2O$ in the presence of lithium and biphenyl.

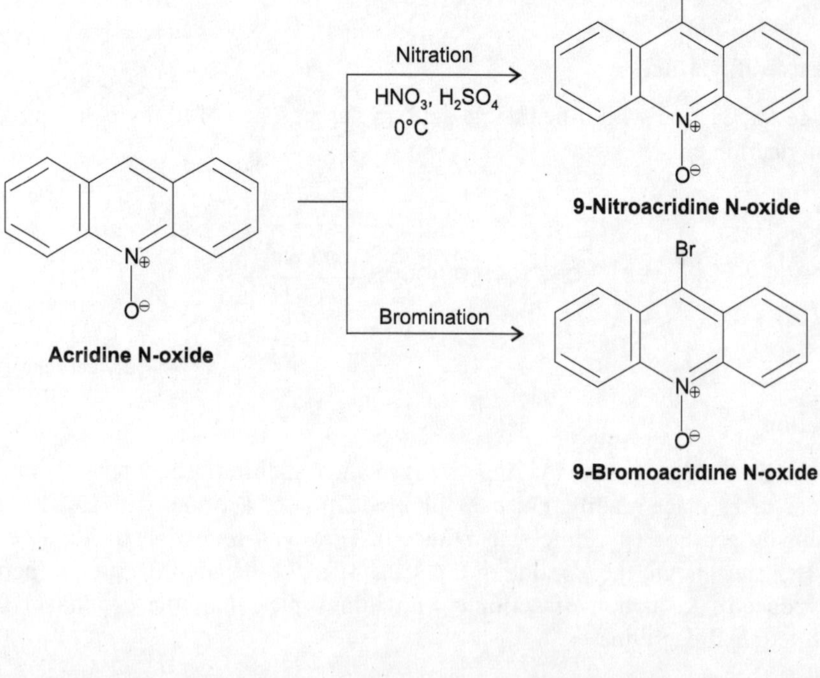

9-Nitroacridine N-oxide

Acridine N-oxide

9-Bromoacridine N-oxide

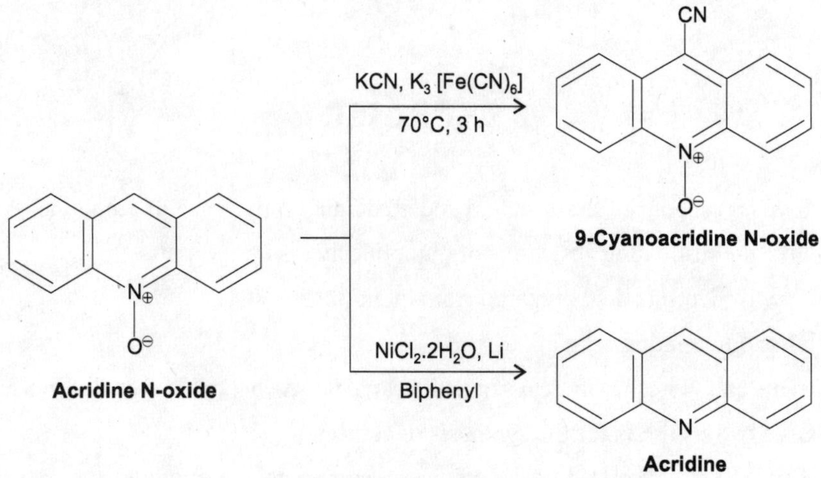

Acridine N-oxide

KCN, K$_3$ [Fe(CN)$_6$]
70°C, 3 h

9-Cyanoacridine N-oxide

NiCl$_2$.2H$_2$O, Li
Biphenyl

Acridine

EXERCISE

Q.1   Give an account of the structure and medicinal importance of quinoline.

Q.2   Give the resonating structures of quinoline and isoquinoline.

Q.3   How is nicotinic acid prepared from quinoline?

[Hint. On treating with alk. $KMnO_4$]

Q.4   Name and write the structures of antimalarial drugs containing quinoline nucleus.

Q.5   Give two methods for the synthesis of acridine.

Q.6   Write the structure and use of medicinal compounds of acridine.

Q.7   Explain:

(a) Electrophilic substitution in isoquinoline takes place at 5-position preferably.

(b) The formation of 3-chloroquinoline when indole is heated with $CHCl_3$ in presence of KOH.

[Hint. (a) pyridine ring is electron deficient and thus electrophilic substitution takes place in benzene ring]

Q.8   Write short note on:

(a) Skruap quinoline synthesis.

(b) Friedlander synthesis of quinoline.

(c) Pomeranz-Fritsch synthesis of isoquinoline.

(d) Bischler-Napieralski synthesis of isoquinoline

Q.9   Carry out the following conversions:-

(a) Indole to quinoline

(b) 9-Acridone to acridine

(c) Aniline to 2-methyl quinoline.

Q.10 Give the product based on Skraup's synthesis from

(a) 3-Bromo-4-aminotoluene and glycerol

(b) 1-Aminonapthalene and glycerol

Q.11 What is the order of reactivity of quinoline, pyridine, quinolone N-oxide towards nitration?

Ans. pyridine < quinoline < quinolone N-oxide

Pyridine is least reactive because it is highly deactivated towards electrophilic attack due to presence of electronegative nitrogen atom. In case of quinoline attack takes place at the benzene ring. Quinolone N-oxide is most reactive amongst these because of the activating effect of the N-oxide group.

Q.12 Complete the following reactions

    (a) Acridine + $SOCl_2$ ⟶

    (b) Acridine + HNO_3/H_2SO_4 ⟶

    (c) Acridine + $NaNH_2$ in liq. NH_3 ⟶

    (d) Quinoline + H_2 + Raney Ni ⟶

    (e) Aniline + acetylacetone + H_2SO_4 ⟶

Ans. (a) 9-Chloroacridine   (b) 2,7-Dinitroacridine

    (c) 9-Aminoacridine   (d) 1,2,3,4-Tetrahydroquinoline

    (e) 2,4-Dimethylquinoline

Q.13 Complete the following reactions:

(a) 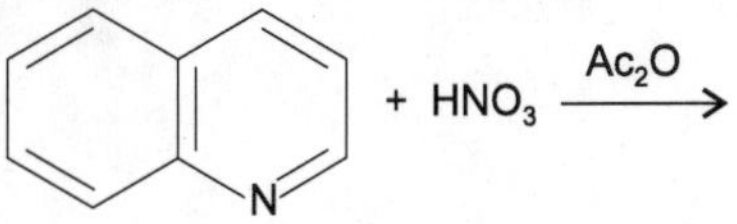
$+ \; HNO_3 \xrightarrow{Ac_2O}$

(b) 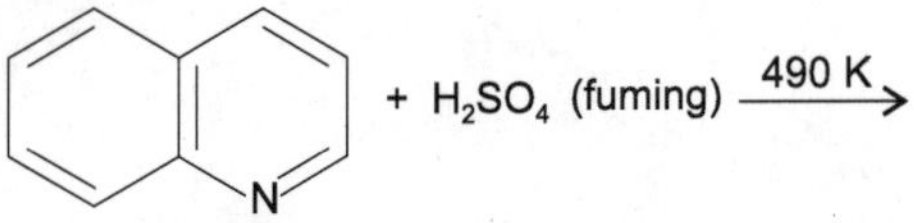
$+ \; H_2SO_4$ (fuming) $\xrightarrow{490\ K}$

(c) 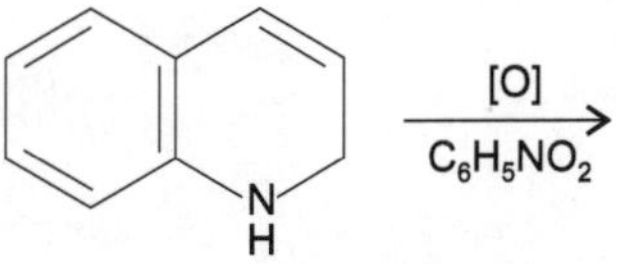
$\xrightarrow[C_6H_5NO_2]{[O]}$

(d) 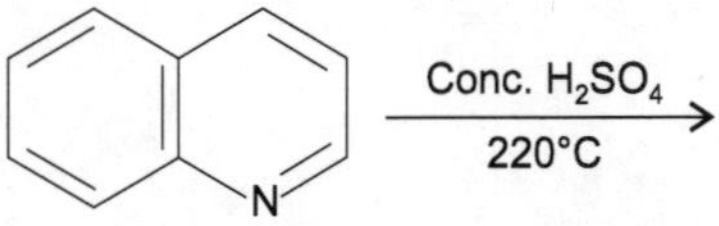
$\xrightarrow[220°C]{Conc.\ H_2SO_4}$

(e) 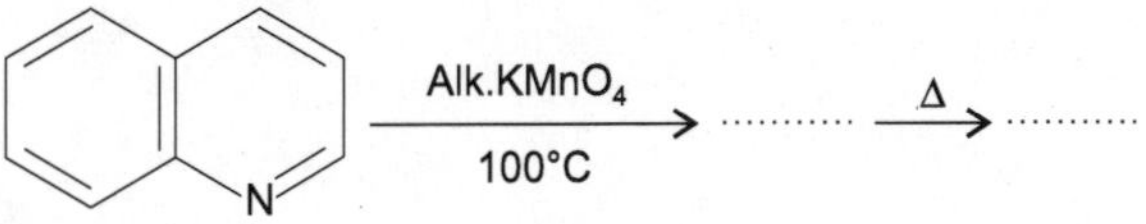
$\xrightarrow[100°C]{Alk.KMnO_4}$ .... $\xrightarrow{\Delta}$ ....

Ans. (a) 3-Nitroquinoline   (b) Quinoline-6-sulfonic acid

    (c) Quinoline   (d) Quinoline-8-sulfonic acid

    (e) Pyridine-2,3-dicarboxylic acid, Nicotinic acid

Q.14 Explain why electrophilic attack takes place in the benzene ring and nucleophilic attack in pyridine ring of quinoline.

Ans. In quinoline during electrophitic attack the pyridine ring gets converted to pyridinium ion, thus undergo electrophitic attack less preferably. However, nucleophilic attack leads to formation of a stable intermediate anion, thus nucleophile attack preferably in pyridine ring.

Six Membered Heterocylic Compounds with Two Nitrogen Atoms

Pyridazine, Pyrimidine, Pyrazine and Barbiturates

11.1 DIAZINES: PYRIDAZINE, PYRIMIDINE AND PYRAZINE

11.1.1 Introduction

The diazines are compounds derived from benzene by the replacement of two of the ring carbon atoms by nitrogen. The three isomeric six membered ring diazines are **pyridazine**, **pyrimidine** and **pyrazine** having two nitrogen atoms in 1-2, 1-3, and 1-4 relationship, respectively.

| Pyridazine | Pyrimidine | Pyrazine |

These three diazines are stable, colourless and water soluble compounds. Pyridazine (bp 208°C) has higher boiling point as compared to pyrimidine (bp 124°C) and pyrazine (bp118°C) due to the polarizability of the N–N unit. It results into extensive dipolar association in the liquid.

11.1.2 Significance of diazines

Pyridazine ring system does not form a part of any natural product and thus it has been less extensively investigated than other diazines. It is mainly used in research and industry as building block for the synthesis of more complex compounds. The pyridazine structure is found within a number of synthetic herbicides such as **pyridafol**. It is also found within the structure of several pharmaceutical drugs such as **cefozopran** (antibiotic), **cadralazine** (antihypertensive), **minaprine** (antidepressant), and **hydralazine** (smooth muscle relaxant).

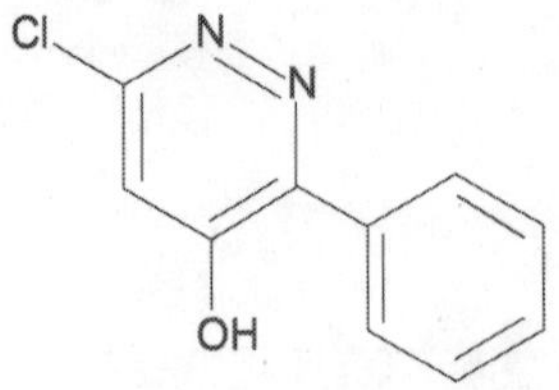

Pyridafol

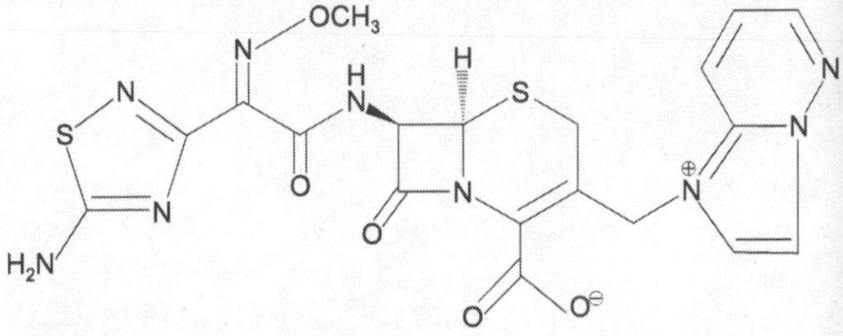

Cefozopran

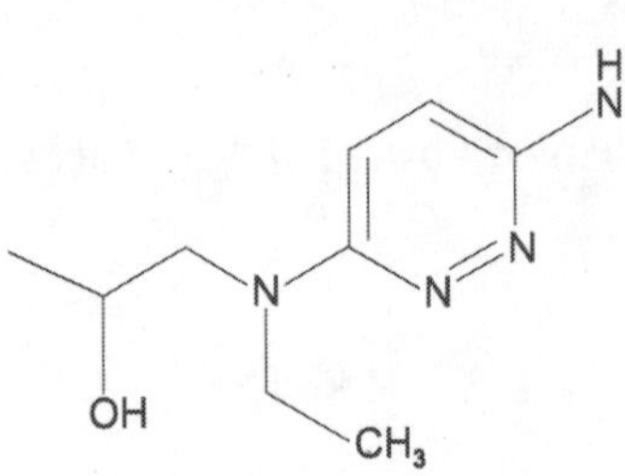

Cadralazine

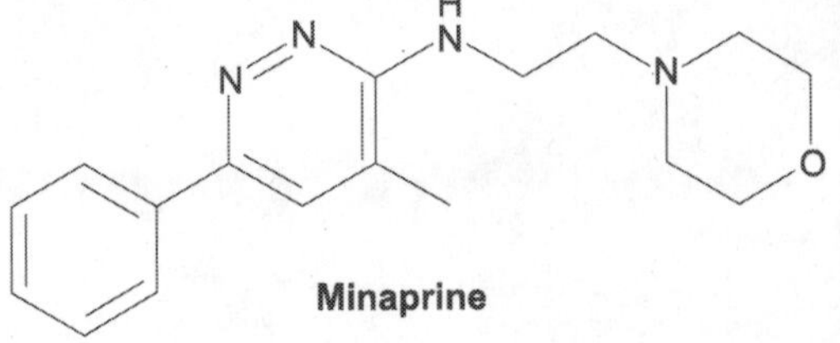

Minaprine

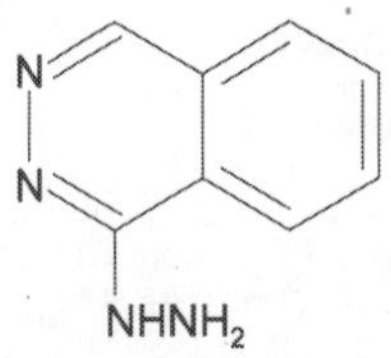

Hydralazine

Pyrimidines are the most important member of all the three diazines. Pyrimidines as such or as part of a molecule are widely distributed in nature. There are innumerable pyrimidine based drugs which are categorized as barbiturates, sulfa drugs, antibiotics and antitumor agents. Several sulfa drugs such as **sulfadiazine, sulfamerazine, sulfametoxydiazine, sulfasomidine** and **sulfametomidine** are pyrimidine derivatives. Purines and pyrimidines are the building blocks of nucleic acids. The **purine** ring system consists of a pyrimidine ring fused to an imidazole ring. The three pyrimidines **cytosine, uracil** and **thymine** occur widely in nucleic acids.

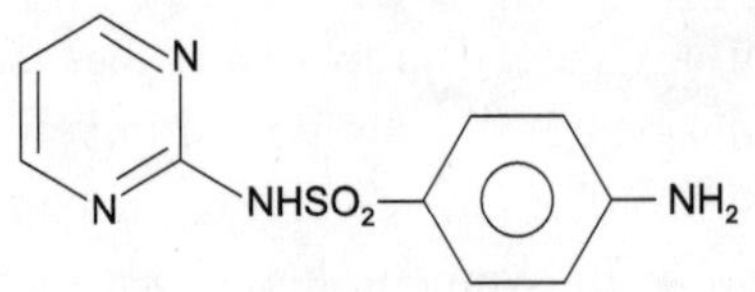

Sulfadiazine

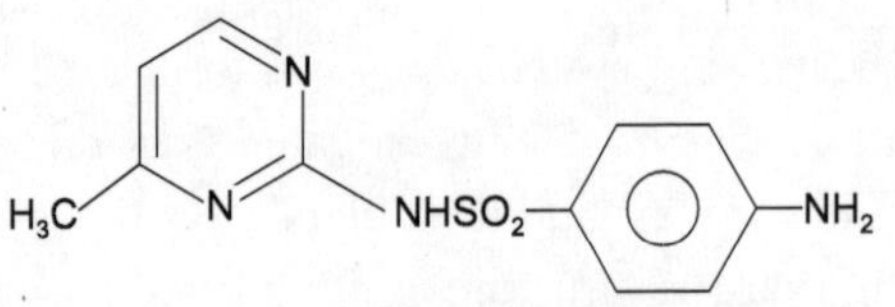

Sulfamerazine

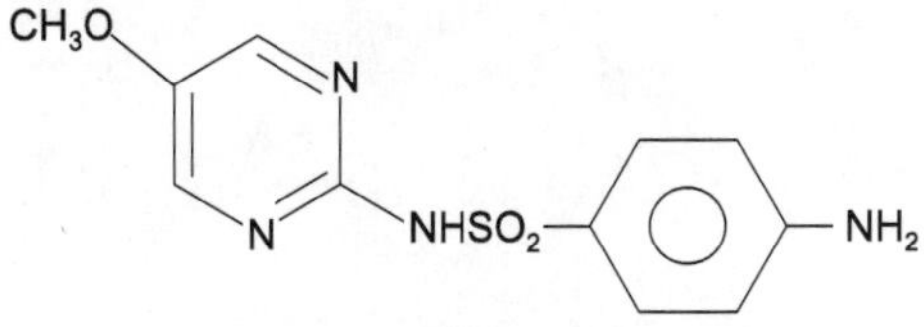

Sulfametoxydiazine

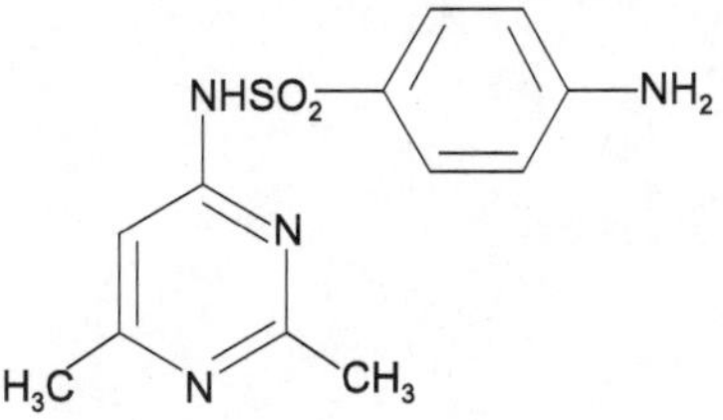

Sulfasomidine

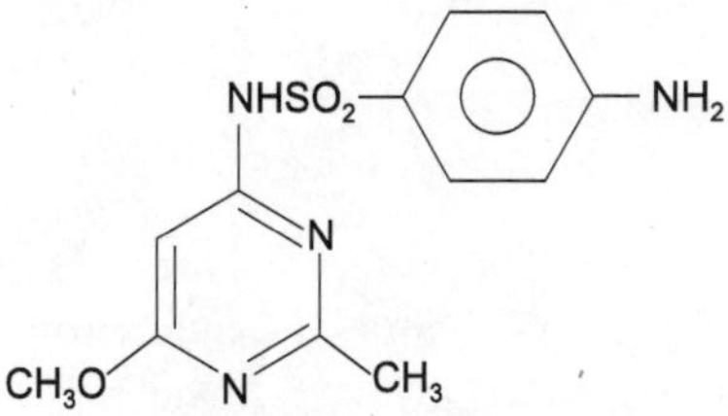

Sulfametomidine

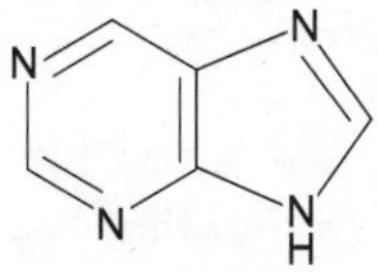

Purine

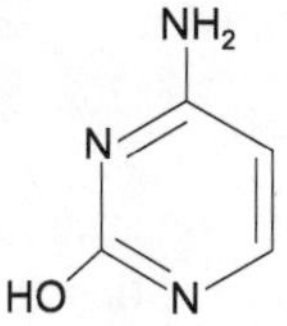

Cytosine

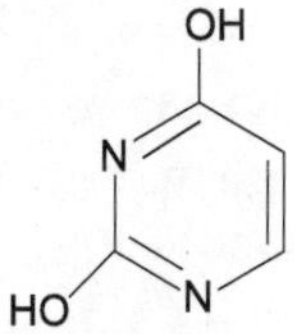

Uracil

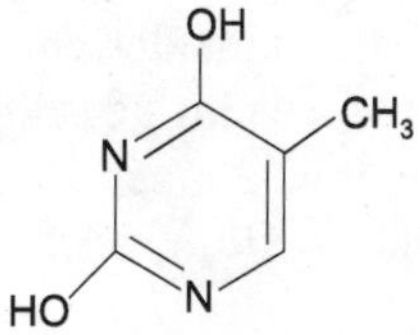

Thymine

Pyrazines such as **pteridine** and **phenazine** also occur in nature. They are useful as antibiotics, diuretics and anti-tumor agents.

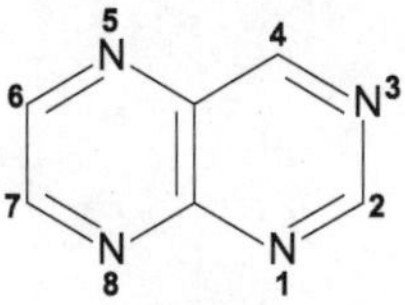

Pteridine

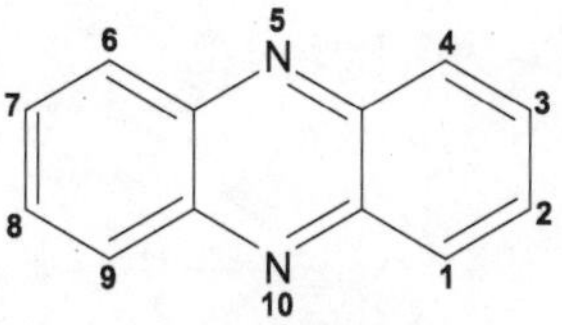

Phenazine

11.1.3 Structure

The three isomeric diazines namely pyridazine, pyrimidines and pyrazine, are aromatic in nature. The resonance energy of benzene, pyridine and diazines is given below:

Benzene > Pyridine > Pyrimidine > Pyrazine > Pyridazine

| **Resonance energy (kJ/mol)** | 151 | 117 | 109 | 101.67 | 92.04 |

Both the nitrogen atoms in diazines are pyridine type, with lone pair of electrons held in sp² hybridized orbitals and not delocalized around the ring. Pyrimidine is resonance hybrid of the following resonating structures:

Resonance in pyrimidine

11.1.4 Basicity

The three diazines are weaker bases than pyridine. The order of base strength is **pyridine > pyridazine > pyrimidine > pyrazine** (pK_a values of their protonated conjugate acids are 5.2, 2.3, 1.3, 0.6, respectively). Diazines have lower pK_a values than pyridine because the second N-atom present in the ring reduces their basicity as compared to pyridine *via* inductive effect. Pyridazine is more basic than pyrimidine and pyrazine because of the effect of adjacent lone pair electrons (α-effect) which repel each other, thus protonation of the nitrogen atom leads to decrease in repulsion between the two lone pairs. Pyrimidine is a much weaker base than pyridine and accepts two protons under strongly acidic conditions (pK_a = 1.3 and -6.9).

Pyridine

pK_a = 5.2

Pyridazine

pK_a = 2.3

Pyrimidine

pK_a = 1.3 pK_a = –6.9

Pyrazine

pK_a = 0.6

11.1.5 Dipole moment

Pyridazine (μ = 3.95 D) has higher dipole moment value than pyrimidine (μ = 2.4 D). The dipole moment value of pyrazine is zero, which supports its symmetrical structure.

Pyridazine   **Pyrimidine**   **Pyrazine**

μ = 3.95 D μ = 2.4 D μ = 0 D

11.1.6 Spectral data

The ^1H NMR of diazines show downfield shift of α-hydrogens. This is due to deshielding caused by the inductive effect of the heteroatom. The symmetrical structure of 1,4-diazine (pyrazine) is reflected in the ^1H NMR and ^{13}C NMR spectra with only one signal for the ring protons at 8.6 ppm and one for the ring C-atoms at 145.2 ppm recorded in $CDCl_3$. The ^1H and ^{13}C NMR (given in brackets) of diazines are shown below:

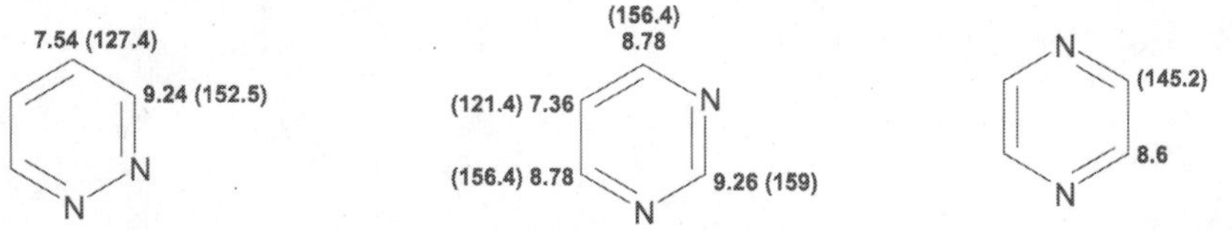
7.54 (127.4)
9.24 (152.5)
N
N
(156.4)
8.78
(121.4) 7.36
N
(156.4) 8.78
9.26 (159)
N
N
(145.2)
8.6
N

Pyridazine — ^1H NMR (δ, ppm, $CDCl_3$); ^{13}C NMR (δ, ppm, DMSO-d_6)

Pyrimidine — ^1H NMR (δ, ppm, $CDCl_3$); ^{13}C NMR (δ, ppm, $CDCl_3$)

Pyrazine — ^1H NMR (δ, ppm, $CDCl_3$); ^{13}C NMR (δ, ppm, $CDCl_3$)

11.1.7 Methods of preparation of diazines

11.1.7.1 Methods of preparation of pyridazine

(*i*) **From 1,4-dicarbonyl compounds:** Alkyl or acyl substituted pyridazines can be prepared by one step cyclization from an unsaturated diketone and hydrazine.

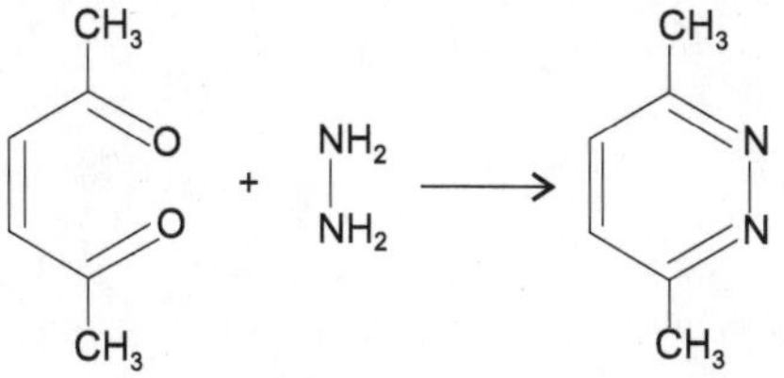
CH3
O
O
CH3
+
NH2
NH2
CH3
N
N
CH3

If the four carbon component is saturated, oxidation is carried out in the final step to get aromatic pyridazine.

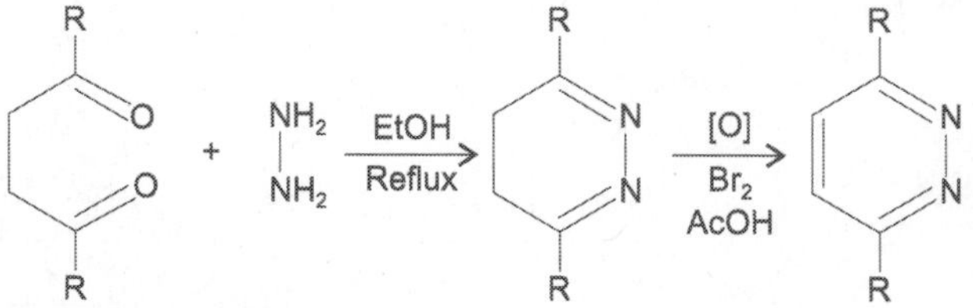
R
O
O
R
+
NH2
NH2
EtOH
Reflux
R
N
N
R
[O]
Br2
AcOH
R
N
N
R

An example is the synthesis of **cotton herbicide** (**4**) by the reaction of hydrazine with γ-keto ester. Hydrazine adds to the keto ester (**1**) to give dihydropyridazolone (**2**). Reaction of dihydropyridazolone (**2**) with Br_2/AcOH followed by dehydrohalogenation yield pyridazolone (**3**). Further reaction of pyridazolone (**3**) with POCl$_3$ and finally nucleophilic substitution reaction leads to the final **cotton herbicide pyridazine derivative (4)**.

(*ii*) **From maleic anhydride:** The condensation of hydrazine or it's derivatives with maleic anhydride also results in the formation of pyridazine ring. For example, reaction of hydrazine with maleic anhydride gives hydroxypyridazinone, which on treatment with POCl$_3$ gives 3,6-dichloropyridazine, reduction of which with H$_2$/Pd, C in aq. NH$_3$/CH$_3$OH forms pyridazine.

Meleic anhydride   **Pyridazine**

11.1.7.2 Methods of preparation of pyrimidine

(*i*) **By cycloaddition:** Cycloaddition of a 1,3,5-triazine with an alkyne followed by subsequent loss of hydrogen cyanide gives pyrimidine derivative.

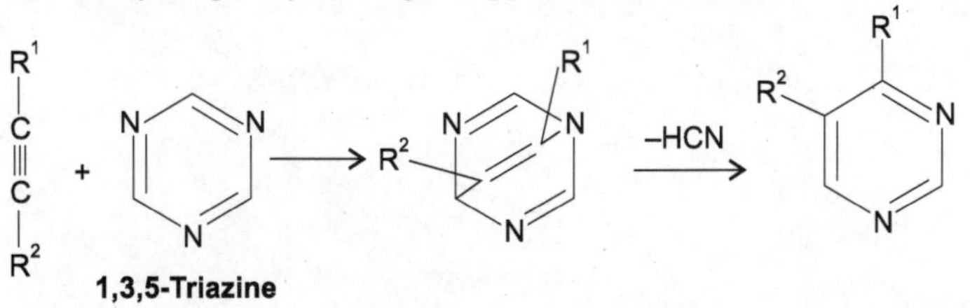

1,3,5-Triazine

(*ii*) **Biginelli reaction:** The dihydropyrimidinones can be prepared by classic Biginelli reaction. The Biginelli reaction is a multiple component reaction (MCR) of ethyl acetoacetate, an aryl aldehyde (such as benzaldehyde) and urea to give 3,4-dihydropyrimidin-2(1*H*)-ones.

Biginelli, P., Gazz. Chim. Ital.1893, 23, 360.

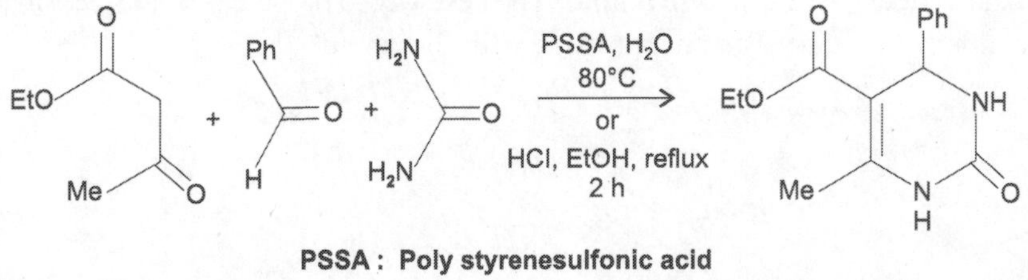

PSSA : Poly styrenesulfonic acid

This reaction was developed by Pietro Biginelli in 1891 and catalyzed by Bronsted acids and/or by Lewis acids such as boron trifluoride. Dihydropyrimidinones, the products of the Biginelli reaction, are widely used in the pharmaceutical industry as calcium channel blockers and antihypertensive agents.

Mechanism:

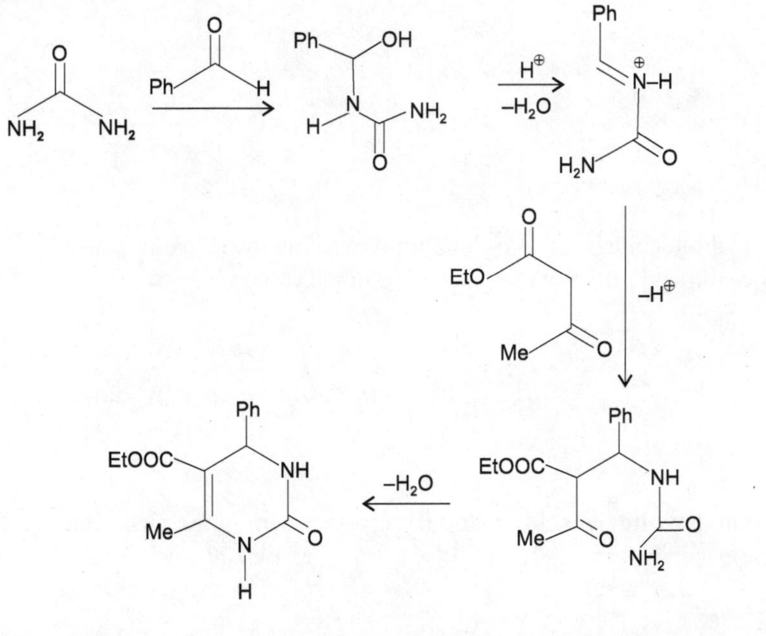

Synthesis of rac-monastrol: Monastrol is a mitosis blocker by kinase Eg5 inhibition. It is synthesized by Biginelli reaction. A one pot reaction of ethyl acetoacetate, 3-hydroxybenzaldehyde and thiourea in presence of $Yb(OTf)_3$ gives a racemic mixture of monastrol.

Dondoni, A.; Massi, A.; Sabbatini, S., Tet. Lett. **2002**, *34, 5913-5916*

(±)-Monastrol

***(iii)* From 1,1,3,3-tetraethoxypropane:** The best way to prepare pyrimidines involves the reaction of 1,1,3,3-tetraethoxypropane with formamide.

***(iv)* From chloro or methyl pyrimidines:** Pyrimidine itself can be prepared from its chloro or methyl derivatives. For example, oxidation of 4,6-dimethylpyrimidine, followed by thermal decarboxylation of resulting pyrimidine-4,6-dicarboxylic acid gives small yield of pyrimidine.

4,6-Dimethylpyrimidine

Also, hydrogenolysis of 4,6-dichloropyrimidine over palladium charcoal in the presence of magnesium oxide gives pyrimidine.

***(v)* From barbituric acid:** Gabriel prepared pyrimidine first time in 1900 from barbituric acid by treating it with $POCl_3$ followed by zinc dust.

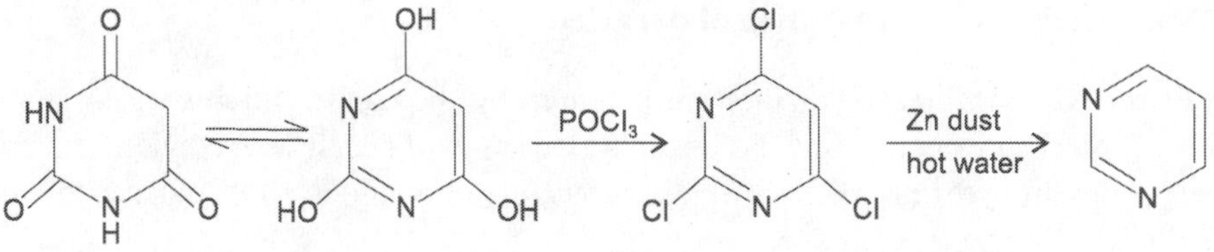

Barbituric acid

(*vi*) **Synthesis of trimethoprim:** Trimethoprim is a pyrimidine based bacteriostatic antibiotic. It inhibits the production of tetrahydrofolic acid, which is necessary for synthesis of protein. It can be synthesized as shown below.

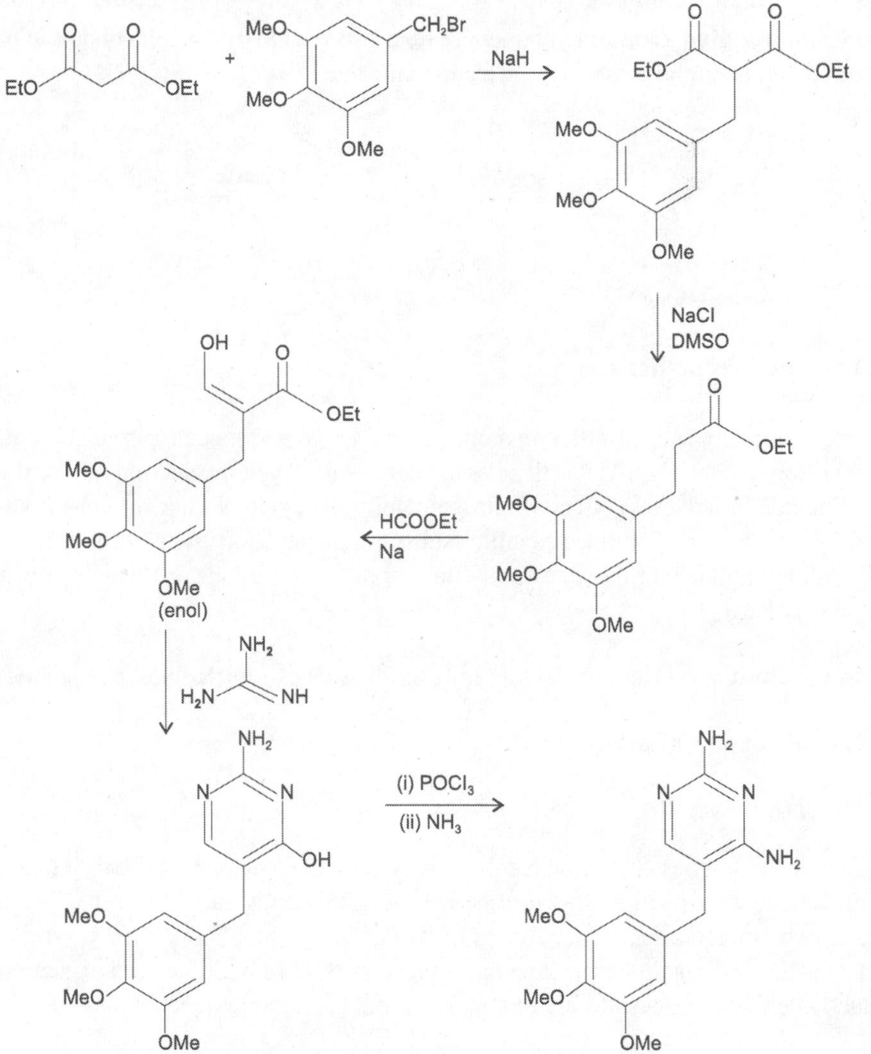

Trimethoprim

11.1.7.3 Methods of preparation of pyrazine

(*i*) **From 1,2-diamines:** Pyrazines are prepared by the condensation of 1,2-diamine with 1,2-dicarbonyl compound. For example, reaction of benzil with ethylene diamine gives a dihydropyrazine which is oxidized by CuO or MnO_2 in KOH/ethanol to yield 2,3-diphenylpyrazine.

Benzil   Ethylene diamine   2,3-Diphenyl pyrazine

(*iii*) **From α-amino ketone:** Pyrazines are readily prepared by the self condensation of α-amino ketone in the presence of mercuric chloride.

α–**Aminoketone**

11.1.8 Reactions of diazines

The presence of a second nitrogen atom in diazines leads to electron depletion at the carbon atoms of the ring. Thus, diazines are comparitively more electron deficient than pyridine due to inductive effect of nitrogen atoms. They do not undergo electrophilic substitution easily. The electrophilic substitution reaction such as nitration or sulfonation on diazine ring occurs only under drastic conditions or if some activating group is present.

However, the diazines are very susceptible to the action of nucleophillic reagents.

11.1.8.1 Reactions of pyridazine

N-Alkylation

Pyridazine undergoes alkylation reactions to reduce the lone pair-lone pair interaction of the adjacent nitrogen atoms (α-effect). Unsymmetrically substituted diazines give rise to two isomeric quaternary salts. For example, alkylation of 3-methylpyridazine and 3-methoxy-6-methylpyridazine takes place at N-1. In both these cases steric and inductive effects decides the site of attack and not mesomeric effect.

3-Methylpyridazine

3-Methoxy-6-methylpyridazine

Electrophilic substitution

The 3, 4, 5 and 6 positions in pyridazine nucleus are electron deficient. No sulfonation or nitration of pyridazine has been reported as such. However, activating groups present on the pyridazine ring makes nitration possible.

Reaction with nucleophilic reagents

The attack of nucleophile generally takes place at 3-position. For example, reaction of pyridazine with n-BuLi in ether followed by oxidation gives 3-n-butylpyridazine.

Oxidation

Pyridazine is resistant to the attack of oxidizing agents because of electron deficiency in the ring. However, with hydrogen peroxide the N-oxide formation takes place but no di-N-oxide is obtained.

Reduction

Pyridazine is fully reduced with Na/MeOH.

Cycloaddition reaction

Pyridazine forms 1:2 adduct with maleic anhydride at room temperature.

11.1.8.2 Reactions of pyrimidine

N-Alkylation reaction

Pyrimidine and its simple derivatives react with alkyl halides to give N-alkylpyrimidinium halide. The pyrimidines like other diazines react with methyl iodide to give monoquaternary salts.

Dialkylation of pyrimidines cannot be carried out with simple alkyl halides, however the more reactive trialkyloxonium tetrafluoroborates can convert pyrimidine and other diazines into diquaternary salts.

Electrophilic substitution

The presence of electron withdrawing nitrogen atoms in the pyrimidines ring causes significant electron depletion at C2, C4 and C6. Thus electrophilic attack is expected to occur at C5 or on the ring nitrogen atoms.

Nitration of pyrimidines and its alkyl derivatives is not possible. However, pyrimidines with electron releasing group undergo nitration at 5-position under vigorous conditions.

Pyrimidine-2(1*H*)-one under nitrating conditions i.e. concentrated sulfuric acid and potassium nitrate gives 5-nitropyrimidine-2-(1*H*)-one.

Sulfonation of pyrimidines bearing at least one electron releasing group is possible by using chlorosulfonic acid.

Vapour phase bromination of unsubstituted pyrimidines gives 5-bromopyrimidine. The 5-chloro, 5-fluoro and 5-iodo pyrimidines are known but they cannot be prepared by direct halogenation. However, direct halogenation of pyrimidines bearing one or more electron releasing groups is very easy. Bromination of pyrimidine hydrochloride also gives 5-bromopyrimidine.

Oxidation

Pyrimidine reacts with peracids to give N-oxides. However, the product formed is relatively less stable under the acidic conditions.

Reactions with nucleophilic reagents

Pyrimidine gets converted to pyrazole by reaction with hot hydrazine.

Pyrazole

11.1.8.3 Reactions of pyrazine

Electrophilic substitution

Chlorination of 2-methylpyrazine occurs under mild conditions and it may involve addition-elimination mechanism.

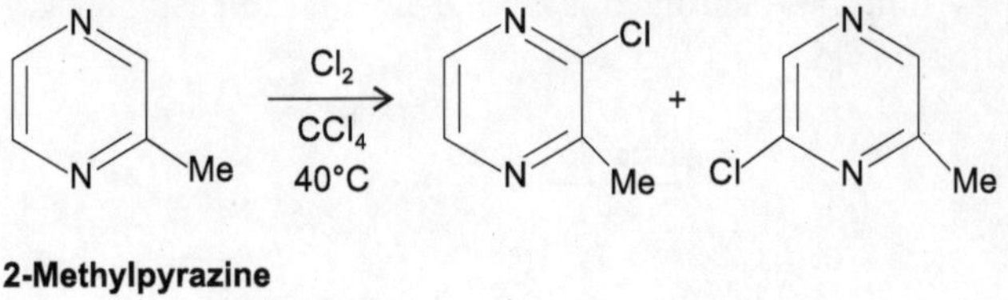

2-Methylpyrazine

Oxidation

2,5-Dimethylpyrazine reacts with hydrogen peroxide to give mixture of mono N-oxide and di N-oxide.

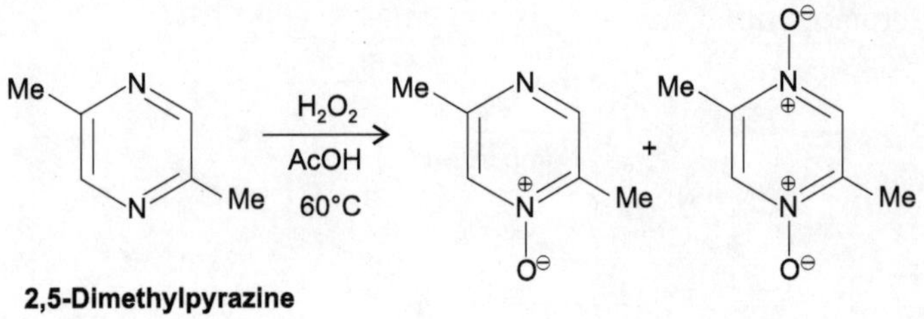

2,5-Dimethylpyrazine

Nucleophilic substitution

Pyrazine also undergoes nucleophilic substitution with great difficulty. For example, reaction of 2,5-dimethylpyrazine with sodamide gives only 35% of 3-amino-2,5-dimethyl pyrazine.

2,5-Dimethylpyrazine   **3-Amino-2,5-dimethylpyrazine**

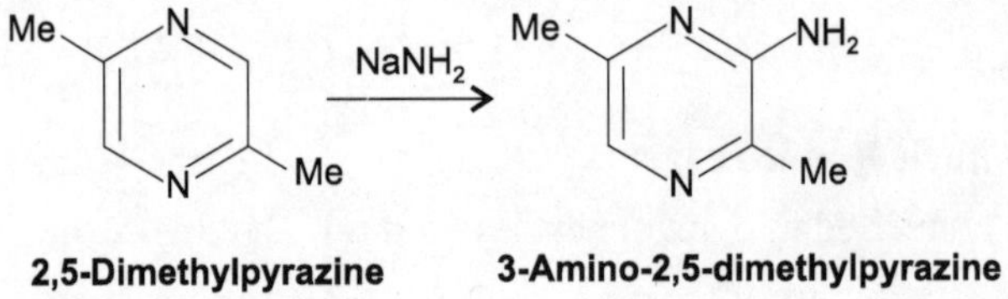

Reduction

Pyrazine ring can be reduced by sodium alcohol to piperazine.

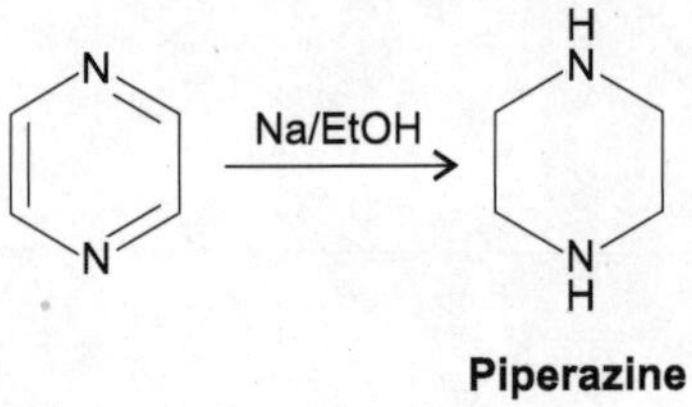

Piperazine

11.2 BARBITURATES

11.2.1 Introduction

Barbituric acid also known as malonyl urea or 6-hydroxyuracil was discovered by the German chemist Adolf von Baeyer in 1864. It is an odourless powder soluble in water. Barbituric acid has two tautomeric forms the trione structure and the 2,4,6-trihydroxypyrimidine structure, however the most probable structure is the trione structure.

Trione   **2,4,6–Trihydroxypyrimidine**

The trione structure of barbituric acid is in agreement with its reactions. For example, the presence of active methylene group is further confirmed by the fact that it readily forms an oximino derivative with nitrous acid.

Barbituric acid

(Violuric acid)
5-Oximinobarbituric acid

Furthermore, the trione structure is supported by the fact that methylation of barbituric acid with MeI/NaOH gives N-methyl derivatives, which indicates the presence of NH groups.

11.2.2 Biological importance

Barbiturates are derivatives of barbituric acid. Barbiturates act as sedatives, anticonvulsant, anesthetic and hypnotics. They are used as sleeping pills. However, barbituric acid itself has no hypnotic properties. They are also effective as anxiolytics. Barbiturates also have analgesic effect, however these effects are somewhat weak, preventing barbiturates from being used in surgery in the absence of other analgesics.

However, barbiturates are still used in general anesthesia, for epilepsy and assisted suicide.

The first barbiturate to be used in medicine in 1903 was 5,5-diethyl barbituric acid (barbital or veronal), and the second, phenobarbital first marketed in 1912. Emil Fischer and Joseph van mering discovered that veronal effectively puts dogs to sleep. Between the 1920s and the mid-1950s, barbiturates were the only drugs used as sedatives and hypnotics.

Barbital has long duration of action. The physiological properties of barbiturates depend upon the substitutions that are present at C5 position. They act by depressing the activity of the central nervous system. Most of the barbiturates are prepared and sold as their sodium salts. Some common derivatives of barbituric acids and their properties are shown below in Table 11.1.

Table 11.1: Derivatives of barbituric acid

Structure	Common Name	Properties
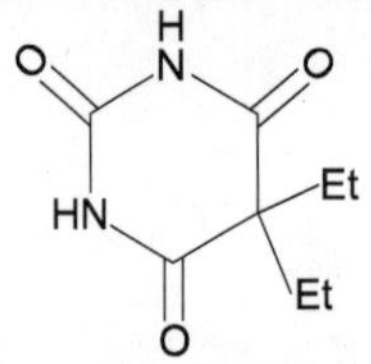	Barbital or Barbitone or Veronal	Hypnotic (Sleeping aid)
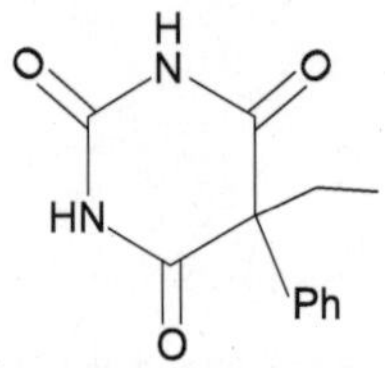	Phenobarbital or Luminal	Long acting CNS depressant, sedative, hypnotic and anticonvulsant
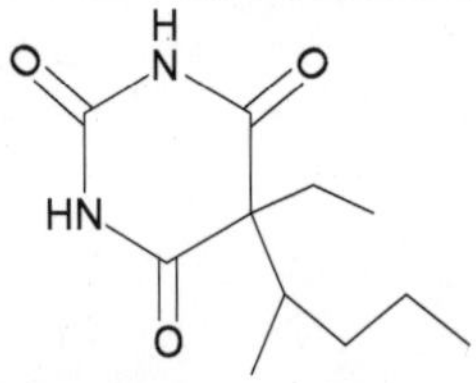	Pentobarbital or Pentobarbitone or Nembutal	Hypnotic for short term, sedative, veterinary anesthetic
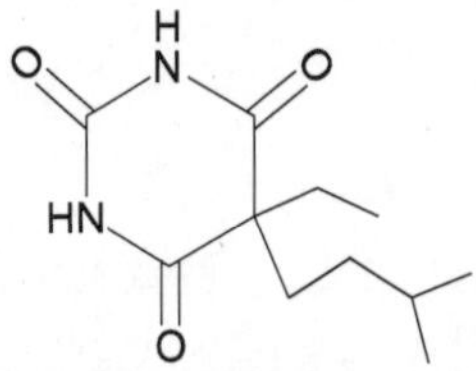	Amobarbital	Sedative and hypnotic

Structure	Common Name	Properties
	Secobarbital or Seconal	Anaesthetic, anxiolytic, anticonvalsant, sedative hypnotic
	Allobarbital	Sedative and hypnotic
	Sodium pentothal or Truth serum or Thiopental	Intravenous anesthetic

11.2.3 Methods of preparation of barbituric acid

Barbituric acid is prepared by condensing urea with malonic acid in the presence of phosphoryl chloride.

Malonic acid **Urea** **Barbituric acid**

Barbituric acid can also be prepared from diethyl malonate and urea in ethanolic sodium ethoxide.

Various 5-amino, alkyl or dialkyl derivatives of barbituric acid can be prepared by condensing amino, alkyl or dialkyl malonate with urea. For example, 5,5-diethyl barbituric acid (barbitone or barbital or veronal) can be prepared from diethyl 5,5-diethylmalonate and urea in the presence of sodium ethoxide.

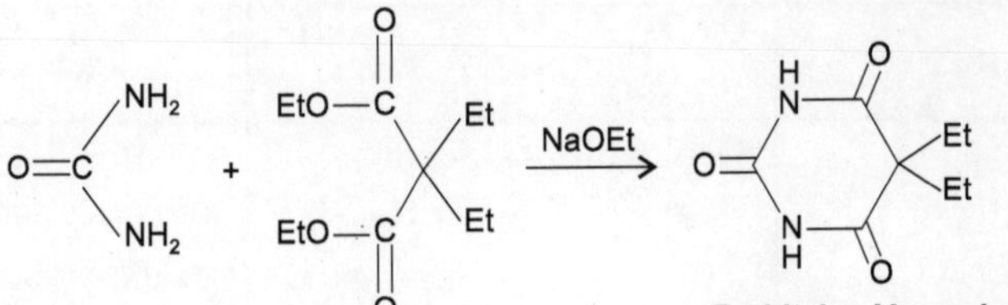

Barbital or Veronal

11.2.4 Reactions of barbituric acid

Barbituric acid has an active methylene group (CH$_2$) and one or two alkyl groups can be introduced at 5-position as shown below.

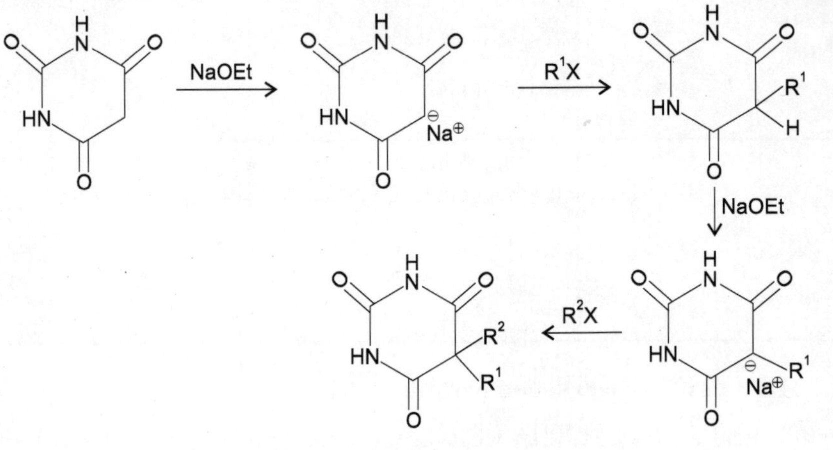

When R^1 = R^2 = Et; 5,5-Diethyl barbituric acid or veronal or barbitone or barbital

By varying R^1 and R^2 all other 5-disubstituted barbiturates can be synthesized

Barbituric acid undergoes nitration at the methylene carbon. It also reacts with diazonium salts and aldehydes.

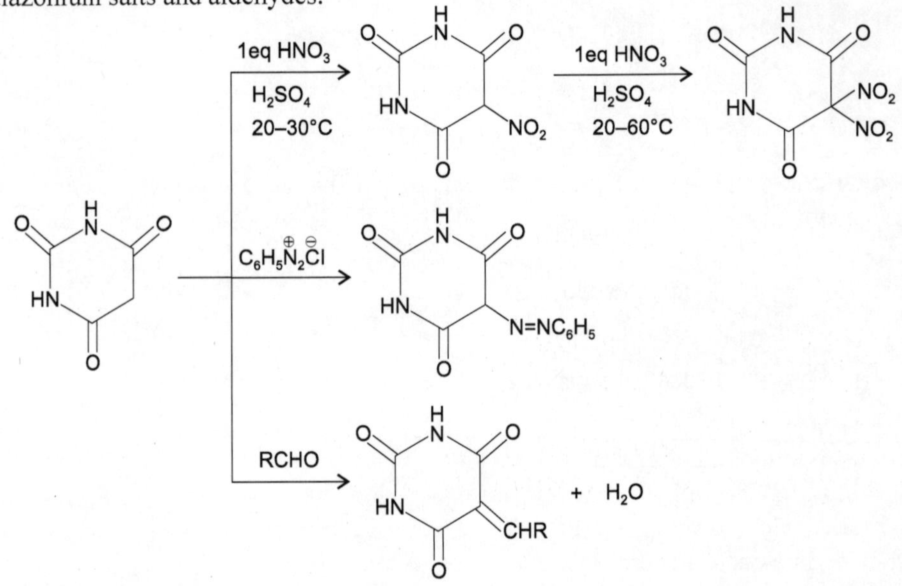

EXERCISE

Q.1 Draw the resonating structures of pyrimidine.

Q.2 How will you account for the trione structure of barbituric acid?

Q.3 Give a method for the preparation of trimethoprim and its use.

Q.4 What happens when pyridazine is treated with maleic anhydride?

[Hint. forms 1:2 adduct with maleic anhydride]

Q.5 Explain:

(a) The three diazines: pyridazine, pyrimidine and pyrazine are weaker bases than pyridine.

(b) The dipole moment of pyrazine is zero.

(c) Only one signal is observed in the ^1H and ^{13}C NMR spectra of pyrazine.

Q.6 Complete the following reactions:-

(a) Pyrimidine + m-CPBA + $CHCl_3$ at rt $\longrightarrow$

(b) Malonic acid + urea + $POCl_3$ $\longrightarrow$

(c) Pyridazine + Na + MeOH $\longrightarrow$

(d) Barbituric acid + CH_3CHO $\longrightarrow$

Ans. (a) Pyrimidine N-oxide   (b) Barbituric acid

(c) 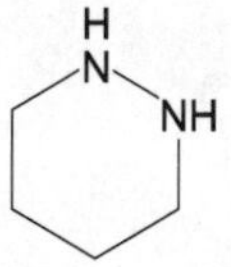   (d)

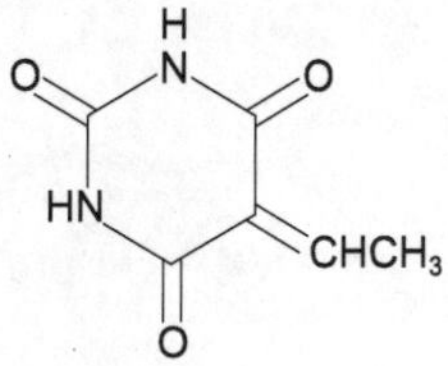

Q.7 Carry out the following conversions:-

(a) Pyrimidine to pyrazole

(b) Barbituric acid to pyrimidine

Q.8 Write a short note on Biginelli reaction

Q.9 Write the structures and give method of preparation of Veronal (barbital or barbitone) and Luminal (phenobarbital).

Bicyclic Ring Systems Derived from Pyridazine, Pyrimidine and Pyrazine

Cinnoline, Phthalizine, Quinazoline, Quinoxaline and Purine

12.1 BENZODIAZINES: CINNOLINE, PHTHALIZINE, QUINAZOLINE AND QUINOXALINE

12.1.1 Introduction

The benzofused derivatives of the three diazines (pyridazine, pyrimidine and pyrazine) are cinnoline, phthalizine, quinazoline and quinoxaline. These naphthyridines (compounds in which two of the carbon atoms of naphthalene ring have been replaced by nitrogen atoms) are isomeric to each other.

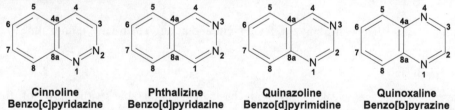

| Cinnoline | Phthalizine | Quinazoline | Quinoxaline |
| Benzo[c]pyridazine | Benzo[d]pyridazine | Benzo[d]pyrimidine | Benzo[b]pyrazine |

Cinnoline (mp 40°C), phthalizine (mp 90°C), quinazoline (mp 48°C) and quinoxaline (mp 31°C) are colourless crystalline solids. The taste of cinnoline resembles that of chloral hydrate.

12.1.2 Biological importance

Cinnoline shows antibacterial activity against *Escherichia coli*. **Luminol** (5-amino-1,2,3,4-tetrahydrophthalazine-1,4-dione) displays an intensely blue chemiluminescence on oxidation with alkaline hydrogen peroxide in the presence of haemin. **Hydralazine** a vasodilator and antihypertensive agent is a derivative of phthalazine.

Luminol

Hydralazine

The quinazoline skeleton appears in many alkaloids, most commonly in the form of 4-quinazolinone moieties. For example, the alkaloid **arborine** obtained from plant family rutaceae contains the quinazoline structure. **Methaqualone** (hypnotic), **quinethazone** (oral diuretic), **proquazone** (analgesic and antirheumatic) and **prazosin** (antihypertensive) are quinazoline based pharmaceuticals.

Arborine   **Methaqualone**   **Quinethazone**

Proquazone   **Prazosin**

Antibiotics such as **echinomycin, levomycin** and **actinoleutin** have quinoxaline ring system.

12.1.3 Basicity

Benzodiazines are weak bases with basicities comparable with those of the corresponding parent heterocycles, with the exception of quinazoline. Quinazoline (pKa of conjugate acid is 3.3) is much more basic than pyrimidine (pKa of conjugate acid is 1.3).

Pyridazine   **Pyrimidine**   **Pyrazine**
pK_a= 2.3 pK_a= 1.3 pK_a= 0.6

Cinnoline   **Phthalazine**   **Quinazoline**   **Quinoxaline**
pK_a= 2.6 pK_a= 3.5 pK_a= 3.3 pK_a= 0.6

(pK_a are of conjugate acids)

12.1.4 Spectral data

The ^{1}H and ^{13}C NMR (given in brackets) spectrum of benzodiazines shows signals at the following chemical shift values:

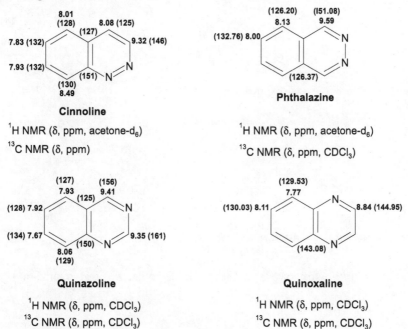

Cinnoline

^{1}H NMR (δ, ppm, acetone-d$_6$)

^{13}C NMR (δ, ppm)

Phthalazine

^{1}H NMR (δ, ppm, acetone-d$_6$)

^{13}C NMR (δ, ppm, CDCl$_3$)

Quinazoline

^{1}H NMR (δ, ppm, CDCl$_3$)

^{13}C NMR (δ, ppm, CDCl$_3$)

Quinoxaline

^{1}H NMR (δ, ppm, CDCl$_3$)

^{13}C NMR (δ, ppm, CDCl$_3$)

12.1.5 Methods of preparation of benzodiazines

12.1.5.1 Methods of preparation of cinnoline

(*i*) **von Richter cinnoline synthesis:** The preparation of cinnoline derivatives involving the diazotization of 3-(ortho-aminoaryl)propiolic acids or ortho-aminoarylacetylenes, and subsequently intramolecular cyclization of resulting diazonium intermediate in the presence of a halide or hydroxide is known as the **von Richter cinnoline synthesis**. For example, diazonium salt of 3-(ortho-aminophenyl)propiolic acid yields 4-hydroxycinnoline. The reaction is temperature dependant and the formation of cinnoline is favoured at high temperature.

von Richter, V., Ber. **1883**, *16, 677.*

Mechanism: Cinnoline derivatives are formed by diazotization of 3-(ortho-aminoaryl) propiolic acids followed by hydration and cyclization. The 3-(ortho-aminophenyl) propiolic acid diazonium chloride on heating in water at 70°C and subsequent cooling separates 4-hydroxycinnoline-3-carboxylic acid. When this acid is heated at 260°C, carbon dioxide is liberated with the formation of 4-hydroxycinnoline.

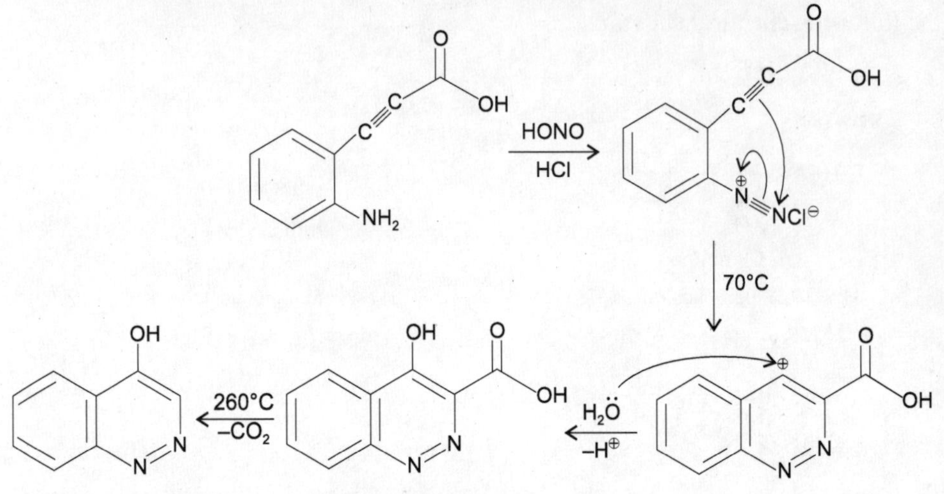

4-Hydroxycinnoline   **4-Hydroxycinnoline-3-carboxylic acid**

Similarly, the ortho-aminoarylacetylenes on diazotization and cyclization yields cinnolines.

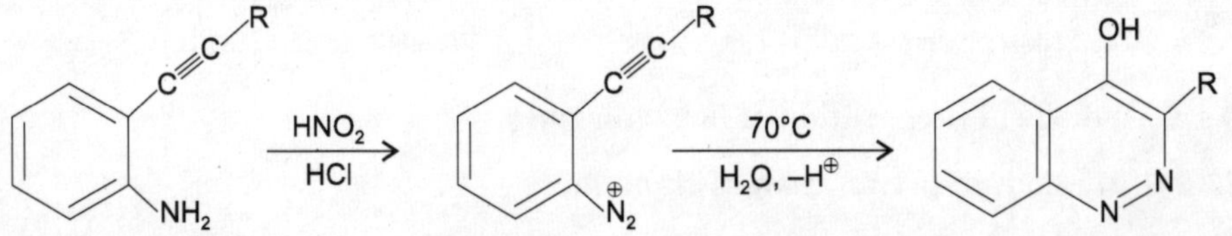

Busch and Rast converted 4-hydroxycinnoline to cinnoline *via* the formation of 4-chlorocinnoline. The 4-chlorocinnoline on treatment with iron/H_2SO_4 and followed by oxidation with mercuric oxide yields cinnoline.

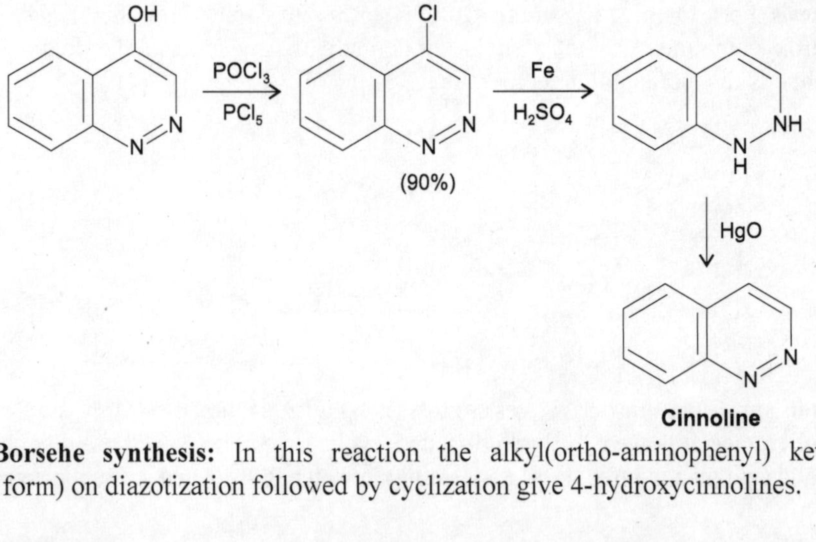

(90%)

Cinnoline

(*ii*) **Borsehe synthesis:** In this reaction the alkyl(ortho-aminophenyl) ketones (enol form) on diazotization followed by cyclization give 4-hydroxycinnolines.

Mechanism: The reaction involves intramolecular coupling of the diazonium cation with the enolate anion.

Another, example of **Borsehe synthesis** is the formation of 4-hydroxy-6-nitrocinnoline from 2-amino-5-nitroactophenone.

(80%)

(*iii*) **Widman-Stoermer synthesis:** The Widman-Stoermer synthesis involves formation of 3,4-disubstituted cinnolines from ortho-aminostyrenes.

Mechanism: It is a ring closing reaction of ortho-aminostyrenes with nitrous acid *via* the formation of diazonium salt.

(*iv*) **From cinnoline-4-carboxylic acid:** Cinnoline itself can be synthesized by decarboxylation of cinnoline-4-carboxylic acid on heating in presence of benzophenone.

12.1.5.2 Methods of preparation of phthalizine

(*i*) **From 1,2-diacyl benzenes:** 1,4-Disubstituted phthalizines can be obtained by cyclocondensation of 1,2-diacyl benzenes with hydrazine.

12.1.5.3 Methods of preparation of quinazoline

In accordance with the significance of quinazoline derivatives, various synthetic methods have been developed for their construction. Typically, quinazolin-4(3*H*)-ones are prepared from anthranilic acid or its derivative.

(*i*) **Niementowski synthesis:** The thermal condensation of anthranilic acid with an amide to 4-oxo-3,4-dihydroquinazoline (also known as benzoylene urea) is referred to as the **Niementowski reaction**. High yield is expected when anthranilic acid is condensed with a lower order of amides such as formamide or acetamide, whereas the yield decreases when a higher order of amide is used in this reaction.

Niementowski, S.V.J., Prakt. Chem. **1895**, *51, 564.*

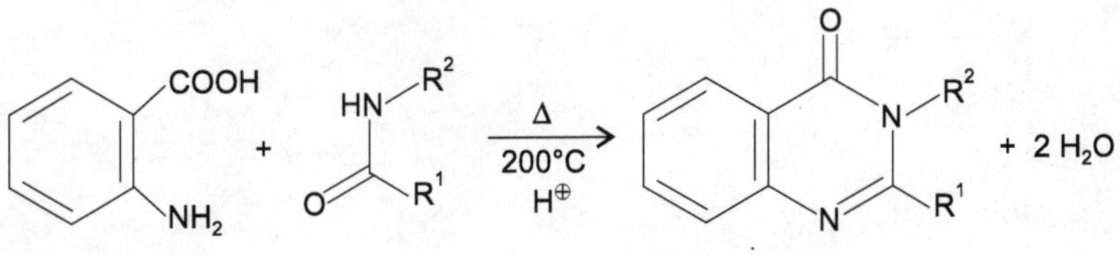

Anthranilic acid

4-Oxo-3,4-dihydroquinazolines
or
Quinazolin-4-(3*H*)ones

Mechanism: The reaction proceeds *via* acid catalyzed formation of ortho-amidine intermediate, followed by the nucleophilic attack of the nitrogen nucleophile at the carboxylic carbonyl group which is activated by protonation to produce the 4-oxo-3,4-dihydroquinazolines or quinazolin-4(3*H*)-ones upon elimination of water.

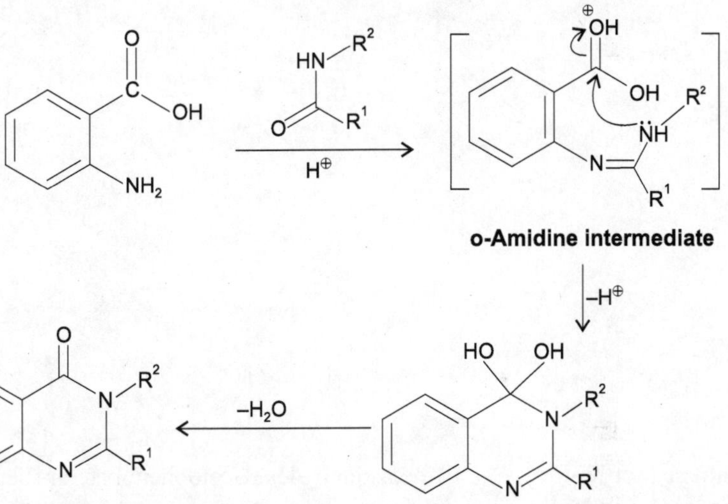

o-Amidine intermediate

Alternatively, 4-oxo-3,4-dihydroquinazolines can be synthesized by cyclization of N-acylanthranilic acid with ammonia or primary amines *via* amide formation.

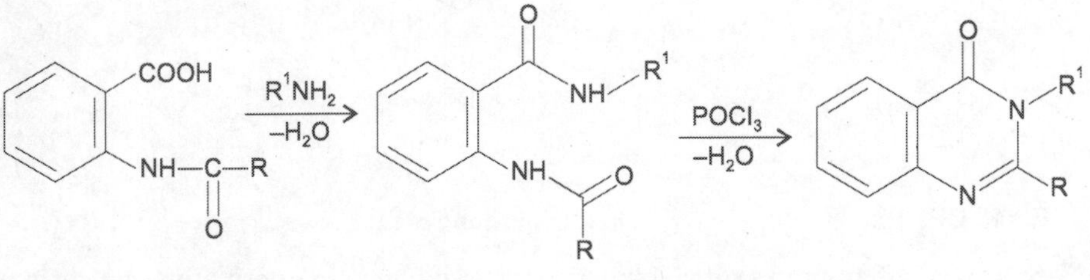

(*ii*) **Bischler synthesis** (first described in 1893): In this reaction ammonium salt of N-acylanthranilic acid undergoes cyclocondensation to generate 2-substituted quinazolinone.

Mechanism: It involves fusion of ammonium N-acetylanthranilate to generate a diamide which undergoes cyclodehydration to form 4-quinazolinone. 4-Quinazolinone tautomerises under reaction conditions to form 4-hydroxyquinazoline.

Alternatively, ortho-(acylamino)-benzaldehydes, acetophenones or benzophenones undergo cyclocondensation with ammonia to produce corresponding quinazoline derivative *via* intermediate imine.

R = H, CH₃, Ph   **Imine intermediate**

(*iii*) **Synthesis of trimetrexate:** Trimetrexate is quinazoline derivative which is an anticancer agent. It is a dihydrofolate reductase inhibitor. The synthesis of trimetrexate reported by Davoll and Johnson, involves an acid catalysed condensation reaction between 1-nitro-4-chloro-ortho-toluonitrile and guanidine to produce 2,4-diamino-5-methyl-6-nitroquinazoline. Nitro group is further reduced using H_2/Pd-C to give 2,4,6-triamino-5-methylquinazoline which is converted to nitrile on reaction with CuCN/HCl. Condensation of nitrile with 2,3,4-trimethoxyaniline produces trimetrexate.

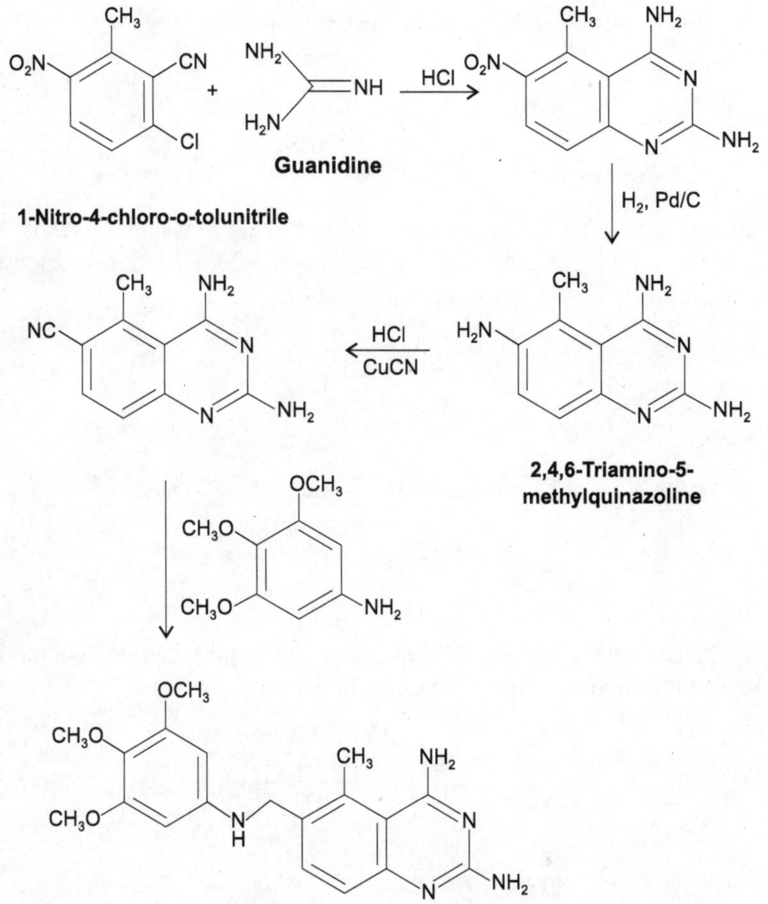
Guanidine
1-Nitro-4-chloro-o-tolunitrile
HCl
H2, Pd/C
HCl
CuCN
2,4,6-Triamino-5-methylquinazoline

Trimetrexate

12.1.5.4 Methods of preparation of quinoxaline

(*i*) **From ortho-diamines and dicarbonyl compound:** The classical synthesis of quinoxalines involves the condensation of an ortho-phenylenediamine and a 1,2-dicarbonyl compound. The reaction is very facile and is most widely used for the synthesis of quinoxaline itself and its alkyl substituted derivatives.

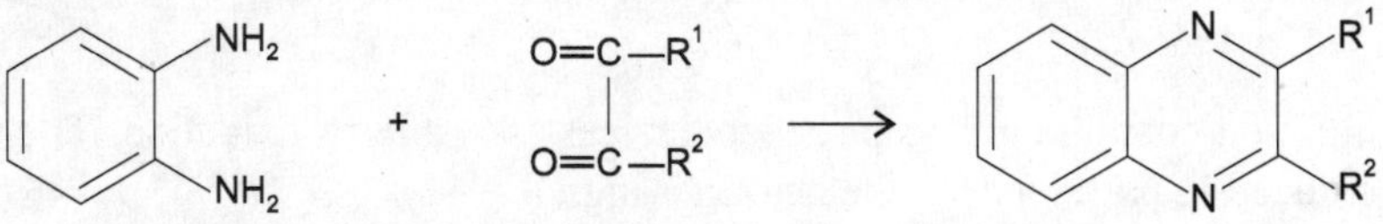

o-Phenylenediamine **α-Dicarbonyl compound**

The condensation of glyoxal with ortho-phenylenediamine yields quinoxaline in almost quantitative yield.

o-Phenylene diamine   Gyoxal   Quinoxaline

Substituted phenylglyoxals are the starting 1,2-dicarbonyl compounds for the synthesis of 2-arylquinoxalines. Similarly, the aryl-2-ketoacid on reaction with o-phenylenediamine yields 3-aryl-2-quinoxalinones.

Reaction of benzil with ortho-phenylenediamine in the presence of 2-iodoxybenzoic acid (IBX) as catalyst also yields substituted qunoxaline.

Benzil   o-Phenylenediamine

12.1.6 Reactions of benzodiazines

The known reluctance of pyridine to take part in electrophilic substitution reaction suggests that the introduction of a second nitrogen atom into the ring would render it even less reactive towards electrophiles. Thus reactivity of benzodiazines towards electrophiles is less than quinoline and isoquinoline. If S_EAr reactions take place, they lead to substitution in the benzene ring.

Nucleophilic substitution of benzodiazines occurs in the diazine ring, particularly if substituted by halogen.

The nitrogen-containing ring in benzodiazines is stable to oxidation. The diazine ring is reduced in preference to the benzene ring.

12.1.6.1 Reactions of cinnoline

Electrophilic substitution

Nitration of cinnoline with HNO_3/H_2SO_4 gives 5-nitrocinnoline and 8-nitrocinnoline.

5-Nitrocinnoline   **8-Nitrocinnoline**

Nucleophilic substitution

Substituted cinnolines undergo nucleophilic substitution reactions more readily. For example, the 4-chloro group of 4-chlorocinnoline can be replaced by a hydroxyl group, ethoxy group or aromatic amino group.

4-Chlorocinnoline

Reduction

4-Phenylcinnoline on treatment with zinc-acetic acid gets converted to 3-phenylindole. However, on reduction with zinc and ethanolic ammonia, 4-phenylcinnoline yields 1,2-dihydro-4-phenylcinnoline.

1,2-Dihydro-4-phenylcinnoline   **3-Phenylindole**

Reduction of 4-hydroxycinnolines with phosphorus and hydriodic acid leads to the formation of the tetrahydro derivative.

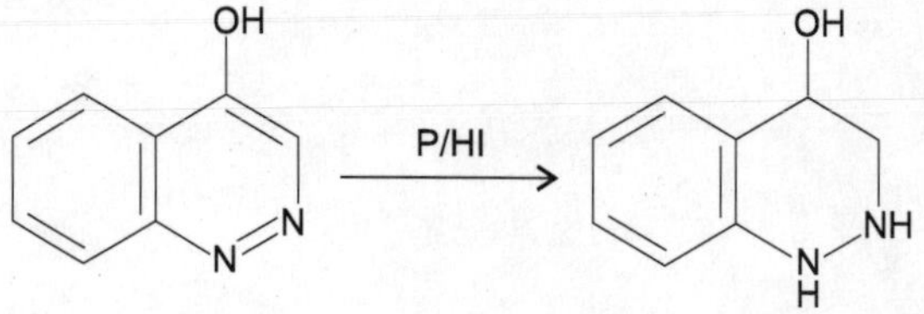

Oxidation

Oxidation of cinnoline with H_2O_2/AcOH or percarboxylic acids (RCOOOH) gives the mixture of cinnoline N-1-oxide, N-2-oxide and N-1,N-2-dioxide.

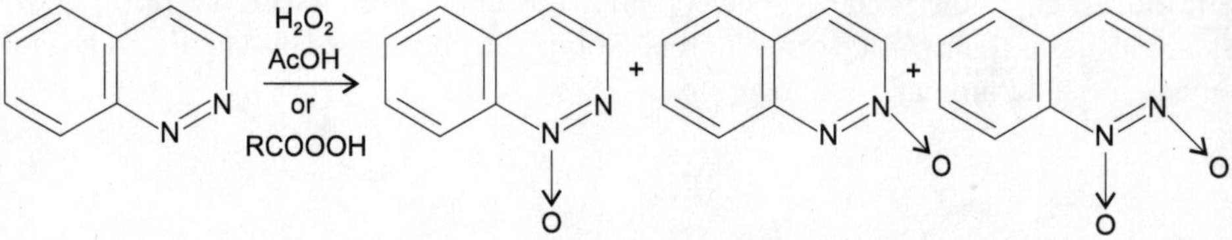

On oxidation with potassium permanganate, 4-phenylcinnoline gets converted to 5-phenylpyridazine-3,4-dicarboxylic acid, which can be decarboxylated to 4-phenylpyridazine.

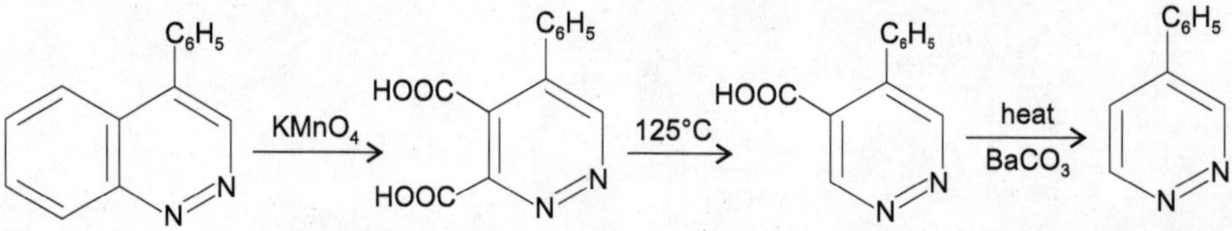

Photolysis

Cinnoline on photolysis with colloidal silver results into formation of N,N'-disilver derivative of ortho-(2-aminovinyl)aniline.

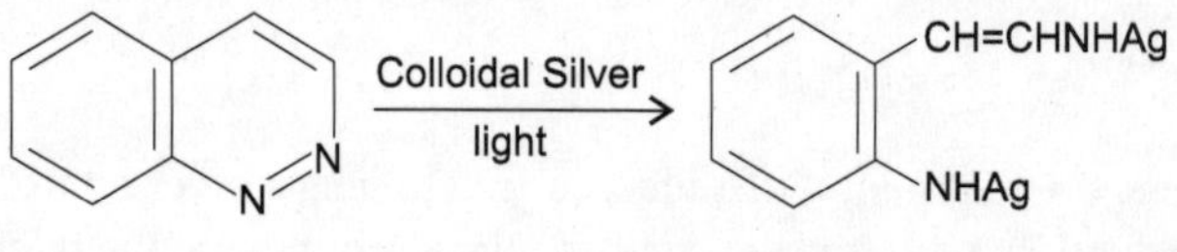

N,N'-Disilver derivative

12.1.6.2 Reactions of phthalizine

1-Methyl and 1-phenylphthalizine on reaction with methyl iodide in benzene gives quarternary salt.

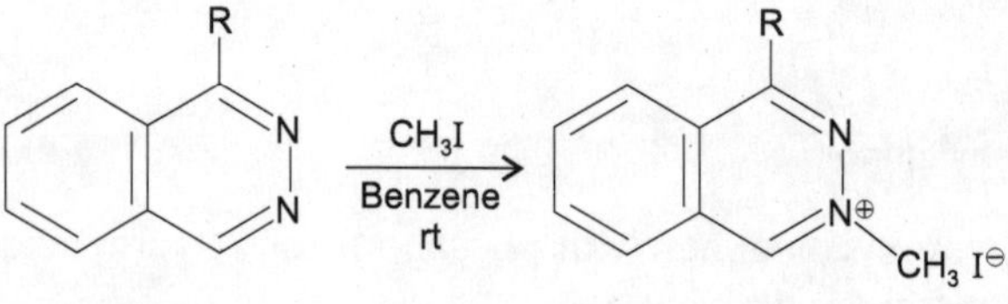

R = CH$_3$, C$_6$H$_5$

Nitration of phthalazine with KNO_3/H_2SO_4 gives mixture of 5-nitrophthalazine and 8-nitro-1(2H)-phthalazinone.

5-Nitrophthalizine   **8-Nitro-1(2H)-phthalazinone**

Reaction of phthalizine with methyl lithium in acidic medium yields 1-methyl-1,2-dihydrophthalizine which undergoes autooxidation to form 4-methylphthalizine and a dimer. Phthalizine also reacts with Grignard reagent to give 4-substituted phthalzine on 1,2-addition followed by autoxidation.

Phthalizine   **1-Methyl-1,2-dihydrophthalizine**

4-Phenylphthalizine   **4-Methyl**   **Peroxyphthalizine compound**
                            phthalizine   **(dimer)**

Phthalizine on reaction with pyridine-borane complex in acetic acid results into the formation of 2-acetyl-1,2,3,4-tetrahydrophthalizine.

2-Acetyl-1,2,3,4-tetrahydrophthalizine
(57%)

Reaction of phthalizine with monoperphthalic acid in ether yields phthalazine N-oxide. Upon oxidation with alkaline potassium permanganate phthalizine yield pyridazine-4,5-dicarboxylic acid.

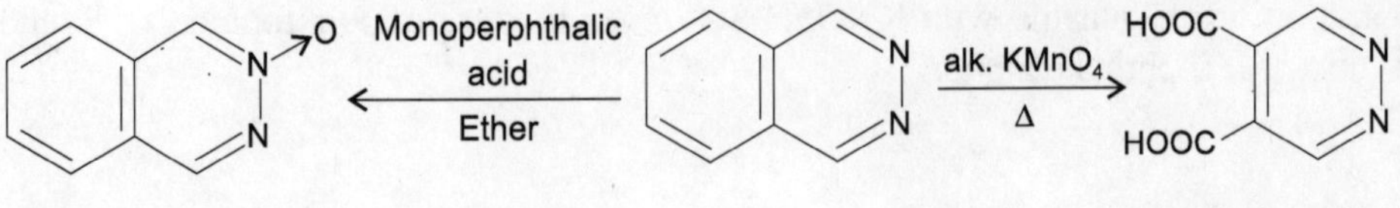

Phthalazine N-oxide

Pyridazine-4,5-dicarboxylic acid

12.1.6.3 Reactions of quinazoline

The reaction of 4-chloroquinazoline with phenyllithium resulted in the formation of 2-(benzylideneamino)benzonitrile as the minor product and 2-(diphenylmethylamino) benzonitrile as the major product.

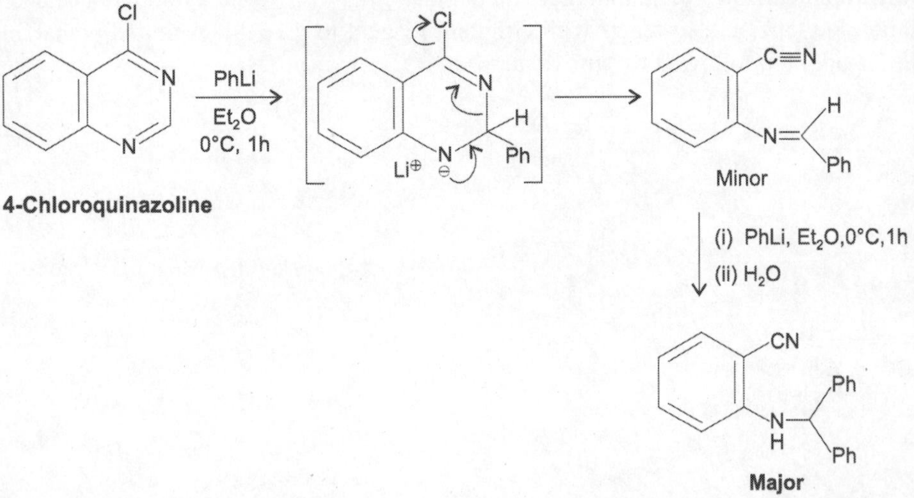

4-Chloroquinazoline

Minor

(i) PhLi, Et₂O,0°C,1h
(ii) H₂O

Major

The reactions of 2,4-dichloroquinazoline with amines, PhLi and alcohols shows C4 regioselectivity. The chlorine present at 4-position is substituted, however chlorine atom at 2-position remain intact.

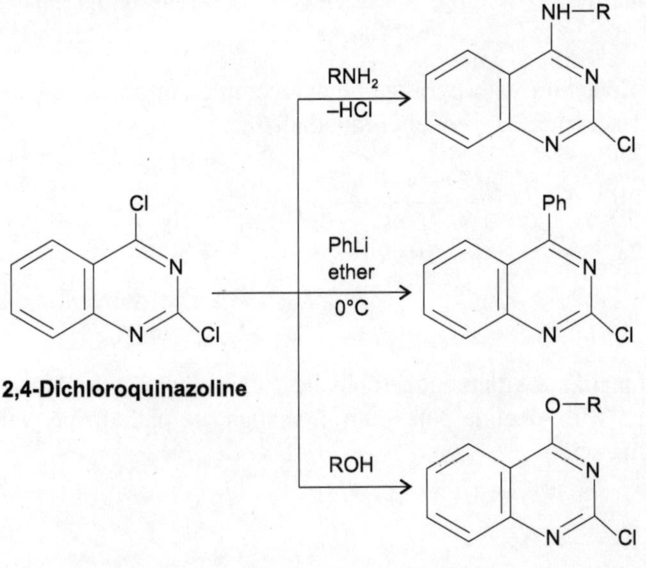

2,4-Dichloroquinazoline

The S_NAr reactivity of the quinazoline system offers synthetic applications for this heterocycle. 4-Chloro-2-phenylquinazoline, on reaction with phenolates, gets converted to the 4-(aryloxy)quinazoline. This undergoes a **Chapman rearrangement** at 300°C with migration of the aryl residue to N-3 forming the 3-arylquinazolin-4(3*H*)-one. The latter hydrolyses with aqueous acid to form benzoxazinone and a primary arylamine.

4-Chloro-2-phenylquinazoline

300°C
(1,3-migration)

$ArNH_2$ +

H_2O
$H^\oplus$

3-Arylquinazolin-4(3*H*)-one

Hence, these heterocycles can be used for the conversion of phenols into the corresponding primary arylamines.

However, 2-Chloro-4-phenylquinazoline under similar conditions i.e., with phenolate ions do not undergo substitution.

$ArO^\ominus$ → No reaction

2-Chloro-4-phenylquinazoline

Side chain reactivity is also observed in benzodiazines. For example, Mannich reaction of 2,4-dimethylquinazoline displays selectivity for the 4-substituted quinazolines.

HCHO
R_2NH

+ H_2O

2,4-Dimethylquinazoline

12.1.6.4 Reactions of quinoxaline

Electrophilic substitution reactions

The symmetry of quinoxaline ring makes the 6- and 7-positions equivalent. When activating substituents are present in the benzene ring substitution usually become more facile. When substitution is in the heterocyclic ring, the situation varies depending on the reaction conditions.

Nitration: Quinoxaline is resistant to nitration under mild conditions. However, on treatment with a mixture of oleum and nitric acid at 90°C for 24 h it gives 1.5% 5-nitroquinoxaline and 24% of 5,6-dinitroquinoxaline.

5-Nitroquinoxaline   **5,6-Dinitroquinoxaline**

(1.5%)   (24%)

Nitration of 6-methoxyquinoxaline and 5-methoxyquinoxaline gives 6-methoxy-5-nitroquinoxaline and 5-methoxy-6,8-dinitroquinoxaline, respectively.

6-Methoxyquinoxaline   **6-Methoxy-5-nitroquinoxaline**

5-Methoxyquinoxaline   **5-Methoxy-6,8-dinitroquinoxaline**

Free radical substitution

Irradiation of quinoxaline in acidified methanol furnishes 2-methylquinoxaline and the reaction is suggested to go through a pathway involving electron transfer from the solvent to an excited state of the protonated quinoxaline.

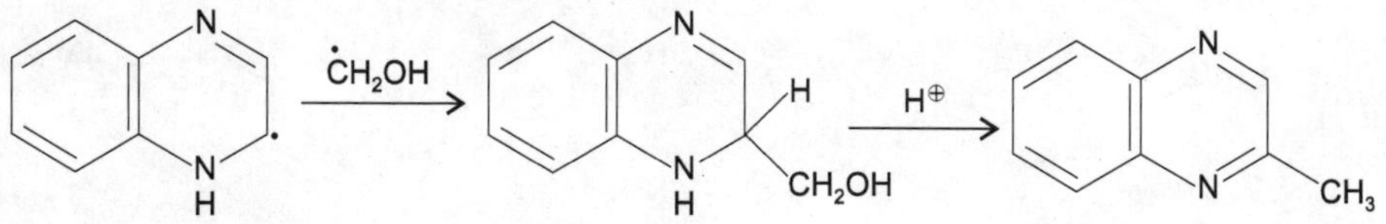

2-Methylquinoxaline

Nucleophilic addition reactions

Quinoxalines undergo facile addition reactions with nucleophilic reagents. Thus two molecular proportions of Grignard reagent can be added across quinoxaline molecule. The reaction of quinoxaline with allyl magnesium bromide gives after hydrolysis of initial adduct 86% of 2,3-diallyl-1,2,3,4-tetrahydroquinoxaline.

86%

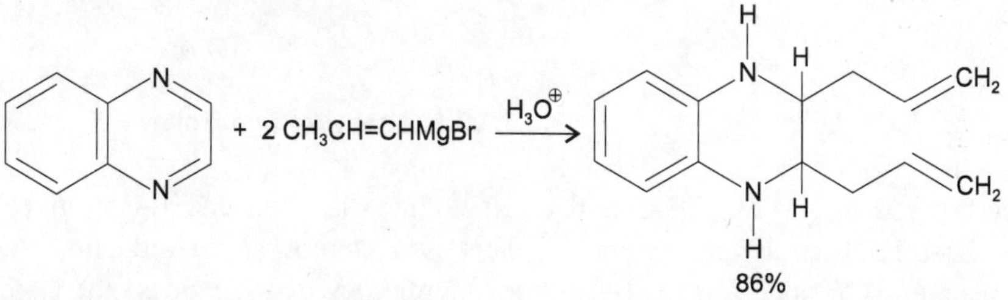

Similarly, the 2,3-bis[3-(dimethylamino)propyl]-1,2,3,4-tetrahydroquinoxaline derivative results from the reaction of quinoxaline and 3-(dimethylamino)propyl magnesium bromide followed by hydrolysis.

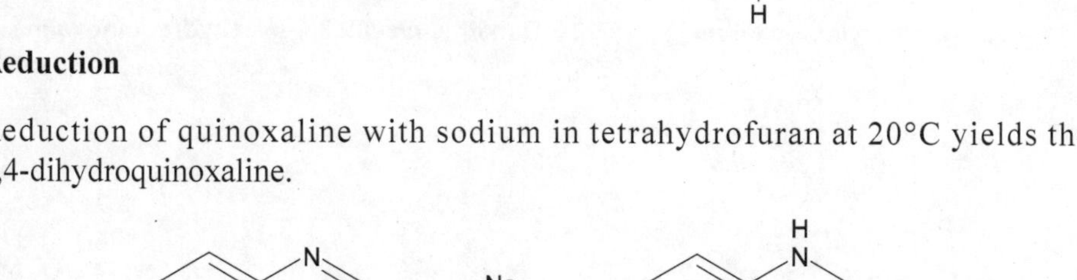

Reduction

Reduction of quinoxaline with sodium in tetrahydrofuran at 20°C yields the 1,4-dihydroquinoxaline.

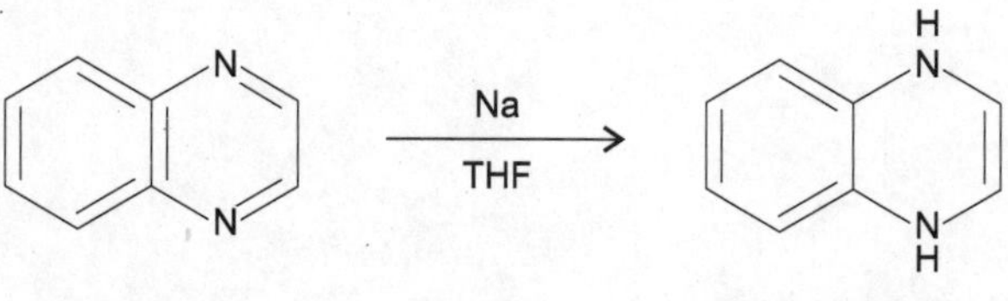

Qinoxaline   **1,4-Dihydroquinoxaline**

The 2-phenylquinoxaline on reduction with Na/THF first gives 2-phenyl-1,4-dihydroquinoxaline which readily rearranges to the thermodynamically more stable 1,2-dihydroquinoxaline.

2-Phenylquinoxaline   **2-Phenyl-1,4-dihydroquinoxaline**   **1,2-Dihydroquinoxaline**

1,2,3,4-Tetrahydroquinoxaline derivatives are formed when quinoxalines are reduced with lithium aluminium hydride in etheral solution.

$$\xrightarrow[\text{Ether}]{\text{LiAlH}_4}$$

1,2,3,4-Tetrahydroquinoxaline

Similarly, reduction of 2,3-dimethylquinoxaline in benzene also gives the meso(*cis*)-1,2,3,4-tetrahydro derivative. This is a stereospecific reduction and no *trans* isomer is formed, since the lithium aluminium hydride does not isomerise the dl-(*trans*) compounds. Low temperature, platinum catalysed hydrogenation of 2,3-dimethylquinoxaline in benzene also gives meso(*cis*)-1,2,3,4-tetrahydro-2,3-dimethylquinoxaline.

De Selms, R.C.; Mosher, H.S., J. Am. Chem. Soc. **1960**, *82, 3762-3765*

$$\xrightarrow[\text{Benzene}]{\text{LiAlH}_4}$$

2,3-Dimethylquinoxaline   **2,3-Dimethyl-*cis*-1,2,3,4-tetrahydroquinoxaline**

Pt/H$_2$
Low temp
Benzene

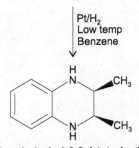

2,3-Dimethyl-*cis*-1,2,3,4-tetrahydroquinoxaline

Hydrogenation of quinoxaline or 1,2,3,4-tetrahydroquinoxaline over a 5% rhodium on alumina catalyst at 100°C and 136 atm. or over freshly prepared Raney nickel gives meso(*cis*) decahydroquinoxaline in high yield.

Quinoxaline

1,2,3,4-Tetrahydroqunoxaline

5% Rh-Al$_2$O$_3$, H$_2$
100°C, 136 atm
or
Raney Ni, H$_2$

***cis*-Decahydroquinoxaline**

Oxidation

Various methods have been used for N-oxidation of quinoxalines. Treatment of quinoxaline with one equivalent of peracetic acid in acetic acid gives quinoxaline-1-oxide and with excess of peracetic acid, quinoxaline-1,4-dioxide is formed.

Peracetic acid
excess

Peracetic acid
Acetic acid

Quinoxaline-1,4-dioxide

Quinoxaline-1-oxide

Reaction of quinoxaline with 30% aqueous hydrogen peroxide in acetic acid gives quinoxaline-2,3-dione. While with alkaline KMnO$_4$ as oxidising agent pyrazine-2,3-dicarboxylic acid is obtained.

alk.KMnO$_4$

30%
aq. H$_2$O$_2$
CH$_3$COOH

Pyrazine-2,3-dicarboxylic acid

Quinoxaline-2,3-dione

12.2 PURINE

12.2.1 Introduction

Purine is a heterocyclic aromatic compound with a pyrimidine ring fused to an imidazole ring. The word **purine** (means pure urine) was given by the German chemist Emil Fischer in 1884. The numbering of the purine ring system is anomalous and starts from the nitrogen of the pyrimidine nucleus.

Purine
(or 9H-Purine)

Four tautomeric forms are possible for the purine containing N-hydrogen. In the crystalline state purine exists predominantly as the 7H-tautomer. In solution both 7H- and 9H- tautomers of purine are present in approximately equal proportion whereas the 1H- and 3H- tautomers are not significant.

1H-Purine   **9H-Purine**   **7H-Purine**   **3H-Purine**

12.2.2 Importance of purines

Purines, substituted purines as well as their tautomers occur widely in nature. Purines and pyrimidine bases are constituents of DNA and RNA, which have great importance in life processes. **Adenine** and **guanine** are the two purine bases present in nucleic acids. Some other important purines are **xanthine, hypoxanthine, caffeine, theobromine, isoguanine** and **uric acid**.

Adenine   **Guanine**   **Xanthine**   **Hypoxanthine**

Caffeine   **Theobromine**   **Isoguanine**   **Uric acid**

Purines are also present in a number of other important biomolecules, such as ATP, GTP, cyclic AMP, NADH and coenzyme A. Meat products contain high concentration of purine but plant-based diets are low in purines. Higher levels of meat and seafood consumption are associated with an increased risk of gout.

12.2.3 Basicity

Purine is a weak base pK_a value of the conjugate acid is 2.5. It gets protonated in acidic medium, with the predominant form being the N-1-protonated cation.

pKₐ = 2.5

In strong acid solution a di-cation is formed by protonation at N-1 and the nitrogen of five membered ring of purine.

pKₐ = 2.5

Predominant cation   **Dication**

The basicity of purine is not much affected by the presence of oxygen functionality (hypoxanthine has a pK_a of 2.0 for conjugate acid). Presence of electron donating groups increase the basicity. Amino groups increase the basicity (pK_a for conjugate acid of adenine is 4.2), and oxo groups reduce the basicity of aminopurines (conjugate acid of guanine has a pK_a of 3.3).

pKₐ of conjugate acid :	2.0	4.2	3.3
	Hypoxanthine	**Adenine**	**Guanine**

12.2.4 Methods of preparation of purine

(*i*) **From uric acid:** Emil Fischer synthesized purine for the first time in 1899. The starting material for the reaction sequence was uric acid (isolated from kidney stones by Scheele in 1776). Uric acid on reactions with PCl_5 gives 2,6,8-trichloropurine, which in the presence of HI and PH_4I gets converted to 2,6-diiodopurine. The 2,6-diiodopurine thus formed can be reduced to purine using zinc dust.

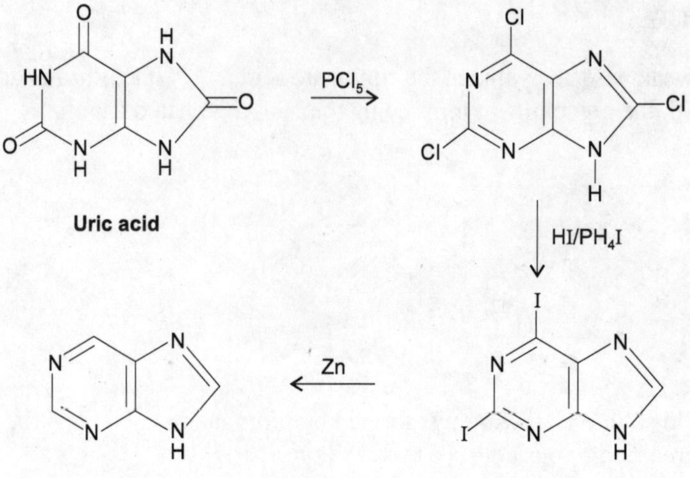

Uric acid

Purine

(ii) From formamide: The 6-aminopurine can be obtained by dehydration of formamide in the presence of $POCl_3$.

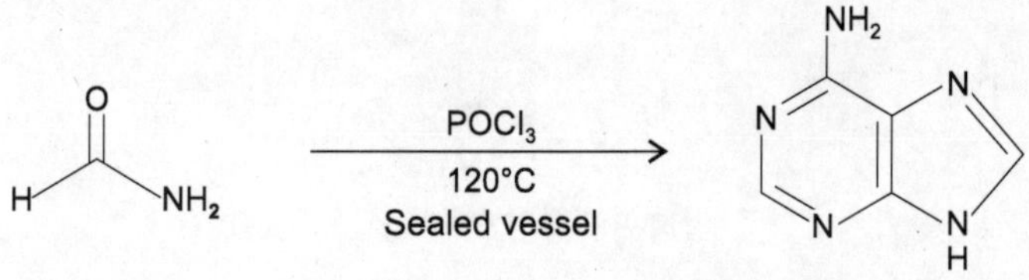

Formamide   **6-Aminopurine**

Purine is obtained in good yield when formamide is heated in an open vessel at 170°C for 28 hour.

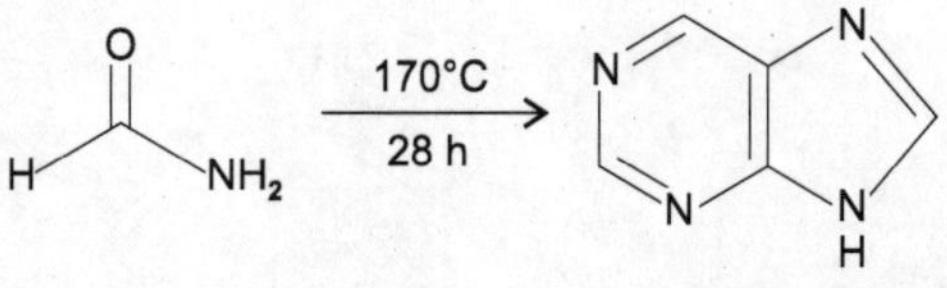

Formamide   **Purine**

(iii) From hydrogen cyanide: Adenine (pentamer of HCN) was first isolated in abiogenic experiments from an aqueaus solution of NH_3 and HCN in 1960 by Oro et al. Four molecules of HCN tetramerize to form diaminomaleodinitrile, which can be converted into almost all natural occurring purines (Oro and Orgel et al). Thus adenine, xanthine, hypoxanthine and guanine can be obtained through polymerisation of HCN under abiogenic conditions.

*Oro, J., Biochem. Biophys. Res. Comm. **1960**, 2, 407; Oro, J., Nature **1961**, 191, 1193.*

(*iv*) **Traube synthesis:** Traube purine synthesis is a classic reaction for the preparation of 8-unsubstituted purines by reaction of an amine substituted pyrimidine with formic acid or formamide on heating.

Traube first synthesised guanine from 2,5,6-triaminopyrimidin-4-ol and formic acid.

(*v*) **By acylation:** The 8-substituted purines are prepared using acylating agents corresponding to higher acids.

11.2.5 Reactions of purines

Alkylation at nitrogen

N-Alkylation of purines is complex and can take place on the neutral molecule or *via* an N-anion. Purine on reaction with iodomethane yields 7,9-dimethylpurinium salt.

Purines can undergo both electrophilic and nucleophilic attack at carbon in the five membered ring and nucleophilic reactions at carbon in the six membered ring.

Halogenation: Purine itself forms an [N–X]$^+$ complex, but does not undergo C-substitution. However, halogenation occur at C8 in the derivatives of adenine and xanthine.

Nitration: Under vigorous conditions, xanthines undergo nitration to form 8-nitro derivative.

Reaction with nucleophiles

The 2- and 6-chloro purines react with hydrazine which is a good nucleophile.

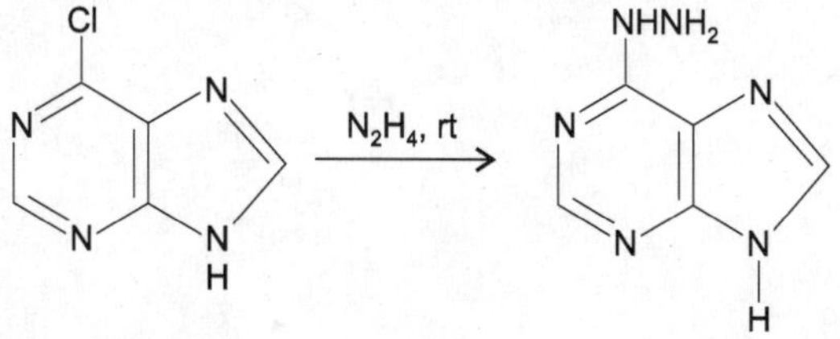

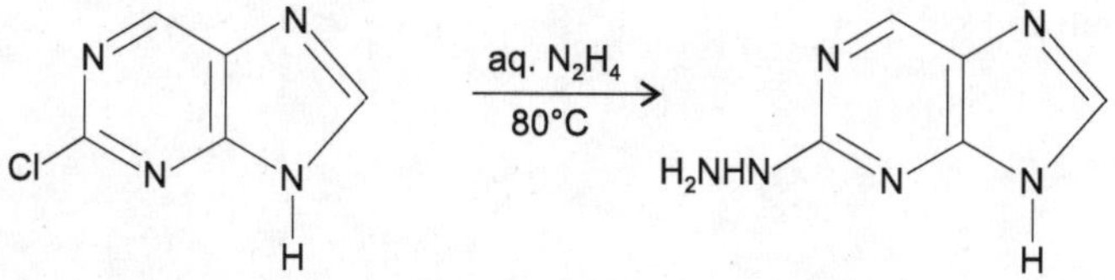

Reduction

The reduction of substituted purines is very complex and lead to the formation of ring-opened products. However, catalytic reduction in the presence of acylating agents yield acylated 1,6-dihydro purines.

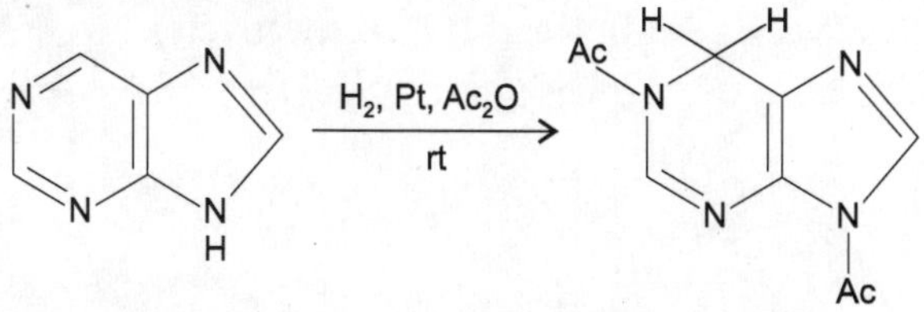

Oxidation

Peracid N-oxidation of purines forms 1- and/or 3-oxides, depending upon the reaction conditions. Adenine and adenosine give 1-oxides, whereas guanine yields the 3-oxide.

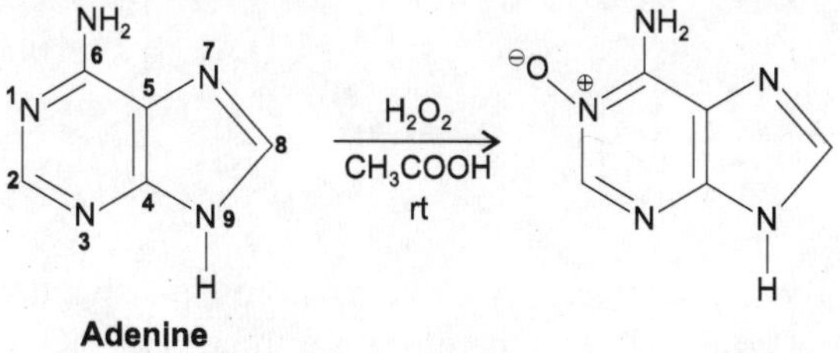

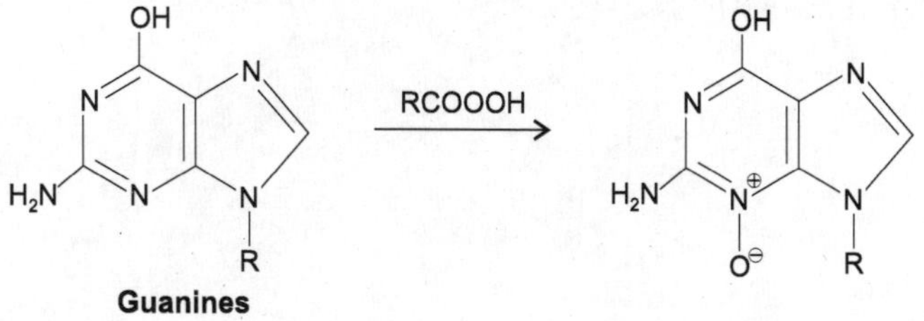

Adenine

Guanines

Coupling with diazonium salts

Amino- and oxy-purines couple with diazonium salts at their 8-position under weakly alkaline medium.

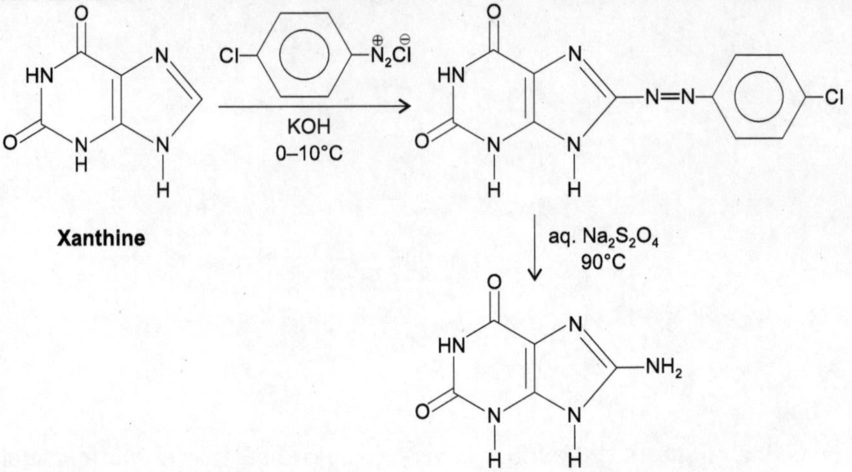

Xanthine

Deprotonation of N-hydrogen

Purine, (pK$_a$ of its conjugate acid is 8.9), is more acidic than imidazole or benzimidazole (pK$_a$ of their conjugate acids are 14.5 and 12.3, respectively).

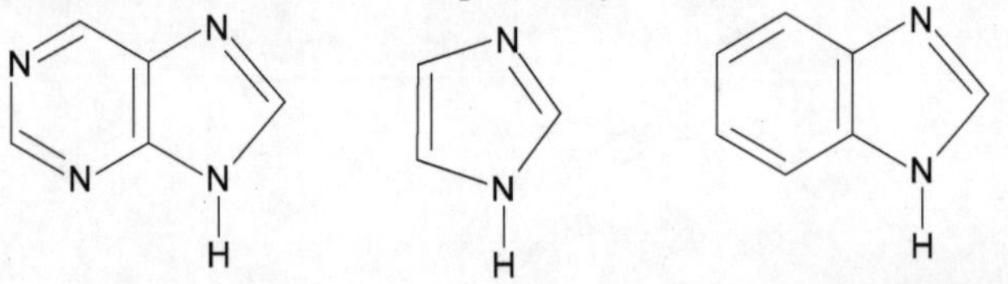

	Purine	**Imidazole**	**Benzimidazole**
pK$_a$ of conjugate acids :-	8.9	14.5	12.3

The relatively high acidity of purine is due to extensive delocalisation of the negative charge over the four nitrogens as shown below.

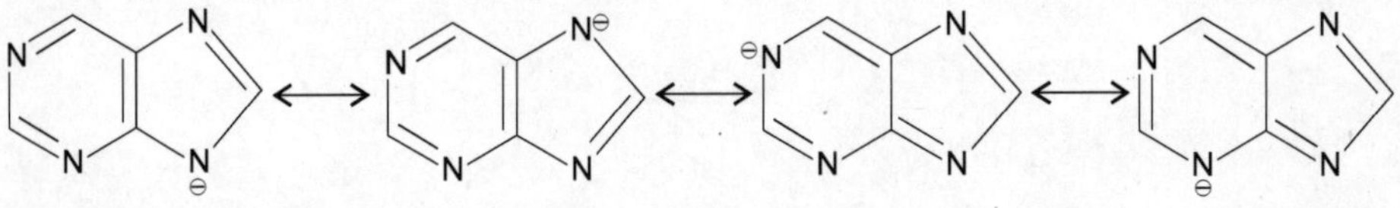

Oxy-purines are more acidic than purine due to more extensive delocalisation which involves the carbonyl groups as well, therefore, xanthine has a pK$_a$ of 7.5 and uric acid, 5.75 (of conjugate acids).

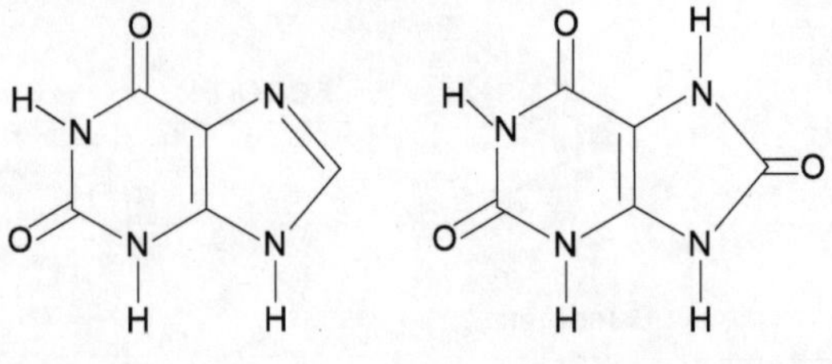

	Xanthine	**Uric acid**
pK$_a$ of conjugate acid :-	7.5	5.75

EXERCISE

Q.1 How will you synthesize purine from uric acid?

Q.2 Discuss Traube's method for the synthesis of guanine.

Q.3 Carry out the following conversions:

(a) 4-Hydroxycinnoline to cinnoline

(b) 4-Phenylcinnoline to 4-phenylpyridazine

Q.4 Write short note on:-

(a) Widman-Stoermer synthesis

(b) Borsehe synthesis

(c) Niementowski synthesis

(d) Bischler synthesis

Q.5 Complete the following reactions:-

(a) Ortho-Phenylenediamine + Glyoxal ⟶

(b) Cinnoline + H_2O_2/AcOH ⟶

(c) Phthalizine + Alk. $KMnO_4$ ⟶

(d) Quinoxaline + H_2/Raney Ni ⟶

Ans. (a) Phthalizine

(b) Mixture of cinnoline N-1-oxide, N-2-oxide and N-1,N-2-dioxide

(c) Pyridazine-4,5-dicarboxylic acid

(d) Decahydroquinoxaline

Q.6 What happens when 4-phenylcinnoline is treated with zinc/acetic acid?

Ans. 3-Phenyl indole is obtained.

Q.7 Explain why purine is more acidic than benzimidazole.

Ans. Due to extensive delocalisation of the negative charge on the four nitrogens.

Seven Membered Heterocyclic Compounds

Azepine, Oxepin, Thiepin and Benzodiazepines

13.1 AZEPINE, OXEPIN AND THIEPIN

13.1.1 Introduction

Azepine, oxepin and thiepin are the three main unsaturated seven membered heterocyclic compounds with one heteroatom.

1*H*-Azepine **Oxepin** **Thiepin**

These seven membered heterocyclic rings containing one heteroatom are not planar and do not comply with Huckel rule. Thus they behave as polyenes and are nonaromatic in nature.

Out of these the nitrogen containing seven membered heterocyclic compound exists in four tautomeric forms these are the 1*H*-, 2*H*-, 3*H*- and 4*H*- azepines. Amongst these four forms the 3*H*-azepine is more stable. The 1*H*-azepine is an unstable red oil at −78°C. The 1*H*-azepine can rearrange to the more stable 3*H*-azepine in the presence of acid or base. The parent 1*H*-azepine is not stable, although many N-substituted derivatives are known and can be synthesized.

1*H*-Azepine **2*H*-Azepine** **3*H*-Azepine** **4*H*-Azepine**

Tautomeric forms of azepine

Oxepin exists as an equilibrium mixture of its two equally contributing valence-tautomeric monocyclic and bicyclic forms. The bicyclic form is a benzene oxide (1,2-epoxybenzene). The energy barrier of the benzene oxide and oxepine tautomerization is 29.4 kJ/mol. The bond parameters of oxepin are not known.

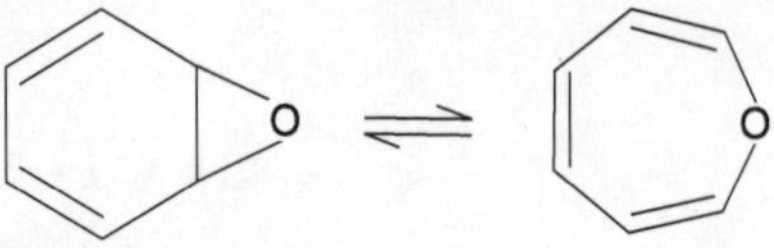

Benzene oxide   **Oxepin**

Spectroscopic data confirms that oxepins have a polyolefinic structure with localized C=C bonds. Oxepin exists in a non planar boat conformation.

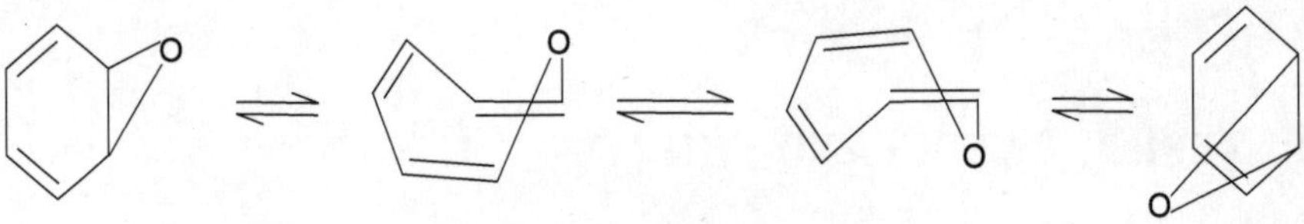

Conformations of oxepin

13.1.2 Biological importance

The seven membered heterocycles are found in many bioactive compounds. An azepine based drug **carbamazepine** is used for the treatment of epilepsy and bipolar disorder. **(±)-Janoxepin** is a tricyclic oxepin containing antiplasmodial natural product isolated from the fungus *Aspergillus janus*. **Zaltoprofen** containing a seven membered ring system with sulfur acts as a nonsteroidal anti-inflammatory drug.

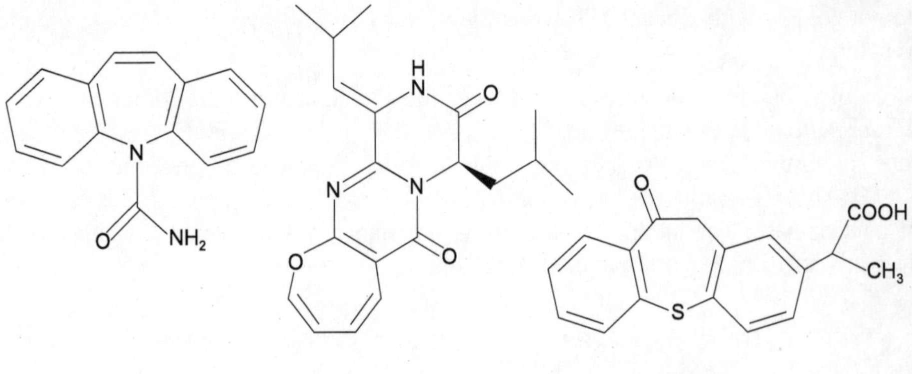

Carbamazepine   **(±)-Janoxepin**   **Zaltoprofen**

13.1.3 Spectral data

The ^1H and ^{13}C NMR (given in brackets) of $1H$-azepine and oxepin are shown below:

1H-Azepine

¹H NMR (δ, ppm, CDCl₃)
at –60°C

Oxepin

¹H NMR (δ, ppm)
¹³C NMR (δ, ppm)

13.1.4 Methods of preparation of azepine, oxepin and thiepin

13.1.4.1 Methods of preparation of azepine

(i) Synthesis of N-ethoxycarbonyl-1H-azepine: The 1H-azepines are generally obtained by valence-bond isomerization of azanorcaradines.

Azanorcaradines   **1H-Azepines**

The azanorcaradines are themselves prepared by the reaction of arenes with nitrene. For example, the reaction of benzene with ethoxycarbonyl nitrene gives N-ethoxycarbonyl-1H-azepine *via* valence bond isomerization. This is a stable compound because of the presence of electron withdrawing groups on the nitrogen atom. However, this on hydrolysis forms 1H-azepine which tautomerizes to 3H-azepine.

N-Ethoxycarbonyl-1H-azepine

(i) $\overset{\ominus}{O}H$
(ii) heat

3H-Azepine   **1H-Azepine**

(*ii*) **Synthesis of 2-alkoxy-3H-azepine:** Nitrobenzene on deoxygenation with tributylphosphine forms intermediate phenyl nitrene which on reaction with primary or secondary alcohols gives 2-alkoxy-3H-azepine.

Phenyl nitrene   2-Alkoxy-3H-azepine

(*iii*)   **Synthesis   of   N-alkoxycarbonyl-4,5-dimethoxycarbonyl-1H-azepine:**
N-Alkoxycarbonyl pyrrole reacts with dimethyl acetylenedicarboxylate (DMAD) in
the presence of aluminium chloride in DCM to give Diels-Alder adduct, which on
photolysis gives corresponding 7-azaquadricyclane. Thermal rearrangement of which
yields N-alkoxycarbonyl-4,5-dimethoxycarbonyl-1H-azepine.

DMAD

Diels-Alder adduct

N-Alkoxycarbonyl-4,5-dimethoxy   **7-Azaquadricyclane**
carbonyl-1H-azepine

(*iv*) **Synthesis of N-carbomethoxyazepine:** Monoaddition of iodine isocyanate
(pseudohalogen) to 1,4-dihydrobenzene (obtained from Birch reduction of
benzene) followed by refluxing with methanol yields   *trans*-iodocarbamate. The
trans-iodocarbamate undergoes cyclization to yield an aziridine derivative in the
presence of sodium methoxide and dry tetrahydrofuran. Bromination of the aziridine
derivative gives a dibromo compound. The dibromo compound gets converted to
N-carbomethoxyazepine under reflux conditions in the presence of NaOMe/THF.

1,4-Dihydrobenzene   ***trans*-Iodocarbamate**
                                      (54%)

N-Carbomethoxyazepine

(*v*) **Synthesis of 1*H*-azepine:** N-carbomethoxyazepine on reaction with trimethylsilane yields trimethylsilyl 1*H*-azepine-N-carboxylate. This when treated with methanol forms the corresponding carbamic acid which on decarboxylation yields 1*H*-azepine in solution at –60°C.

(*vi*) **Synthesis of 2*H*-azepine:** The 2*H*-azepine could be prepared for the first time in only 1% yield. The reaction proceeds *via* a ring construction involving N-Boc deprotection, followed by treatment with strong base (DMAP or DABCO). The 2*H*-azepine was obtained after an intramolecular imine formation and base-induced elimination of acetate. The 2*H*-azepine formed was stable at 25°C for 48 h.

2H-Azepine
(1%)

13.1.4.2 Method of preparation of oxepin

(*i*) **From cyclohexa-1,4-dienes:** The oxepins are synthesized from benzene monoxide in a manner similar to the synthesis of azepine *via* valence-bond isomerization of azanorcaradines.

The cyclohexa-1,4-dienes can be readily converted to monoepoxide, which undergoes bromine addition to the other double bond giving 3,4-dibromo-7-oxabicyclo[4.1.0] heptane. Double dehydrobromination of this compound with sodium methoxide or DBU leads to the formation of benzene oxide which is in equilibrium with monocyclic form, oxepin.

Oxepin

For example 3,7-dimethyloxepin can be synthesized by the above method as shown.

3,7-Dimethyloxepin

(*ii*) **From 2,5-disubstituted furans:** Thermal reaction of 2,5-disubstituted furans with dimethyl acetylenedicarboxylate (DMAD) gives Diels-Alder adduct, which on photolysis gives corresponding 7-oxaquadricyclanes. Thermal rearrangement of which yields 2,7-disubstituted-4,5-dimethoxycarbonyloxepin.

(*iii*) **From Dewar benzene:** Epoxidation of Dewar benzene with m-CPBA followed by photolysis or pyrolysis of the monoepoxide yields oxepin.

Dewar benzene

Oxepin

13.1.4.3 Methods of preparation of substituted thiepin

The seven membered sulfur containing thiepin has not been synthesized till now. However, substituted thiepins are known. These are stable compounds and can be synthesized.

(*i*) **From 3-aminothiophenes:** Substituted thiepin can be synthesized by the reaction of 3-aminothiophenes and activated alkyne *via* [2+2] cycloaddition followed by electrocyclic cyclobutene fission.

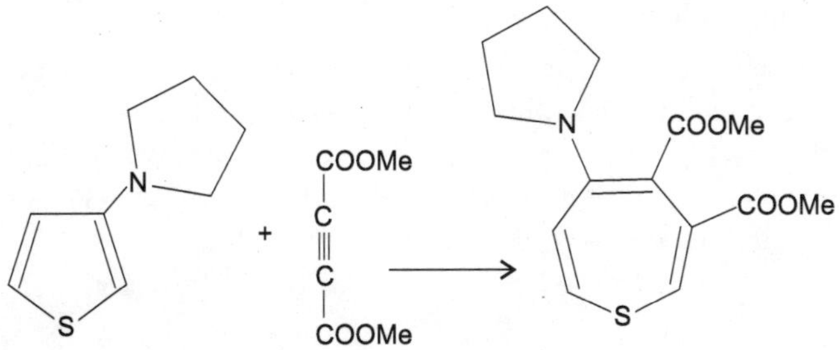

13.1.5 Reactions of azepine, oxepin and thiepin

13.1.5.1 Reactions of azepine

2-Methyl-N-carbomethoxyazepine undergoes photo ring contraction to give 2-carbomethoxy-3-methyl-2-azabicyclo[3.2.0]hepta-3,6-diene.

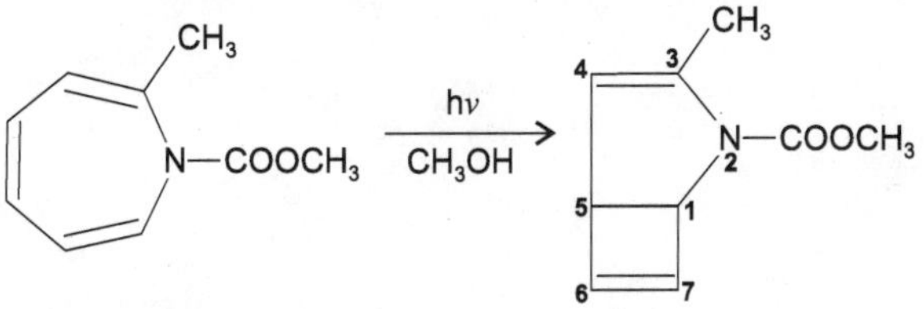

13.1.5.2 Reactions of oxepin

Reaction with acids

Oxepins undergo thermal or acid catalyzed rearrangement to phenols *via* their arene oxides. For example 4-methyloxepin gives para-cresol on treatment with acids.

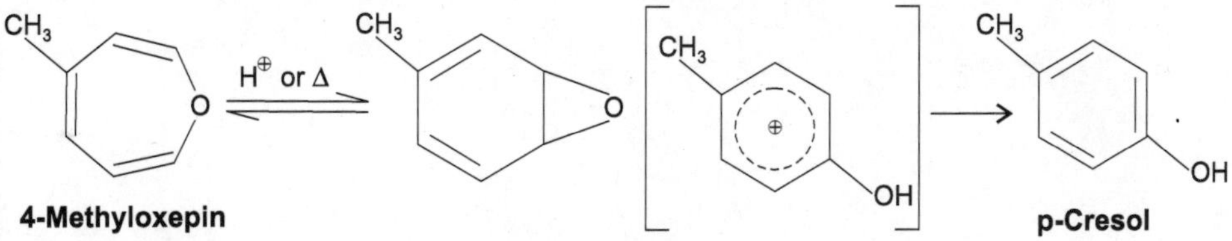

4-Methyloxepin   **p-Cresol**

Bromination

Bromination of oxepin gives 2,7-dibromooxepin, which reacts with cyclopentadienyl magnesium bromide to give 2,7-dicyclopentadienylyloxepin. Which further on treatment with triethylamine give fulvalene derivatives.

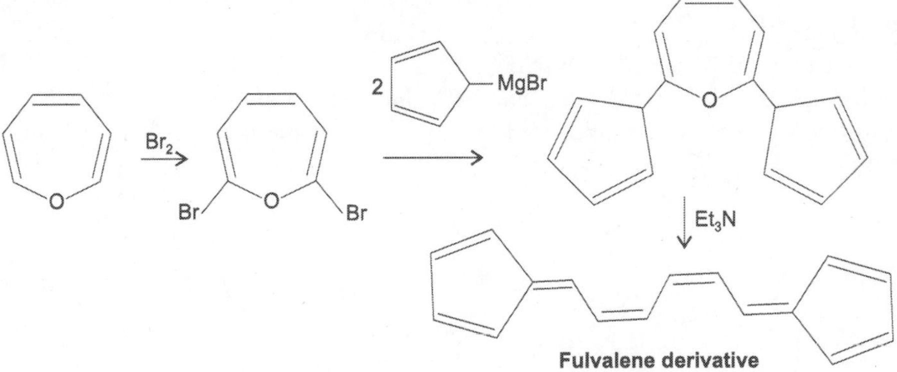

Fulvalene derivative

Reaction with nucleophiles

The arene oxide, tautomeric form of oxepin, reacts with various nucleophiles such as azide anion, methyl lithium, dimethyl magnesium, methanol/methoxide, NaSR, Na_2S to give the corresponding cyclohexadiene derivative. Water adds to oxepin in the presence of *Epoxide hydrolase* to yield trans diol. However, it does not react with nitrogen nucleophiles such as NH_3, NH_2^-, RNH_2.

13.1.5.3 Reactions of thiepin

The oxidation of 2,7-di-tert-butylthiepin gives 2,7-di-tert-butylthiepin-1-oxide which at a temperature above -15°C gets converted into ortho-di-tert-butylbenzene. In this case the 1-oxide is not oxidized to dioxide because of the bulky groups at 2 and 7 position.

ortho-di-tert-butylbenzene

Thiepin can undergo desulfurization to give arenes *via* formation of thiirane.

13.2 BENZODIAZEPINES

13.2.1 Introduction

Benzodiazepines are heterocyclic ring systems formed by the fusion between benzene and a diazepine ring system. Diazepine is a heterocycle with two nitrogen atoms, five carbon atoms and the maximum possible number of cumulative double bonds. Benzodiazepines are a large class of compounds, used as sedatives, anxiolytics and skeletal muscle relaxants. The first synthesized benzodiazepine was **chlordiazepoxide** (Librium), and it was discovered serendipitously by L. Sternbach and E. Reeder. **Diazepam** (Valium) is another benzodiazepine based drug. Diazepam is often used before a surgery to sedate the patient and remove anxiety. It causes temporary loss of memory. Diazepam is also used for treatment of anxiety, insomnia and symptoms of alcohol withdrawal.

Librium (Chlordiazepoxide)   **Diazepam (Valium)**

13.2.2 Synthesis of diazepam or valium

The synthesis of diazepam from N-methyl-4-chloroaniline is shown below.

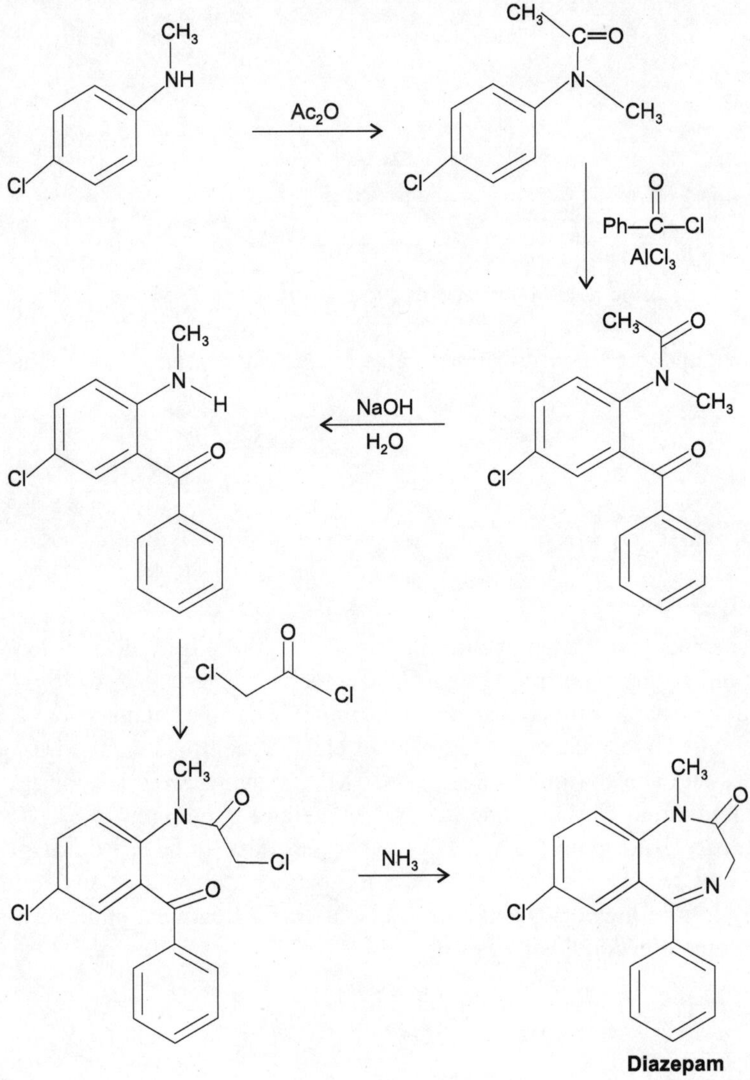

Diazepam

EXERCISE

Q.1   Draw the non-planar boat conformation of oxepin.

Q.2   Discuss the synthesis of diazepam.

Q.3   How will you convert Dewar benzene into oxepin?

Q.4   Give a method of synthesis of oxepin from cyclohexa-1,4-dienes.

Q.5   Draw the structures of the four tautomeric forms of azepines.

Q.6   How will you synthesize 3H-azepine from benzene?

Q.7   What product is obtained on reaction of oxepin with azide anion.

Q.8   Give the synthesis of valium. What are its uses.

Q. 9   Complete the following reactions:

(a) Benzene $+ N_3COOC_2H_5 + hv \longrightarrow$

(b) Oxepin + $Br_2 \longrightarrow$

Ans. (a) N-Ethoxy carbonyl-1H-azepine

(b) 2,7-Dibromooxepin

Q.10 What happens when oxepin is treated with $NaSCH_3$.

Ans.

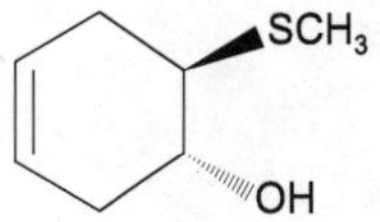

Porphyrins

14.1 PORPHYRINS

14.1.1 Introduction

Porphyrins are naturally occurring nitrogenous biological pigments. The name of porphyrin is derived from the Greek word 'porphyra' which means purple. Porphyrins play essential roles in living organisms. They form the base structure of chlorophyll a, which is involved in photosynthesis as well as a variety of other hemoproteins consisting of porphyrin combined with metals and proteins. Some iron-containing porphyrins are called **hemes**. In animals, iron protoporphyrin IX performs the oxygen carrying and release functions in hemoglobin. Many of the proteins in the mitochondrial electron-transport chain also contain porphyrins. Two Nobel prizes have been awarded for work related to porphyrins. One to Richard Willstatter in 1915 for his research on plant pigments, especially chlorophyll, and the other to Hans Fischer in 1930 for his research on constitution of haemin and chlorophyll and especially for his synthesis of haemin ($C_{34}H_{32}O_4N_4FeCl$).

14.1.2 Structure of porphyrins

A porphyrin ring system consists of four pyrrole rings joined at their α-carbon atoms *via* methine bridges (=CH−). The porphyrin macrocycles are highly conjugated systems, thus have intense absorption bands in the visible region and deep colour. This macrocycle has 26 π-electrons in total. The parent porphyrin is **porphin**, and substituted **porphins** are called porphyrins. Substituted porphyrins are important naturally occurring heterocyclic compounds.

Porphin ring system

5,10,15,20-Tetraphenylporphyrin

There are two distinct substitution patterns namely, β and **meso** in porphyrins. The β-substituted porphyrins resemble naturally occurring porphyrins. However, the meso-substituted porphyrins have no direct biological counterparts. They are found to have application in biometric models and in material chemistry. The substituents at the meso-position can include alkyl, aryl, heterocyclic or organometallic groups, or other porphyrins. The meso-substituted tetra-aryl porphyrins provide versatile building blocks for creating three-dimensional structural patterns.

β-Substituted porphyrin = 2,3,7,8,12,13,17,18   *meso*-Substituted porphyrin = 5,10,15,20

Porphin ring system is highly delocalized as there are nine double bonds around the periphery and hence 18-electron system. Thus these compounds are aromatic having (4n + 2) π-electrons, as in case of [18]annulenes.

18e around the periphery of porphin ring

In porphyrins the two protons on nitrogens are mobile and jump freely among the four nitrogens. Thus the porphin ring system can exist in two tautomeric forms. They are weak bases and they are protonated to form dications. The extensive conjugation in porphyrin and delocalization of π-electrons within the ring makes them intensely colored compound. It is this extensive conjugated system of porphyrin which gives blood its characteristic red color.

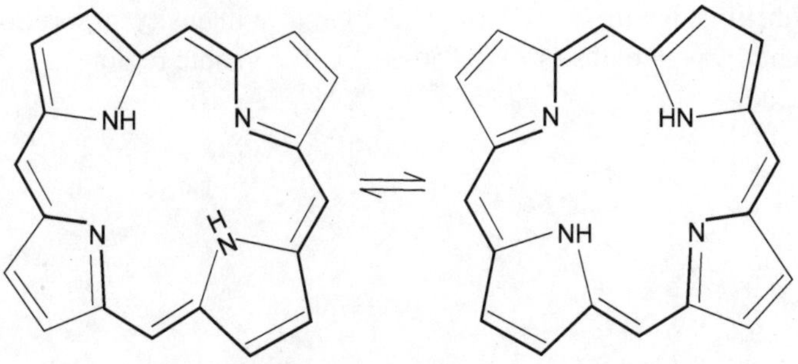

Tautomerism in porphyrin

Porphyrins absorb light in the visible spectrum and have been used as dyes. Because of their unique chemistry, porphyrins are involved in metal binding (ligands), solar cells (convert light or chemical energy), oxygen transport medium (hemoglobin), electron transfer medium (conducting polymers), iron metabolism, hormone synthesis, gene regulation and drug metabolism.

14.1.3 Spectral data

The ^{1}H NMR chemical shift values for the three types of protons in porphin and tetraphenylporphyrin are shown below.

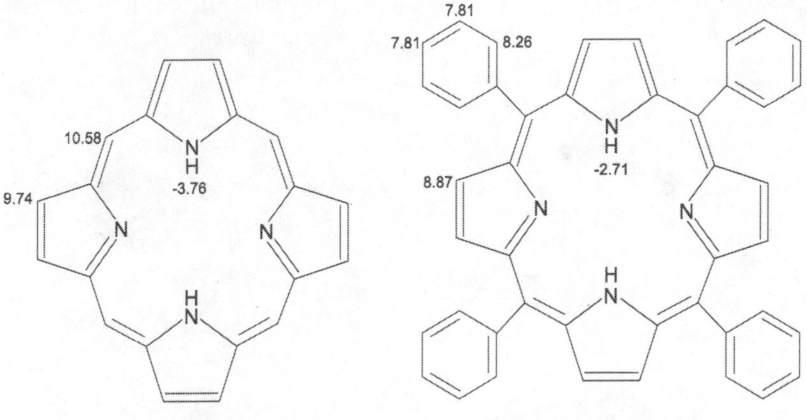

Porphin
^{1}H NMR (δ ppm, CDCl$_3$)

Tetraphenylporphyrin
^{1}H NMR (δ ppm, CDCl$_3$)

The extended delocalization of π-electron system in porphyrins gives rise to a strong ring current. Due to the anisotropic effect from the porphyrin ring current, the ^{1}H NMR signals for the deshielded meso protons (protons on the bridging methine carbons) appear downfield, whereas the signals for the shielded protons on the inner nitrogen atoms appear upfield.

The high molar absorptivity (about 160,000) of porphyrins allows even very low concentrations to be detected by UV spectroscopy. They show characteristic electronic absorption spectra, with one intense band (**Soret band** or **B band**) in the near-ultraviolet region of the spectrum around 390-425 nm depending on whether the porphyrin is β-or meso-substituted with ε > 2×10⁵ and four low-intensity absorption bands (**Q bands**) at higher wavelengths (480 to 650 nm) in the visible region.

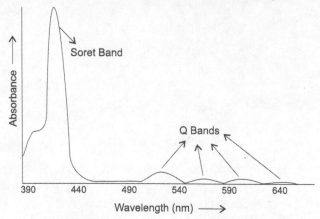

Typical UV-Vis absorption spectrum of porphyrin

14.1.4 Methods of preparation of porphyrins

(*i*) **Rothemund synthesis**: One of the earliest methods of synthesis of porphyrin was developed by Paul Rothemund in 1935. Rothemund first investigated the synthesis of meso-tetramethylporphyrin by the reaction of acetaldehyde and pyrrole in the presence of methanol at high temperatures.

Rothemund, P., J. Am. Chem. Soc. **1935**, *57, 2010.*

In 1941, Rothemund described the preparation of meso-tetraphenylporphyrin [H₂(TPP)], by heating pyrrole and benzaldehyde in presence of pyridine in a sealed vessel at 220°C for 48 h, followed by slow cooling over 10-18 h.

In the laboratory, porphyrins are synthesized by the condensation of aldehyde with pyrrole.

Mechanism: The reaction involves an electrophilic substitution to form rapidly equilibrating mixture of short pyrrole-aldehyde chains on repeated condensations. At the end ring closes to yield **porphyrinogen**, which on spontaneous oxidation upon exposure to air gives corresponding porphyrins.

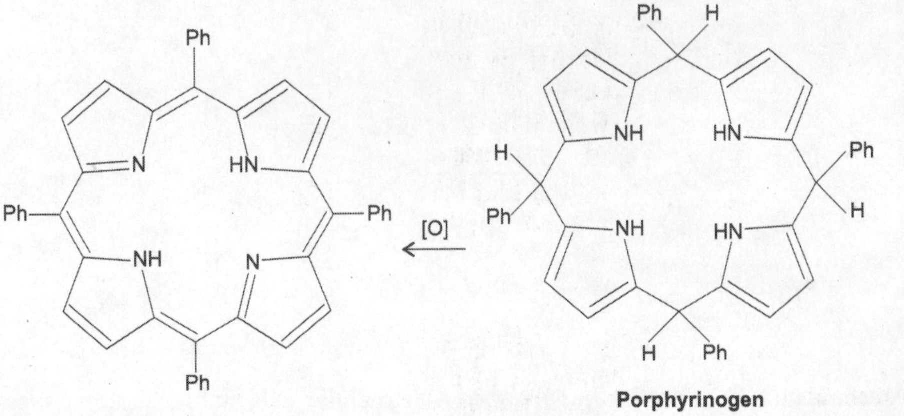

Porphyrinogen

The typical approaches for the synthesis of porphyrins are based on Rothemund synthesis. However, formation of other analogues and the difficulty in isolation of porphyrin are two main drawbacks in these synthesis.

(*ii*) **Adler and Longo synthesis:** Adler and Longo modified the Rothemund reaction by allowing benzaldehyde and pyrrole to react for 30 min in refluxing propionic acid at 141°C under aerobic conditions.

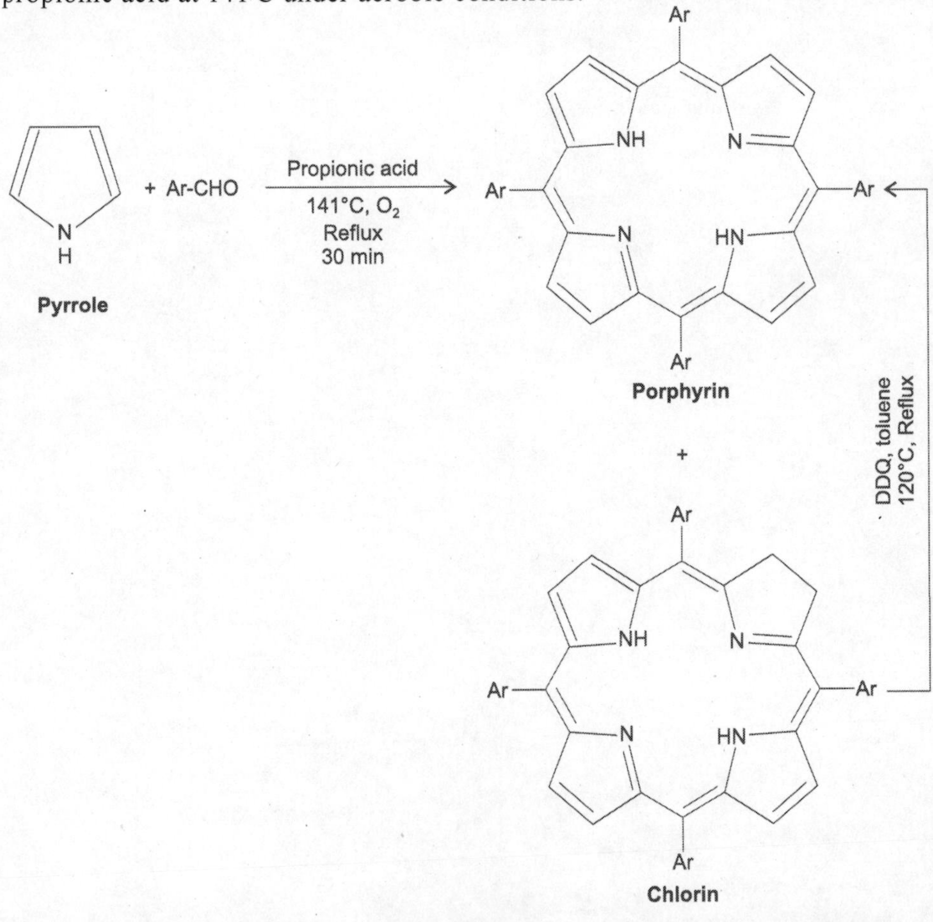

Porphyrin

+

Chlorin

Under these reaction conditions substituted benzaldehydes can be converted to the coresponding porphyrins up to 20% yields. Other acid catalysts which can be used are acetic acid, chloroacetic acid and trifluoroacetic acid. The tetra-arylporphyrin is obtained in crystalline form directly from the reaction medium. In this method about 2-10% of **chlorin** is formed, which can be easily oxidized to the corresponding porphyrin by treatment with DDQ in refluxing toluene. Chlorin constitute three pyrrole and one pyrroline nucleus connected through =CH– linkages.

Adler, A.D.; Longo, F.R.; Shergalis, W., J. Am. Chem. Soc. 1964, 86, 3145.

(*iii*) **Lindsey synthesis:** Lindsey gave the synthesis of meso-substituted porphyrins by the condensation of pyrrole and aldehydes in dicloromethane, at room temperature, using trifuoroacetic acid, BCl_3 or BF_3-etherate as Lewis acid catalysts. It was followed by the addition of a stoichiometric quantity of DDQ or para-chloranil to oxidize the porphyrinogen into the corresponding chlorin-free porphyrin. In this methodology the overall yields can reach upto 50% depending on the aldehydes used.

Lindsey, J.S.; Hsu, H.C.; Schreiman, I.C.; Tetrahedron Lett. 1986, 27, 4969.

Lindsey, J.S.; Schreiman, I.C.; Hsu, H.C.; Kearney, P.C.; Marguerettaz, A.M.; Tetrahedron Lett. 1986, 27, 4969.

Recently, two greener methods have been developed for the synthesis of pophyrin derivatives, one using microwave oven to heat the liquid reagents, and the other by performing the synthesis in the gas phase at high temperatures.

14.1.5 Biosynthesis of porphyrin

The biosynthesis of porphyrin involves the formation of **porphobilinogen** from two molecules of δ-aminolevulinic acid. The mechanism involves the formation of an imine between δ-aminolevulinic acid and the enzyme that catalyzes the reaction. An aldol-type condensation occurs between the imine formed and another molecule of δ-aminolevulinic acid. Nucleophilic attack by the amino group on the imine closes the ring. The enzyme is then eliminated, and removal of a proton creates the aromatic ring of porphobilinogen.

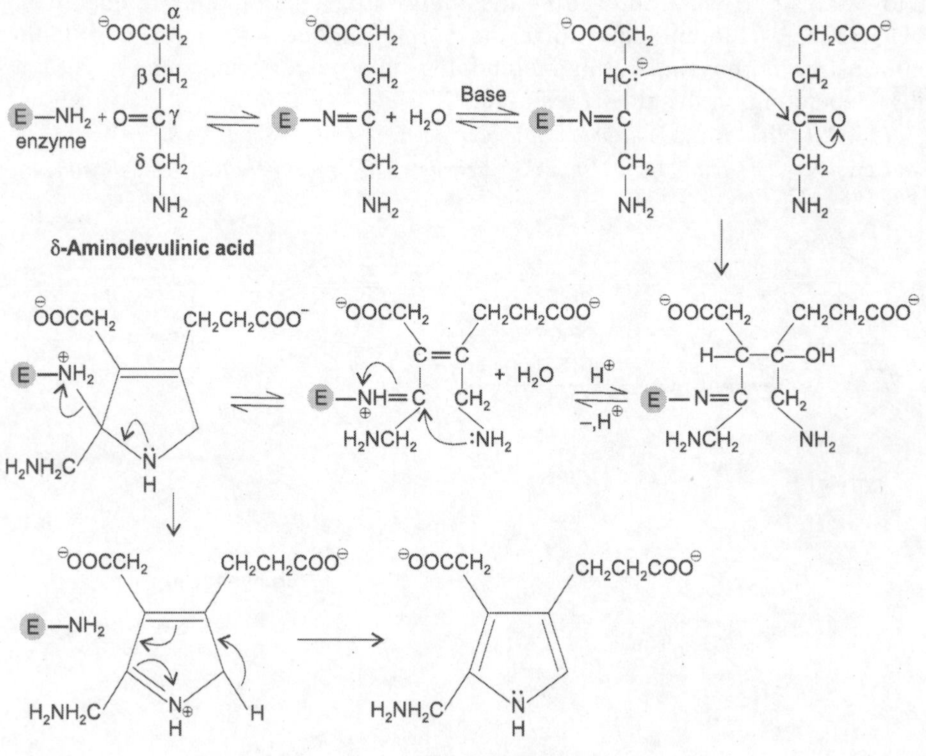

δ-**Aminolevulinic acid**

Porphobilinogen

14.1.6 Complexes of porphyrins

Porphyrins bind metals to form complexes. The metal ion usually has a charge of 2+ or 3+. The binding takes place as shown below:

$$H_2\text{porphyrin} + [ML_n]^{2+} \rightarrow M(\text{porphyrinate})L_{n-4} + 4L + 2H^+$$

where, M = **metal ion** and L = **ligand**

The space in the middle with the four inward pointing nitrogen atoms of porphyrins is just right for complex formation with metal ions such as Fe(II) and Mg(II). Heme, which is found in hemoglobin and myoglobin, contains an iron ion ligated (sharing of nonbonding electrons with a metal ion) by the four nitrogens of a porphyrin ring system. The porphyrin ring system of heme is known as protoporphyrin IX. The ring system plus the iron atom is called **iron protoporphyrin IX**. Heme-metal complexes are strongly coloured, the iron complex is blood red color. The **phthalocyanine-metal complex** based blue and green pigments are used to color plastic shopping bags. The green coloring pigment in leaves of plants is **chlorophyll a**. It contains a porphyrin moiety which is bound to magnesium metal. **Vitamin B12** also has a ring system similar to porphyrin, but one of the methine bridges is missing. The ring system of vitamin B12 is known as a **corrin ring system**. The metal atom in vitamin B12 is cobalt.

Iron protoporphyrin IX
Heme

Phthalocyanine copper complex

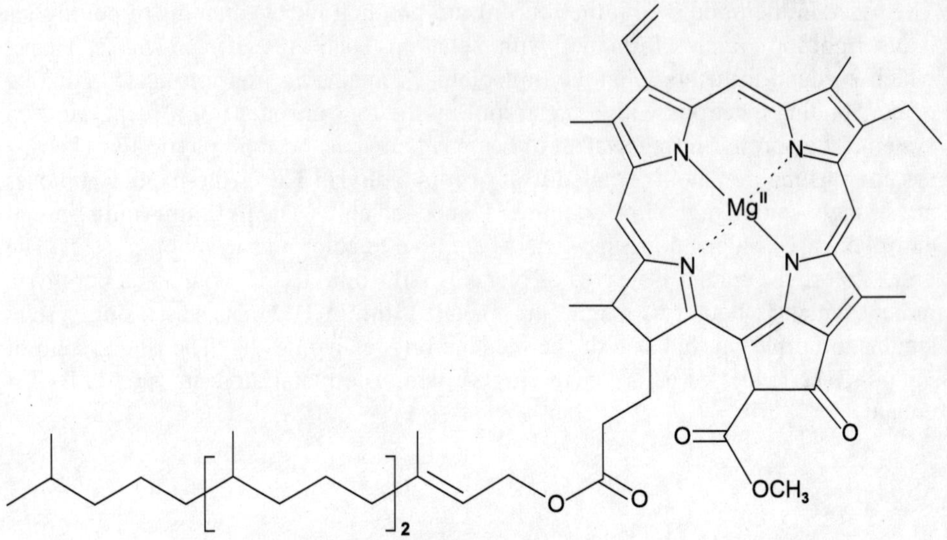

Chlorophyll a

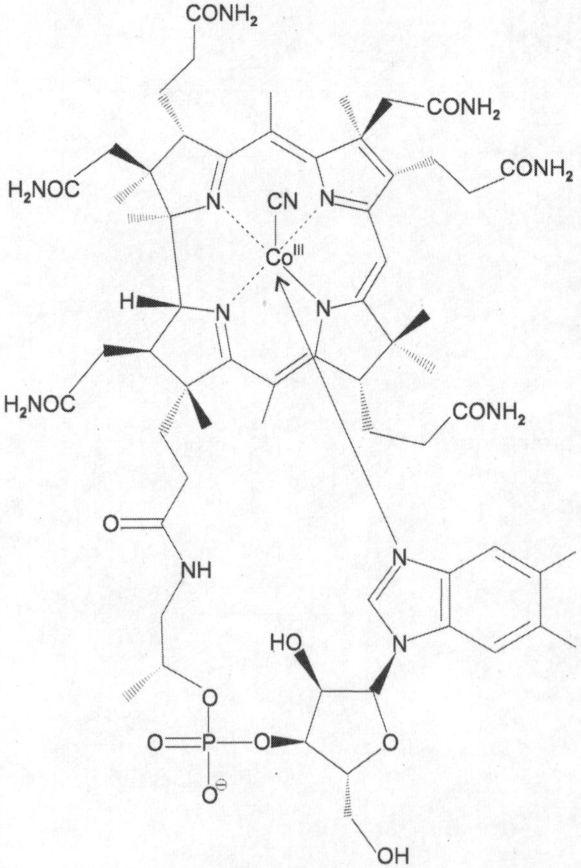

Vitamin B12

EXERCISE

Q.1 What is porphin?

Q.2 What do you understand by β and meso substituted porphyrins?

Q.3 The reaction of pyrrole with HCHO in the presence of a base yields a compound $C_{20}H_{14}N_4$, Which is a structural unit in heme and chlorophyll. Give its structural formula and discuss its aromaticity.

Q.4 Give a comparative account of Rothemund synthesis, Adler and Longo synthesis and Lindsey synthesis of porphyrins.

Q.5 What is the role of DDQ in Lindsey synthesis of porphyrin?

Q.6 Explain the formation of porphobilinogen during the biosynthesis of porphyrins.

Q.7 Name the metal atoms present in the following:

(a) Protoporphyrin IX

(b) Chlorophyll a

(c) Vitamin B12

Ans. (a) Iron (II); (b) Magnesium (II); (c) Cobalt (III)

Q.8 Draw the structure of chlorin.

Index